Fundamentals of Spatial Analysis and Modelling

T0260395

Fundamentals of Spatial Analysis and Modelling

Jay Gao

CRC Press
Taylor & Francis Group
Boca Raton London New York

CRC Press is an imprint of the
Taylor & Francis Group, an **informa** business

MATLAB® is a trademark of The MathWorks, Inc. and is used with permission. The MathWorks does not warrant the accuracy of the text or exercises in this book. This book's use or discussion of MATLAB® software or related products does not constitute endorsement or sponsorship by The MathWorks of a particular pedagogical approach or particular use of the MATLAB® software.

First edition published 2022
by CRC Press
6000 Broken Sound Parkway NW, Suite 300, Boca Raton, FL 33487-2742

and by CRC Press
2 Park Square, Milton Park, Abingdon, Oxon, OX14 4RN

© 2022 Taylor & Francis Group, LLC

CRC Press is an imprint of Taylor & Francis Group, LLC

Library of Congress Cataloging-in-Publication Data

Names: Gao, Jay, author.
Title: Fundamentals of spatial analysis and modelling / Jay Gao.
Description: First edition. | Boca Raton : CRC Press, 2022. | Includes bibliographical references and index.
Identifiers: LCCN 2021035702 (print) | LCCN 2021035703 (ebook) | ISBN 9781032115757 (hardback) | ISBN 9781032115764 (paperback) | ISBN 9781003220527 (ebook)
Subjects: LCSH: Spatial analysis (Statistics) | Spatial analysis (Statistics)--Mathematical models. | Geology--Statistical methods--Data processing.
Classification: LCC QA278.2 .G36 2022 (print) | LCC QA278.2 (ebook) | DDC 001.4/22--dc23/eng/20211021
LC record available at https://lccn.loc.gov/2021035702
LC ebook record available at https://lccn.loc.gov/2021035703

ISBN: 9781032115757 (hbk)
ISBN: 9781032115764 pbk)
ISBN: 9781003220527 (ebk)

DOI: 10.1201/9781003220527

Typeset in Times
by Deanta Global Publishing Services, Chennai, India

To my younger sister, who never ceases to amaze me with her boundless love and care for me.

Contents

Preface

Spurred by the rapid developments in computing technology, geographic information system (GIS) software has experienced phenomenal evolution over the last three decades, transformed from being mostly spatial data manipulation to integrated analytical systems. Such developments have popularised GIS teaching and learning at higher education institutions and enhanced the spatial literacy of the general public. Nevertheless, the GIS approach is not synonymous with spatial analysis, even though it has tremendously facilitated the undertaking of spatial analysis. While closely resembling spatial analysis, GIS analysis is by no means identical to spatial analysis and modelling, let alone having replaced it. Owing to the wide availability of spatial data acquired by location-aware devices, spatial analysis and spatiotemporal modelling (simulation) have been applied to an ever-widening span of fields, creating both new opportunities in understanding how they can be accomplished effectively and raising challenges as to how to meet the ever-expanding demands of users.

Despite the large number of textbooks that have been written about GIS analysis, the same cannot be said of spatial analysis. Although several books have been published on various topics in spatial analysis and modelling so far, they all cover only certain aspects of spatial analysis or modelling, and none of them treats this newly emerging and fast-evolving field of study systematically and comprehensively. Even fewer are suitable as textbooks that treat the subject holistically to cater to the needs of upper-level undergraduate and postgraduate students, especially those who specialise in GIS technology. In this sense, this book serves as a welcome addition to the large number of texts in this area. Dissimilar to the existing ones, this book methodically approaches the topic of spatial analysis in its entirety. It introduces a wide range of ways in which spatial data can be analysed and simulated. By reading this book, students will be exposed to the whole range of analytical methods of geospatial data and can expect to gain a sound understanding of how to carry out spatial analysis/modelling in different computing environments and platforms.

All topics related to spatial analysis and spatiotemporal simulation are logically examined in-depth in this book, including how spatial data are acquired, represented digitally, and spatially aggregated. Also featured prominently in the book are the nature of space and how it is measured. Descriptive, explanatory, and inferential analyses of point, linear, and areal spatial entities are all presented in detail. Predictive spatial analysis is covered under spatial interpolation at the local and global scales. A substantiate portion of the book is devoted to spatial modelling, including how to select variables for a model, how to weigh them and test their importance, how to construct and validate a model, and how to evaluate the accuracy of modelling. The book also captures the latest developments in spatiotemporal simulation, with cellular automata and agent-based modelling extensively expounded. Also covered in the book are the temporal component of geospatial data, their representation, and analysis. Apart from theories, the book also explores how spatial analysis and modelling can be implemented in different computing platforms, supplemented with ample examples. This book is lavishly annotated with tables and diagrams to simplify concepts and contrast disparities of analytical and modelling methods to facilitate deep understanding.

Comprising eight chapters, this book first presents the fundamental concepts related to spatial analyses, such as coordinate systems, measures of space, how spatial data are aggregated, and how space is partitioned. After laying a solid foundation of spatial measures, the book describes spatial analysis from Chapter 3 onward. The first section of Chapter 3 focuses on the acquisition of spatial data, followed by spatial association. Chapter 4 covers the descriptive (and inferential) spatial analysis of point, line, and area data, both individually and collectively as a group. Chapter 5 expounds on spatial interpolation, in which inverse distance weighting and kriging are discussed at length, in addition to minimum curvature and trend surface analysis. Chapter 6 addresses spatial modelling. All elements related to spatial modelling, from model development to model validation, are explained in depth. Chapter 7 covers spatiotemporal simulation, in which cellular automata and agent-based modelling feature prominently. In most cases, examples are provided to illustrate how spatiotemporal simulation is implemented for different applications. Chapter 8 is devoted to time-based spatial analysis, both descriptive and modelling.

This book can serve as the primary textbook for an upper-level undergraduate or postgraduate course on GIS spatial analysis, or just spatial analysis and modelling. Some topics in the second chapter can be skipped for those GIS students who have already gained the background knowledge in a lower-level course. The entire materials can be delivered in about 40–50 hours of face-to-face contact in the classroom and the lab (extra time may be required if the students are expected to undertake an independent project). The students may wish to review the content of each chapter by browsing the online compendium resources at https://drive.google.com/drive/folders/1pd18jGObirpyOymaQeU6MyN8sM_463gG?usp=sharing. The course instructor may wish to refer to the online PowerPoint resources in preparing lecture materials. The sample analytical examples are provided as references only. They are meant to demonstrate possible ways by which practical skills can be gained from hands-on experience in the lab. The supplied data may be substituted with locally sourced data or data collected by the students themselves to make the analyses more relevant to home.

MATLAB® is a registered trademark of The MathWorks, Inc. For product information, please contact:

The MathWorks, Inc.
3 Apple Hill Drive
Natick, MA 01760-2098 USA
Tel: 508 647 7000
Fax: 508-647-7001
E-mail: info@mathworks.com
Web: www.mathworks.com

Author Biography

Dr Jay Gao is an Associate Professor at the Faculty of Science, University of Auckland, New Zealand. He received his BE in photogrammetric engineering from Wuhan Technical University of Surveying and Mapping, China, his MS from the University of Toronto, Canada, and his PhD from the University of Georgia in the US in 1992. Over his academic career, he has carried out many research projects on remote sensing, geographic information systems, the Global Positioning System and their integrated applications in natural resource management and environmental monitoring, focusing on quantitative remote sensing and integration of geo-computational methods. His recent research is focused on applying spatial modelling to studying environmental issues . He is an Associate Editor of the International Society for Photogrammetry and Remote Sensing's *Journal of Photogrammetry and Remote Sensing* and has published more than 160 research articles. His textbook *Digital Analysis of Remotely Sensed Imagery* was published by McGraw-Hill Education in 2009.

Acknowledgements

Although this book represents the ultimate fruition of decades of teaching and research in the geospatial field of study, dating back to my PhD era in the early 1990s, it would not have been possible without the generosity of many others. The contribution of these people towards this book is hereby gratefully acknowledged. I am deeply indebted to the University of Auckland, which granted me the opportunity to teach a course in spatial analysis by myself. In particular, I would like to thank my former colleague, David O'Sullivan, who generously shared his lecture and lab materials with me. Such a collegial spirit is sincerely appreciated and forever cherished. This book could not have been completed so soon without the university granting me a sabbatical that coincided mostly with the COVID-19–triggered lockdowns and border closure, during which the bulk of the book was written. My gratitude towards my former cartographic colleagues, Jan Kelly and Jonette Surridge, is expressed here. They showed me how to use CorelDraw, with which the majority of illustrations contained in this book were drawn.

Next, I would like to acknowledge a few publishers and professional organisations for permitting me to reuse their illustrations in this book. They are: SAGE (Table 4.3), Springer (Figure 6.14), Elsevier (Tables 5.5 and 6.10, and Figures 6.2, 6.3, 6.6a, 8.12, and 8.13), and Taylor & Francis (Figures 6.21, 6.22, 7.14, 7.15, and 8.11, and Table 7.1), the Association of American Geographers (Figure 8.9), and the Society of Exploration Geophysicist (Figure 5.8). I am also deeply indebted to the following individuals and organisations for granting me permission to reuse their diagrams: Tom Patterson at the US National Park Service (Figure 1.3), Visual Capitalist (Figure 1.5), Veros (Figure 2.20b), Jason Brownlee (Figure 6.10), and Keith Clarke (Figure 7.3). Kitchin et al. (Figure 2.21), Joseph Claghorn (Figure 3.7a), Lim et al. (Figure 6.6c), Jaseela et al. (Figure 6.8), Siddayao et al. (Table 6.9), Keith Woodford (Figure 6.16), Thomas Bandholtz (Figure 8.6), and Meyer et al. (Table 8.5) are thanked for granting me the rights to reuse their materials in this book or making their materials openly accessible.

I am especially appreciative of the contributions of my postgraduate students to this book. They have carried out cutting-edge research on spatial modelling (simulation) dating back to as early as the 1990s, including Graeme Aggett on the earthquake damage modelling of Wellington, Sesa Wiguna on modelling landslide hazards in Sukabumi, Indonesia, Xuying Ma on the modelling of air pollutant distribution in Auckland, Tingting Xu on the urban expansion modelling of Auckland, and Xilai Li on the simulation of grassland degradation. Their research results have considerably enriched the content of this book. In addition, Xuying also supplied additional results about trend surface analysis.

Finally, I owe a special thank you to my parents for bringing me into this world with observant eyes and an inquisitive mind that is perpetually ready to be stretched.

Jay Gao
Auckland, New Zealand
June 2021

1 Introduction

1.1 WHAT IS SPATIAL ANALYSIS?

1.1.1 DEFINITION

Although *spatial analysis* has been widely practised in the geographic information system (GIS) community for decades, it is not so easy to define it precisely. In the literature, spatial data analysis has been defined differently by various authors. Goodchild (1987) defined it as a set of techniques devised to generate a spatial perspective on data. It is distinctive from other forms of analysis in that the analytical results are dependent on the locations of features or events being analysed. It yields value-added products from existing datasets, which is not possible otherwise. The analysis involves both locations and the attributes of spatial entities, or just the locations themselves. This definition leans heavily towards the technical perspective, involving both locational and attribute information. Haining (1990) defined it as "a body of methods and techniques for analyzing 'events' at a variety of spatial scales, the results of which depend upon the spatial arrangement of the 'events'". Featured prominently in this definition are techniques that address scale and spatial patterns. Although it is more encompassing than Goodchild's definition by incorporating scale in the analysis, this definition still fails to capture the new developments in spatial analysis, such as spatial modelling and simulation which are one step further than simple descriptive analysis.

Longley et al. (2005) considered spatial analysis as a set of methods that can produce results changing with the locations of the features under analysis. This definition is almost identical to that of Haining (1990). Basically, this definition reduces spatial analysis to a set of analytical tools that have a spatial component. The Environmental System Research Institute (Esri) *GIS Dictionary* expands the definition of spatial analysis as the "process of examining the locations, attributes, and relationships of features in spatial data through overlay and other analytical techniques in order to address a question or gain useful knowledge". This rather narrowly focused definition emphasises the content of spatial analysis and its objectives. However, the field of spatial analysis has evolved to such a degree that spatial modelling and simulation have become the integral components of spatial analysis, but they are excluded from this definition.

All of the aforementioned definitions from different perspectives with different emphases and annotations are decades old. They can no longer reflect the field of spatial data analysis adequately. In this book, contemporary spatial analysis is defined as a series of techniques for describing, analysing, simulating, and predicting the spatial patterns and/or processes of geographic phenomena that may involve mobile agents. This analysis may include spatial statistic indices, spatial regression and adaptive modelling, spatial dynamic modelling, integrated spatial statistic and

DOI: 10.1201/9781003220527-1

spatial mechanism modelling, and spatial complex system modelling. The spatial data can be either geo-referenced points, lines, or areas to identify their patterns and associations, and to predict unknown values of the concerned attribute in relation to its affecting variables. This definition has three unique characteristics: (1) it is rather comprehensive, in that it includes the nature of spatial data that must have geographic coordinates through which observations are associated with their locations on the ground; (2) it captures the spatial dimension of the entity; and (3) it includes the purposes of spatial analysis, namely, to characterise the observed spatial patterns and to predict their values/patterns in the future. What is missing from this definition is the temporal component. This is because time is treated as invariant in descriptive spatial analysis, and is always implicitly dealt with in spatial modelling in which the behaviour of variables in the model varies with time. What is unclear in this definition is the dimension of the geographic entity. At present, it is confined to only two-dimensional phenomena. In reality, some geographic phenomena such as air pollution are inherently three-dimensional. Such phenomena can still be studied by slicing the vertical dimension into certain height bands and then treating each as a single two-dimensional layer. It can be studied using the methods described in this book. This attribute can be expressed as a function of location and time, namely, Z(easting, northing, t).

Implicit in the above definition of spatial data analysis is the development of mathematical models of spatial distributions, the analysis of locational patterns, and the investigation and prediction of space–time dynamics. Spatial analysis embraces a wide cluster of techniques that apply formal, usually quantitative, structures to systems in which the prime variables of interest vary across space. Traditionally, spatial data analysis falls into the domain of quantitative geography, although ecology, urban studies, transportation, and a host of cognate disciplines draw from and have played an instrumental role in the development of this field. In turn, the applications of spatial analysis to some of these fields have enriched and expanded spatial analysis.

1.1.2 Spatial Statistics

Spatial analysis has a number of closely related but not exactly meaningful terms, one of which is *spatial statistics*. In the Esri *GIS Dictionary*, it is defined as the:

> *study of statistical methods that use space and spatial properties (such as distance, area, volume, length, height, orientation, centrality and/or other spatial characteristics of data) directly in their mathematical computations. Spatial statistics may include a variety of analyses, such as pattern analysis, shape analysis, surface modeling and surface prediction, spatial regression, statistical comparisons of spatial datasets, statistical modeling and prediction of spatial interaction.*

This definition encompasses spatial analysis. Besides, it goes one step further by spelling out the exact tasks of spatial analysis. According to this definition, spatial statistics applies statistical methods to the analysis of spatially referenced data. The topological, geometric, or geographic properties of spatial entities are the objects of study. Thus, all tools of traditional statistical analysis can be used to analyse spatial distributions, patterns, processes, and relationships of these entities without any modification. Spatial statistics differ from spatial analysis in that it focuses on the statistical properties of the attribute of a geographic phenomenon. These analyses can be descriptive, inferential, exploratory, and even geostatistical. What is missing from this definition is spatial modelling and simulation. Thus, spatial analysis has a much broader connotation than spatial statistics, which focuses on the quantitative description and exploration of spatial phenomena.

1.1.3 GEOCOMPUTATION

Coined by Openshaw and Abrahart (1996), *geocomputation* started to appear in the literature in the mid-1990s at a specially themed conference dedicated to it. It refers to the application of computing technologies to solving spatial problems, including storage, analysis, and visualisation of spatial data and spatially modelled results (Esri *GIS Dictionary*). This definition emphasises the use of computers and the computing environment for running spatial analyses. It represents a marriage between computer science and spatial analysis. This definition is particularly suited to those spatial analyses that are memory- and CPU-hungry. Thus, geocomputation differs from spatial data analysis in that the latter has a much broader scope, including spatial modelling and simulation, whereas geocomputation aims at developing methods to analyse and model a range of highly complex, often non-deterministic problems with or without relying on GIS. As the field of spatial data analysis evolves, geocomputation has expanded to include principal component analysis, *k*-means clustering analysis, and maximum likelihood classification of satellite imagery, machine learning (e.g., artificial intelligence) decision trees, neural networks in identifying patterns and classifying satellite imagery data, and in mining the sheer volume of geo-referenced data for knowledge discovery. Some of these do not even have a spatial component. The analyses are considered spatial simply because the inputs are two-dimensional layers. Therefore, the spatial component of geographic data is not featured prominently in geocomputation.

1.1.4 GEOSTATISTICS

Geostatistics is a field of study applying statistical methods to analysing spatial data. It is a special branch of applied statistics initially developed by Georges Matheron, a French mathematician and geologist, in mining in the 1970s. This subset of spatial analytical methods aims at spatially depicting the observed attribute of a spatial entity quantitatively or statistically. When it was first developed, geostatistics found applications exclusively in the mining industry (e.g., to estimate ore reserves). It did not experience rapid development and wide uses until the 1990s, when advanced spatial analysis systems such as GIS became available, and the easy and wide availability of spatially referenced data in the digital format. In geography in general, and spatial analysis in particular, geostatistics can be regarded as a special branch of statistics applied to geographic data (e.g., data that have a spatial component attached to them). Since its invention, geostatistics has been widely used to analyse a wide range of spatial data in diverse disciplines, including petroleum geology, hydrogeology, oceanography, geochemistry, geometallurgy, geography, forestry, environmental control, landscape ecology, agriculture (especially in precision farming), and even medicine. These days, it has found even wider applications in more fields in natural sciences, including soil science, hydrology, meteorology, and environmental science. Common to all of these fields is that the data used have a spatial component, and the variable being analysed is predictable to a certain spatial extent. It is called quasi-statistics by Davis (1986) as some requirements of classical statistics are not met with the data. For instance, data independence is impossible to achieve with petroleum drilling holes that are expensive and highly limited in their spatial distribution.

1.2   WHAT IS SO SPECIAL ABOUT SPATIAL ANALYSIS?

1.2.1   A Historic Event

Spatial analysis was practised as early as the mid-19th century. A well-known case of spatial analysis was undertaken by a medical professional named John Snow who was apparently not even a geographer, let alone a spatial analyst. The analysis of spatial data was triggered by a cholera outbreak in Soho, London in 1854. Mysteriously, hundreds of people caught cholera and died during the outbreak. Dr Snow was not just an ordinary physician. Driven by his suspicion that these people caught the disease by drinking contaminated water, he interviewed his patients about the address of their abode during the medical consultation. Afterwards, he plotted out their residential addresses on a paper map (Figure 1.1). According to this map, the majority of the cases were all located in close proximity to the Broad Street pump, spatially clustered around this well. He then advised the local residents in the neighbourhood not to drink the water from this well as he theorised that water from it was contaminated by the human waste from the nearby cholera-infected residents. Soon the plague was brought under control. Unintentionally, Dr Snow became a pioneering epidemiologist, the first known person who manually mapped the distribution of infection cases and exploited spatial analysis to save lives, even though the method he used was rather primitive by today's standards. Nevertheless, the essence of spatial analysis has remained hardly changed over the years. In this case, it is a simple overlay analysis of two sets of spatially referenced data: a map of infected patients with a map of the well distribution. Although we no longer face the risk of the same cholera contamination now, spatial analysis is just as important, relevant, and potent as ever, albeit for different reasons.

1.2.2   Value of the Spatial Perspective

Humans live in space, and every move is spatial in nature. As all phenomena are spatially related, a spatial perspective on them is more insightful than otherwise. As a matter of fact, society cannot function smoothly without spatial analysis. For instance, a traffic accident on a major road can grind an entire city's traffic to a halt.

FIGURE 1.1   Spatial distribution map of cholera cases in Soho, London in the 1854 cholera outbreak in relation to the location of wells, manually drawn by Dr John Snow.

We must know the location of this spot and find alternative routes to divert traffic away from the affected areas and alert pending motorists to the unforeseen changes to avoid traffic congestion and delay. The spatial distribution of traffic accidents in relation to the road network can pinpoint highly dangerous and accident-prone spots. A spatial perspective on these blind spots can help us pin down the causes of crashes, implement remedial measures at these places to reduce human causality and loss to properties, and rationalise the design of safer transportation infrastructure in the future. In addition to spatial location, spatial relationship is also critical to our well-being, and even happiness. The association between living in close proximity to a power transmission line and cancer can be established through buffering and correlation analysis of a large number of cases. In light of this association, caution can be exercised to minimise the risk of exposure to electromagnetic radioactivity by zoning residential areas further away from the transmission corridor.

We are living in an era of constant changes; one of which impacts all humanity is climate. As the global temperature becomes warmer, sea levels will rise, which will threaten coastal settlements and inundate low-lying infrastructure. Inevitably, some coastal settlements lying at an elevation just above the current sea level face the risk of being submerged under seawater in the near future. Spatial analysis can help us identify those at-risk regions so that proactive counter-measures such as sea walls can be put in place to mitigate the foreseeable adverse influences. Such earlier planning actions can save lives and reduce costs later on.

1.2.3   UBIQUITOUS APPLICATIONS

By default, wherever we live has a spatial component to it. Space is intrinsically linked to our daily routine. Spatial decision-making is embedded in every mobility-related activity we engage in. Whether it is driving to work or attending an important business meeting, we need to navigate the complex cityscape. Analysis of our spatial mobility can reveal our collective behaviour (such as the timing of morning rush-hour commuting) and interactions (e.g., going to work by different modes of transport). Spatial knowledge on our collective behaviour can minimise negative consequences of our actions, such as by avoiding traffic jams. It is not just transport that can benefit from spatial analysis. A wide range of scientists rely on spatial analysis to yield solutions to problems faced in diverse disciplines, including social scientists, forensic scientists, traffic engineers, epidemiologists, hydrologists, urban planners, and even landscape ecologists. When a flu epidemic or a highly contagious virus (e.g., COVID-19) emerges, analysis of the spatial pattern of early cases can reveal how it is transmitted from person to person and along which route. Such information can help us to identify where the infected dwell and with whom they have interacted. Such knowledge enables us to better pinpoint the origin of infection and trace the movement of the infected and their social contacts to bring the outbreak under control quickly. Forensic scientists may benefit from the analysis of the spatial pattern of crime and its relation to urban design. A better urban design can deter the potential perpetrators and reduce the likelihood of common crimes, such as burglaries and petty thefts. An urban planner may need to know the spatial pattern of vehicle-originated air pollutants. Any areas that suffer from an unusually high concentration of vehicle-induced air pollutants may indicate a poor urban design that discourages their dispersal, or a poorly laid out road network that is conducive to perennial traffic congestion. Similarly, landscape ecologists may be interested in the spatial variation of street trees and their relationship with other environmental variables such as atmospheric carbon dioxide levels. The correlation of these two variables can reveal the ability of the trees in sinking vehicle-yielded carbon dioxide.

In summary, a large number of fields can benefit from spatial analysis that can either yield new insights into an existing problem or provide better solutions to it. As more location-aware devices such as Global Positioning System (GPS)-enabled mobile phones find increasing uses in our lives, more spatial data are being collected with ease and at a fast pace. Such a colossal volume of data will fuel spatial analysis and open up new applications to previously unknown territories.

1.2.4 Ability to Predict and Explore Alternatives

With the assistance of spatial analysis, it is possible to predict the attribute value of a spatial entity at unsampled sites from neighbouring observations based on the spatial relationship between them. Such predictions are highly prized as they can minimise the cost of sample collection. Alternatively, it is also possible to predict the value at a given site from a number of attributes at the same location, a practice commonly known as spatial modelling. Through this modelling approach, it is possible to explore the influence of a given input variable through sensitivity analysis. Such a predictive method of analysis can yield information on what is going to happen to the target of study in the future based on what has occurred in the past. For instance, it is possible to demarcate the coastal areas that will likely be inundated in the next 50 years based on the magnitude of climate warming in the past. Such knowledge can be used proactively to enact land use plans in coastal areas and to properly lay out different land use zones to minimise the impacts of the anticipated sea-level rise.

The world we live in is rather complex, involving different processes and interactions. In each of these processes, whether social, geographic, or environmental, a plethora of factors are at play. In most circumstances, it is impossible to know the exact effect of an individual variable in the process as it is inevitably intertwined with the effect of another factor or a number of factors. For instance, the extent to which a landsliding event is triggered by a heavy storm can rarely be quantified. However, with the assistance of spatial simulation, we are able to isolate the effect of this variable by allowing it to vary while keeping all other variables unchanged. Through alternating the variables and iteratively changing their values in multiple simulations, it is possible to identify the exact role these variables play in the landslide process. This predictive power is almost impossible to achieve via any other means. The gained knowledge enables us to develop rational hypotheses about the observations. Through spatial modelling, we are able to simulate the spatial interactions of an active agent with the environment to assess its potential impact on the latter. Furthermore, spatial simulation can predict future changes to the variable of interest in different scenarios. For instance, whether a grassland is degraded or not depends on the level of grazing intensity and the degree of external disturbances. Through modifying the intensity of grazing, it is possible to run scenario analysis in the form of "what … if …" and produce results that are able to shed light on the optimal grazing intensity.

1.3 A BRIEF HISTORY

Although spatial analysis came into existence less than 160 years ago, it can be divided into four distinct periods or stages in its evolution (Table 1.1). At every stage, the advancements in computer technology have tremendously facilitated and eased spatial analysis.

1.3.1 Pre-digital and Early Digital Era

This stage took place mainly prior to the mid-1980s, when desktop computers were still in their infancy and the digital revolution starting from the 1960s had not noticeably impacted data analysis in general, and spatial data analysis in particular. Back

TABLE 1.1

Four Main Stages in the Evolution of Spatial Data Analysis and Their Main Properties

Stages	Eras	Main features
Prior to 1980s	Pre-digital and early digital era	Pen and paper, and mainframe computers; primitive, slow; spatial interpolation; development of theories on the spatial behaviour of variables (e.g., theory of regionalised variables)
1980s–1990s	Desktop computing era	Spatial statistics possible with SAS and SPSS; the advent of sophisticated spatial interpolation algorithms (e.g., co-kriging)
1990s–2000s	GIS and GPS era	An explosion of spatial data; study of complex relationships; spatial coding and relationship; simple models; advent of commercial packages for spatial analysis; easy and wide applications
2000s–present	Big data era	Ubiquitous use of positioning-enabled gadgets; crowd-sourced social media data; object-oriented scripting; sophisticated modelling; process-oriented simulation

then, spatial analysis was rather basic, primitive, and cumbersome to undertake as it had to be carried out manually using the pen and paper method (e.g., analogue maps). This situation changed during the quantitative revolution in geography in the late 1960s and early 1970s, when statistics was applied to analysing census data. As early as the late 1980s, with the increasing acceptance of GIS, spatial analysis became possible in the digital environment. For the first time in history, demographers were able to analyse the distribution of income and life expectancy by suburbs using computers. Nevertheless, crude information was produced at a painstakingly slow pace because of the need to punch cards to implement the computing codes. Apart from computing technology, the late 1970s and early 1980s also witnessed the rising popularity of databases and the query language accompanying them. Databases made it possible to build complex models about some geographic phenomena. It was during this period that Tobler (1970) came up with the First Law of Geography, which states: "everything is related to everything else, but near things are more related than distant things". This law firmly lays the theoretical foundation of spatial dependency and spatial auto-correlation. It provides the theoretical backing for distance-based spatial analysis and supports the regionalised variable theory proposed by Matheron (1971). Since this theory required sophisticated computation that was extremely difficult to implement digitally, it did not find wide applications until the next stage, towards the end of the 20th century.

1.3.2 DESKTOP ERA

This period started in the mid-1980s and lasted until the mid-1990s, during which spatial analysis experienced explosive growth, boosted by tremendous advancements in computer technology. Although mainframe computers had been in existence for decades, they were not accessible to most people. Besides, computing in this environment was cumbersome and inefficient. Starting in the late 1980s, micro- and mini-computers, and later desktop personal computers, were gaining increasing popularity in academic circles. The emergence of computer-based statistical analysis packages in the mid-1980s, such as SAS and later SPSS, enabled the analysis of socioeconomic data digitally. Since these systems were designed to analyse primarily socioeconomic data collected via census statistically, analysis of spatially explicit data was highly limited. In other words, aspatial analysis of geographic data was the norm, while the spatial component of the data was mostly neglected.

Even with desktop computers, it was not so easy to carry out statistical analysis as the computer operating system was command-driven and the computer packages were not user-friendly. Even simple analysis had to be implemented via scripting due to the absence of commercially available, generic spatial analysis packages. The coding process was understandably slow, error-prone, and frustrating, and the analysis results could not be visualised graphically. This pre-Microsoft Windows era of data analysis was mostly statistical in nature (e.g., non-spatial), with some hypothesis testing possible. This situation changed with the advent of the Microsoft Windows operating system in the mid-1980s, which allows the user to interact with the computer graphically. Subsequently, Windows-based computer software packages eased the execution of data analysis. A wide range of analyses can be performed by merely clicking a few buttons on the screen. Nevertheless, spatial analysis did not flourish because of the lack of spatially explicit data. This situation did not improve dramatically until the advent of location-aware devices and services such as the Global Positioning System. Apart from the advancements in computer technology, new theories about spatial analysis, such as robust kriging algorithms (Hawkins and Cressie, 1984), were also developed during this period.

1.3.3   GPS AND GIS ERA

This era spanned from the mid-1990s to the mid-2000s, during which GPS and GIS experienced phenomenal growth and development to play a decisive role in spatial analysis. GPS is a positioning system comprising a constellation of 24 satellites revolving around the Earth to triangulate the ground position of a receiver that simultaneously tracks the signals from at least four of the satellites. Although GPS satellites were launched by the US government as early as the 1960s, GPS did not find wide civilian applications until the early 1990s when it was demilitarised. The tracking of features using GPS receivers mounted on vehicles and attached to moving objects (e.g., animal tracking) generated a colossal volume of spatially referenced data. With the ever-widening applications of GPS in navigation and fleet management, spatial data analysis has been broadened to include the temporal component. The wide and easy availability of geospatial data from GPS in the digital format extends spatial analysis to be synonymous with spatiotemporal analysis, especially in transport and mobility data analysis.

In spite of the drastically improved availability of spatially referenced data from GPS, however, spatial analysis did not take off until the mid-1990s, when

Windows-based desktop GIS systems became much more user-friendly and powerful. GIS represents a natural progression from conventional paper-based maps to digital maps with numerous added advantages over analogue maps in spatial data analysis. There is no definite record of exactly when the term GIS came into existence or who coined it. It is generally acknowledged that Dr Roger Tomlinson is the "father of GIS" who guided the development of the Canada Geographic Information System (CGIS) for storing, analysing, and manipulating data collected for the Canada Land Inventory in the 1960s using mainframe IBM computers (Tomlinson, 1970; Goodchild, 2018). This system was designed to assess the land capability of rural Canada from soils, agriculture, recreation, wildlife, waterfowl, forestry, and land use. Otherwise referred to as the spatial data system, the term GIS was finally accepted in academic circles. The predecessors of modern-day GISs included Dual Independent Map Encoding (DIME) in conjunction with geographic base files (GBFs: digital maps of a city's streets, address ranges, and geostatistical codes) in the 1970s and 1980s by the US Census Bureau, and its subsequent successor of the Topologically Integrated Geographic Encoding and Referencing (TIGER) system in 1990 (Broome and Meixler, 1990).

Apart from operational GISs based on mini-computers developed by government agencies, commercial GIS packages had also been released by private companies and universities, such as ARC/Info by the Environmental System Research Institute and IDRISI by Clarke University. Both systems ran on PCs and were able to perform a variety of spatial analyses. Such GIS systems, different from the popular computer-aided design (CAD) systems exclusively for graphic purposes, made GIS accessible to an expanding range of spatial analysts and professionals, and catapulted spatial analysis and modelling to the next level.

1.3.4 Big Data Era

From the early 21st century, spatial analysis has evolved into the big data era owing to the increasing popularity of GPS-enabled devices, especially smart phones and cameras. Their ubiquitous use and auto-tracking have yielded "big" data about people's behaviour and movements and opened up new avenues of spatial analysis and applications, such as fleet management, animal tracking, and human mobility studies (Jiang et al., 2017). Detailed information about where the tracked features are located at what time can be acquired almost instantaneously and continuously. Palm-top mobile devices not only yield spatial data about their carriers, but also allow access to and searches for special information on the World Wide Web that used to be done exclusively from desktop computers. Smart phones provide not only spatiotemporal information about animals and humans, but also about the information their owners have searched for online. Further information about special topics can be easily sourced from popular social media sites such as Twitter and Facebook. They can yield crowd-sourced data. Such big data have opened a floodgate of applications that were unimaginable just years ago, such as the identification of spots most prone to car crashes and their timing in a city, and the hot spots of car crash sites that deserve special attention while driving at a particular time of the day. They also provide an excellent opportunity to harvest and mine them to yield new insights into the collective behaviour of the human population. For instance, it is possible to analyse how people react to a natural disaster, such as the Christchurch earthquakes in 2011, through analysis of the tweets they sent and the location from where they sent them. However, purposely mining such data may require sophisticated scripting. In law and enforcement, home detainees who are forced to wear GPS-enabled bracelets can be tracked and their whereabouts can be monitored remotely in real time. Analysis of the signal transmitted by the wearer's monitor can easily reveal the spatiotemporal behaviour of the wearer and detect whether bail conditions have been breached.

It is also during this era that specially tailored computing platforms and software have been developed and gained popularity, such as the R scripting language (and package) for analysing spatial data, and the NetLogo package for spatial simulation (see Section 1.7.5). Although neither are linked directly to big data, they do allow spatial analysis to be undertaken with greater ease. In particular, they enable inexperienced analysts to learn how to simulate spatial processes with little difficulty owing to their user-friendliness. The advent of such systems has meant that spatial analysis is practised by an ever-expanding field of scientists.

1.4   RELATIONSHIP WITH OTHER PERTINENT DISCIPLINES

The aforementioned history of spatial analysis indicates that it never evolved in isolation. Whether we can perform certain kinds of spatial analysis and at what level of ease all depends ultimately on other pertinent disciplines that either supply the data needed for spatial analysis, or provide a medium for spatial analysis, or offer the theoretical grounding of spatial analysis. These disciplines mainly include, but are not limited to, GIS, remote sensing, GPS, computer science, and statistics, in order of relevance (Table 1.2).

1.4.1   GIS

GIS is a powerful system to store, retrieve, analyse, collate data, and visualise results graphically. As a computer system, GIS is good at managing, manipulating, and analysing spatially referenced data. Naturally, GIS is aligned closely with spatial analysis. In fact, what kind of spatial analysis we can undertake and how easily it can be implemented are subject to GIS functionality and capability. It is not an exaggeration to state that without GIS there would be virtually no spatial analysis, for three reasons:

(1) GIS packages enable spatial data to be efficiently represented in the map format. Initially, most of the spatial data were secondary in nature, digitised from pre-existing analogue maps. In contrast, limited point data were collected using GPS receivers. Later, the GIS database was augmented with easily available satellite images that can be analysed in the spatial domain using a GIS.

TABLE 1.2

Relationship between Spatial Analysis and Five Pertinent Disciplines in Order of Relevance

Discipline	Relationship with spatial analysis
GIS	Supplier of spatial data; functions for routine vector data manipulation and analysis; tools for building spatial models
Remote sensing/ image analysis	Supplier of raster data (e.g., land cover and digital elevation model data); platform of raster data analysis and modelling
GPS	Determinant of the spatial component of spatial data, fast updating of existing spatial data
Computer science	Sophisticated software and algorithms for easing spatial analysis and modelling; powerful machines for implementing spatial analysis
Statistics	Calculation of spatial entities' attributes; hypothesis tests of spatial distribution patterns

(2) GIS provides an environment for integrating spatially referenced data from diverse sources, and for providing analytical functions by which sophisticated analyses and complex modelling can be implemented. In the early days, most of the spatial analysis was restricted to monotonous types of data, with data from different sources seldom integrated. With the use of GIS, data from multiple sources can be easily integrated and exploited in a spatial analysis.

(3) GIS can instantly visualise the results of analysis. It is even possible to produce animations to illustrate the spatial process of a spatial phenomenon, such as landsliding and the spread of wild fires over the landscape. Apart from being visualised graphically in map form in a GIS, the results of analysis and modelling can also be overlaid with other features to explore their potential relationships prior to the actual analysis. The graphic display of the results allows assessment of their reasonableness and provides clues as to how some parameterisation in the model or analysis can be improved to yield a more realistic and desirable outcome.

All sorts of spatial data can be stored in a GIS database and retrieved at will, and queried if the topological relationship among the entities is also stored. GIS is the ideal platform for implementing most spatial analyses and modelling, for three reasons:

(1) GIS databases contain all the data needed for spatial analysis. These data can exist in raster or vector form, and be transformed from one form to another. This platform is useful and important for organising data from different sources logically into layers that are essential in some spatial analysis. Apart from storing and integrating data, GIS offers efficient means of data input, including importation from other digital systems. GIS is also

good at preparing data for spatial analysis, such as data (re)projection, rescaling (dissolving), and buffering. Common to all of these processings is the change in the spatial extent of the data while minimal changes are made to their attributes. Therefore, these kinds of pure spatial operations without involving corresponding analysis of their attributes fall outside the scope of this book and will not be discussed further.

(2) GIS possesses strong and powerful analytical capabilities that can ease certain kinds of spatial analysis. For instance, GIS contains common spatial analysis tools that allow spatial analysis to be executed by clicking a few buttons. A wide range of functions are available for performing all kinds of spatial analysis, including spatial interpolation, neighbourhood, zonal, and surface analyses. Of these analyses, only interpolation will be covered in this book as the other analyses do not involve both the spatial and thematic components simultaneously. Of particular note is ArcGIS's spatial modelling capability. Multiple layers can be input into an operation to model the predictive variable without the intermediate results. Since all layers have been geo-referenced to the same coordinate system, ArcGIS allows the modelling to be done with maximal ease and speed. The operation can be scripted to implement repeat runs with different files or different parameterisation. Naturally, specific tasks of analysis can be implemented in some stand-alone packages. It is troublesome to run analyses involving multiple systems and to change data formats back and forth to circumvent data compatibility problems. In particular, GIS plays an important role in handling very large, complex data and in providing the integration of computer mapping and modelling approaches.

(3) GIS also contains a suite of powerful tools for statistical analysis of data in deriving the relationship between the dependent variable and its explanatory variables to be used in spatial modelling. Furthermore, it is possible to incorporate some stand-alone software into ArcGIS as an extension, for example Patch Analyst as an ArcGIS extension (see Chapter 6). This incorporation eradicates the need for data conversion as all the GIS-produced results can be analysed directly.

1.4.2   REMOTE SENSING

As a means of data acquisition, remote sensing serves as an important source of imagery and non-imagery raster data that have been properly geo-referenced. These data can be airborne or spaceborne, but are always current up to days, or even hours. They are essential to the periodic updating of the GIS database as sensors aboard satellites and aircraft are excellent at acquiring up-to-date imagery data. Through further processing, these data can be converted into useful information needed in some spatial analyses. For instance, detailed land cover information essential in certain spatial analysis and modelling can be produced quickly from remote sensing data. Current and detailed land cover information is vital in a number of applications, such as modelling forest fire risks, watershed hydraulic modelling, and assessing habitat quality in landscape ecology. In fact, remotely sensed data are indispensable in all fields of application that require land cover information. Additionally, remotely sensed imagery can also serve as the ground truth against which the accuracy of modelled or simulated outcomes can be compared to yield an accuracy of modelling, or in case of unreliable results, can guide the modification of model parameterisation. Non-imagery remote sensing data such as light detection and ranging (LiDAR) are also indispensable in all kinds of spatial analysis and modelling that require elevation-related information, such as modelling of sediment yield in an earthquake in relation to surface morphology. Finally, some remotely sensed data processing and analysis systems such as ERDAS Imagine have special built-in functions that enable certain kinds of spatial analysis and modelling to be undertaken in the raster environment with relative ease. For instance, the ModelBuilder in ERDAS allows cartographic modelling to be undertaken easily via overlay analysis of raster layers derived from respective types of remotely sensed data.

1.4.3 GPS

GPS is able to position features anywhere on the Earth's surface or even in the air at every given moment. This system is excellent at collecting data at fixed points and along lines, but physical accessibility to the site to be positioned is essential. Although GPS cannot perform any spatial analysis itself, it is still critically important to spatial analysis in three valuable aspects:

(1) This primarily positioning technology enables data from different sources to be geo-referenced to the same system, without which it is impossible to analyse them spatially when multi-source data are involved. GPS can generate universal coordinates for ground features, through which the geo-datum and coordinate systems adopted by data from different sources can be unified.

(2) GPS is essential in field data collection, although it is utterly inadequate to collect the attribute value of the variable of interest by itself. GPS can guide us to the site where samples are to be collected, and inform us whether the samples have been collected from the pre-designated positions. The coordinates of the collected samples can be determined by placing a GPS receiver near them. Through these coordinates, we are able to link the *in situ* observed properties to the values of other variables at the same location. Via the collected locational data, it is possible to associate the attribute values at the sampling sites with their corresponding spectral or radiometric properties on satellite imagery. If a sufficient number of such data points are collected in the field, then a statistical relationship between the two variables can be established. With the assistance of this relationship, it is possible to convert the satellite image to a map showing the spatial distribution of the sampled variable, thereby extending the point-based observations to the whole study area. This spatial perspective can reveal the spatial pattern of the variable under study (e.g., grassland biomass) and allow the

estimation of its attribute value quantitatively (e.g., total biomass on the ground).

(3) GPS is useful in validating analysis results in the field. Guided by a GPS, we can navigate to the check points in the field and compare the modelled results with the reality observed at the same or corresponding location in the field. Their degree of agreement can indicate the reliability of modelling.

1.4.4 COMPUTER SCIENCE

These days, it is unimaginable to manually perform some mathematical calculations that are indispensable in spatial analysis, such as the solution of multi-linear equations in determining the weights in spatial interpolation, let alone sophisticated spatial modelling and simulation. The alternative is to use computers that have found ubiquitous applications in every scientific endeavour, including spatial analysis. Nowadays, computers have permeated all aspects of our lives to such a degree that we cannot function properly without them. The same is also true with spatial analysis. It is not an exaggeration to claim that without computer science, there would be no spatial analysis. Certain types of computation are so complex that it is impossible to undertake them without the assistance of computing scripts. The type of spatial analysis we can perform digitally and the ease of analysis depend completely on the computer package we are using and its functionality. It is not surprising that spatial analysis did not really take off until the digital era, as discussed in Section 1.3. The advancements in computer technology have revolutionised spatial analysis. Intuitive, interactive graphics-driven computer operating systems have been developed to ease the implementation of spatial analysis and modelling. Powerful computer packages have become available to perform nearly all kinds of spatial analyses. Apart from computer hardware and packages, several scripting languages (e.g., JavaScript, Python, and R) have been developed for running tailored and niche types of spatial analysis in pioneering domains, such as diffusion modelling of contagious diseases in epidemiology, forest fire modelling in landscape ecology, and crowd behaviour modelling in sociology. Scripting languages are valuable and essential in running sophisticated modelling and in implementing transition rules absent from existing systems. Powerful computing software has enabled real-time visualisation of the analytical and spatiotemporal simulation results using fancy designs, and even as animations, which are essential to effectively communicate the outcome of the analysis and modelling to the wider audience and stakeholders.

1.4.5 STATISTICS

Statistics differ from the above disciplines in that it neither supplies the data needed in spatial analysis nor serves as a platform for performing or easing spatial analysis and modelling. Instead, statistical information and theories can guide certain types of spatial analysis (e.g., inferential analysis), dictate the type of sensible analysis (e.g., univariate or multi-variate regression), and provide theoretical backing for the appropriateness of some analyses, such as data independency and randomness. Without a good understanding of basic statistics, or guidance by the underpinning theories, some spatial analyses could be flawed, and the results generated misleading. Thus, statistics can be regarded as the core or soul of (inferential) spatial analysis. This is because, to a large degree, spatial analysis just represents the extension of statistics to the spatial domain. Therefore, any requirements in statistical analysis are equally relevant in and applicable to spatial analysis. Statistics is particularly indispensable to inferential spatial analysis, in which hypotheses can be either rejected or accepted at certain confidence levels following some kind of statistical test. Knowledge of probability and data distribution is valuable in carrying out inferential tests on the spatial properties of data, such as the randomness of point patterns. Finally, rudimentary statistic knowledge is vital in understanding common terms used in spatial analysis. Possession of sound knowledge in statistics is important to interpret the analytical results properly, especially with regard to the generalisation of the generated findings. Therefore, competent statistical knowledge and numeracy skills are imperative to carry out sound and sensible spatial analysis.

1.5 TARGETS OF ANALYSIS

In spite of the wide range of fields to which spatial analysis has been applied, as discussed in Section 1.2.3, surprisingly, the targets of spatial analysis are rather limited in their scope and quantity. In fact, there are only three primary targets or elements of spatial analysis, namely, spatial relationship, spatial pattern, and spatial process.

1.5.1 SPATIAL RELATIONSHIP

Spatial relationship refers to the positioning of one spatial entity in relation to another, or the spatial arrangement of a group of entities in the same geo-referencing system. They may be of the same nature or different types. Such entities can be point, linear, or areal features (Figure 1.2), or a combination of them. When a large group of spatial entities exist in space, they can form various spatial relationships among them, such as proximity to each other and adjacency. One particular proximity relationship is the nearest neighbour, which could be the nearest highway (line) from a hospital (point). Apart from adjacency, the spatial relationship can be bordering, falling within (e.g., an island in a lake), overlapping, or belonging (e.g., a forest boundary is made up of a road) (Figure 1.2). In the last instance, the spatial relationship is formed between two types of spatial entities: a linear feature comprising an areal feature. A spatial relationship can also exist between points and areas, such as the number of COVID-19 infection cases in a suburb. The spatial relationship among features of the same type can be described as dispersed or compact, even, clustered, or random. For linear features, the relationship can be intersecting and parallel (Figure 1.2). Typical spatial relationships of polygons may be described as intersecting, overlapping, enclosing, and adjacent. Such diverse kinds of spatial relationship require a large variety of methods of spatial analysis to study them, and they can produce a whole range of spatial analysis metrics. Which of these spatial relationships should be the target of analysis depends on the purpose of undertaking spatial analysis in the first place.

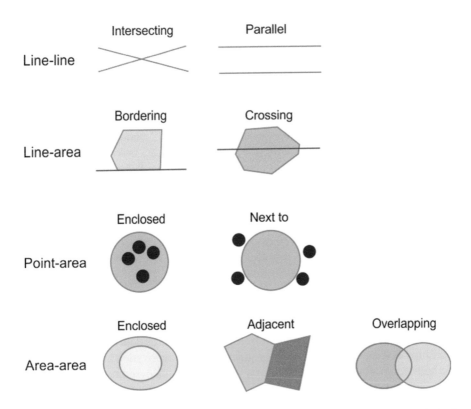

FIGURE 1.2 Typical spatial relationships between spatial entities of different typological dimensions.

Typical spatial relationships among spatial entities of different topological dimensions also exert an influence on how they should be analysed spatially (Table 1.3). Not all spatial relationships are equally common in reality. Some of them will be analysed more frequently than others. Of the six possible types of relationship, point–point, point–area, and area–area relationships are the most commonly analysed. For instance, the point–point relationship is analysed to determine commuting length and duration, such as from a residence to the nearest bus stop. It is also analysed to study the degree of spatial dispersion of residents surrounding a school so as to minimise the commuting of students attending it. In contrast, line–line and line–area relationships are the least commonly analysed because they are relevant to only a narrow range of application fields (e.g., hydrology) that require different methods of analysis (e.g., network analysis).

The point-in-area relationship is commonly analysed to determine density, such as the number of houses in a suburb, population density, and the distribution of schools in different suburbs of a city. Such information can show whether schools have a spatially balanced distribution and whether there is any gap in the distribution caused by population growth. Virtually an extension of the point–point relationship,

TABLE 1.3

Major Relationships between Spatial Entities of Different Topological Dimensions and the Types of Spatial Analysis that Can Be Performed on Them

Entity 1	Entity 2	Examples	Objects of spatial analysis
Point	Point	Distance between a residential dwelling and a school	Nearest neighbour, central tendency, clustering
Point	Line	Distance between a petrol station and the nearest motorway	Proximity
Point	Area	Number of schools in a suburb; the number of COVID-19 cases in a city	Enclosure, density, evenness, point pattern and distribution
Line	Line	Relationship between a river and a motorway	Intersection, connectivity
Line	Area	Road segments within a buffer zone, length of a bridge over a lake	Passing, intersection, length, sinuosity
Area	Area	A greenfield site in an urban area, a forest adjoining a lake	Overlap, within, adjacency

the area–area relationship is commonly analysed for census data to identify the relationship between two factors, such as socioeconomic deprivation and physical well-being/life expectancy. In epidemiology, it can also be analysed to study the spatial association between the proportion of green fields and respiratory illness cases that must be enumerated over a spatial extent.

It must be noted that all the aforementioned spatial relationships and their applications are valid for spatial entities represented in the vector format. In the raster format, spatial relationship does not exist at the object level as all spatial entities are invariably represented as grid cells of uniform size and shape. Thus, it can occur only between two variables, such as vegetative cover and elevation (Figure 1.3). This kind of spatial relationship is global and spatially varying, and has to be analysed statistically. Spatial relationships between entities (or objects) and those between spatial variables require different methods of analysis.

1.5.2 SPATIAL PATTERN AND STRUCTURE

Spatial pattern refers to the spatial arrangement or distribution of a group of spatial entities of either identical or different identities. Pattern implies some kind of unique and regular repetitions or occurrences of the same entities (Figure 1.4). Spatial patterns can be analysed for point, linear, or areal features. Common point patterns are described as clustered, random, or dispersed. Spatial patterns are not so easy to analyse quantitatively as all patterns involve some kinds of geometric properties that are not easily quantifiable. In contrast, it is relatively easy to compare an observed pattern with some standard patterns to assess its nature (e.g., clustered or dispersed). It is also possible to test whether the observed pattern conforms to a standard one statistically.

FIGURE 1.3   Spatial relationship between one variable (vegetative cover) and another (elevation). It is a global relationship between the two variables over the entire extent of study (source: Tom Patterson, with permission).

FIGURE 1.4   Spatial patterns of settlements in the north-western US as seen from a night-time satellite image (source: screen capture from https://worldview.earthdata.nasa.gov/).

Spatial structure refers to the spatial organisation of entities or features, or how they are distributed spatially. Most spatial phenomena do not have a definitive structure, except urban areas where functional entities are connected to each other. In ecology, it refers to the spatial organisation and distribution of communities. In spatially structured communities, sites near each other are compositionally more similar than more distant ones. Spatial structure differs from spatial relationship in that spatial entities are related to each other in one way or another. Spatial structure is seldom studied in spatial analysis.

1.5.3 SPATIAL PROCESS

Spatial process is defined as the temporal evolution or spatial movement of a phenomenon or spatial entity over an area or from a point, such as the spread of bushfires, the dispersal of seeds in the landscape, urban sprawl, and the spread of a plague (Figure 1.5). An epidemic plague can spread from one suburb to another in the same city. Vehicle-sourced air pollutants can be dispersed along artery roads to nearby densely populated residential areas. A wild fire can spread from the ignition point to an entire hillside. Common to all these spatial processes is the temporal component or the variation of the attribute with time. Such processes can be studied at a temporal increment ranging from seconds (e.g., forest fires) to decades (e.g., urban sprawl). The targeted spatial process of study may be pertinent to the same variable whose spatial extent is continuously changing over time (e.g., spatial diffusion), or it may be about the attribute of the same location at multiple times (e.g., spatial change). The former involves only one spatial variable (attribute) whose spatial extent may

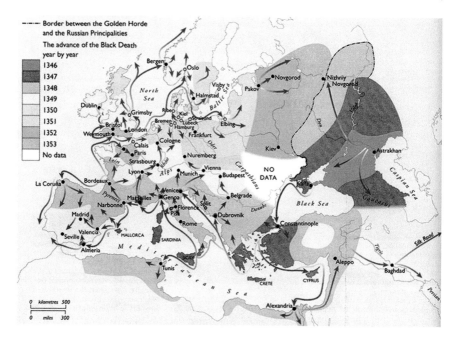

FIGURE 1.5 Spatial process involves time-dependent changes in spatial extent, such as the spread of the second pandemic of the Black Death in Europe (1346–1353). In this diagram, colour is used to illustrate the spatial extent in a given year, and arrows are used to indicate the direction of spatial propagation (e.g., spread by merchants and travellers) (source: Visual Capitalist, 2019, with permission).

expand or contract, such as urban sprawl. The latter implies mobility and transition between multiple states. For instance, the dispersal of seeds to a vacant spot can cause its colonisation by weeds, turning the bare ground into a vegetative cover in landscape ecology modelling. Spatially, the newly colonised spot may not be continuously adjoining an existing vegetation patch. A communicable disease may spread out spatially from one suburb to another in the same city. It is similar to fire spread in the landscape, except that all the new infectious cases are point entities, while the fire-burned areas are areal in nature. The colonisation of habitat by a certain species lies in between, in that it can be treated as the dispersal of the same seeds over some distance, or it can be perceived as changes in the areal extent if all the seeds are treated collectively and indiscriminately. In this case, the end result may be polygons representing the changed spatial extent of the habitat for a given type of invasive species (e.g., gorse).

Compared with spatial relationships and spatial patterns, spatial processes are much more challenging to understand and study, especially if they take place over a very long time (e.g., centuries), as it is almost impossible to verify the results of analysis and modelling. Therefore, how a spatial process is evolving is best studied via simulation, in which certain parameters or variables can be held constant so as to examine the effects of other variables. If some of the required inputs are unknown or impossible to obtain in the field (e.g., the clonal growth rate of sedges), then processes can only be studied via scenario analysis by assuming a value for them.

1.6 TYPES OF SPATIAL ANALYSIS

In terms of the nature of analysis, spatial data analysis and modelling fall into three categories – descriptive, inferential, and predictive – each having its own objectives, suitable methods of analysis, characteristics, and applicability to a unique type of data (Table 1.4).

1.6.1 DESCRIPTIVE SPATIAL ANALYSIS

Descriptive spatial analysis aims to come up with a quantitative measure or produce metrics using analytical tools to depict a spatial entity or its relationship with other entities, such as the shape of a polygon and its distance from the nearest neighbouring polygon. It is also possible to quantify the spatial pattern of a group of entities. The results from descriptive spatial analysis are all factual and replicable. They can help answer questions about a spatial entity. Take forest fire as an example. Descriptive analysis results may pertain to the portion of the burned area in the landscape, mean and standard deviation of the burned patch size, ratio of the burned patches to the total forest area, the shape of burned patches, and their distance from other burned patches. These results provide just a numerical value to indicate a special quality of the burned patches. Jointly, they can paint a general picture about the burning outcome and intensity. Through spatial quantification, it is possible to compare the effects of different forest fires.

TABLE 1.4
Common Approaches of Spatial Data Analysis and Their Main Features

Type of analysis	Main features/tasks	Examples
Descriptive analysis	Generation of quantitative metrics to indicate a degree or state	Habitat shape, tree density, and predominant wind directions
Exploratory analysis	Formulation and testing of hypotheses about the properties of a population from samples	Spatial distribution of trees in a watershed; dependency of crimes on urban design
Predictive analysis	Estimation of values at unsampled locations from neighbouring observations	Spatial distribution of air pollutant concentration predicted from a limited number of observations
Spatial modelling	Estimation of a quality or an attribute value across space from a number of inputs and/or nearby influences statistically	Monetary damage/loss to properties caused by an earthquake; risk of landslides; likelihood of forest fires
Spatial (temporal) simulation	Exploration of spatial processes under the influence of mobile agents	The propensity of grassland degradation and recovery under different grazing schemes; urban sprawl; bushfire spread over the landscape

Common descriptive spatial analyses can be carried out easily using existing Microsoft Windows-based computing systems by clicking on a few buttons after the spatial data have been input into the right format. A large number of metrics can be calculated in one session. In spite of this ease, however, caution must be exercised in interpreting the results of descriptive spatial analyses, especially when there is no one-to-one correspondence between a given quantitative measure and a spatial property. For instance, the same quantitative value of fractal dimension can be associated with numerous shapes. The main limitation of descriptive spatial analysis is that a single quantitative value is unable to fully capture the whole spatial property of an entity that is inherently spatial (e.g., two-dimensional) in nature. Some spatial entities are so complex spatially or geometrically that they cannot be realistically or accurately measured by a single number.

1.6.2 EXPLANATORY/INFERENTIAL SPATIAL ANALYSIS

Explanatory spatial analysis attempts to explain the occurrence of an event in which a number of factors may be at play, or how one variable affects another spatially. Closely related to explanatory spatial analysis is *exploratory spatial data analysis*, which aims to detect spatial properties of data and spatial patterns in them (Haining et al., 1998). Exploratory spatial analysis can reveal something about the population from which the samples of analysis are collected. Two very significant components of explanatory spatial analysis are the formulation of hypotheses about the spatial properties of an event and the assessment of spatial models to fit the observations. For instance, the importance of the considered factors or variables to the spread of a bushfire can be studied via explanatory spatial analysis. Assessment of spatial models can be achieved via *inferential spatial analysis*. It attempts to gauge the nature of distribution or deviation of the observations from some kinds of standard ones. The analysis focuses on the spatial patterns and geometric properties of spatial entities, especially a group of points or polygons, not so much on their attribute values. It can shed light on the spatial arrangement of observations by testing whether they conform to a standard pattern of distribution, such as random or clustered. Testing hypotheses regarding spatial patterns and relationships is the core of explanatory or confirmatory spatial analysis. How to formulate a hypothesis about a spatial pattern or process depends on its nature. For instance, hypotheses about a forest fire can be formulated from the perspective of fuel, climatic conditions, land use, and topography. In the test, the observed pattern is compared with some kind of standard distribution that is regarded as the benchmark. The hypothesis is either accepted or rejected following the test outcome. Inferential spatial analysis can be carried out for spatial relationships, spatial patterns, and spatial processes. For instance, we can test whether a point pattern is random or whether an observed spatial relationship is significant. Having said this, however, it must be borne in mind that the distinction between explanatory and exploratory analysis is not always so clear-cut. As an example, the study of the spatial pattern of burglary incidents can be either explanatory or exploratory, and the two terms can be used interchangeably in this context.

Both explanatory analysis and inferential spatial analysis face the same issue associated with data sampling. For explanatory spatial analysis, data have to be sampled spatially. In reality, the collected samples may be spatially correlated, which violates the assumption underlying traditional non-spatial data analysis. Another hurdle in inferential spatial analysis is that the data may actually represent the entire population. Samples themselves are also the population, so there is no need to infer about the properties of the population from which these samples are drawn. Similar to descriptive spatial analysis, both exploratory and explanatory approaches can only explore or describe spatial patterns and relationships, and are unable to study spatial processes. This task must rely on predictive spatial analysis, which can provide predictions about what is going to happen in the future under a given set of conditions.

1.6.3 PREDICTIVE SPATIAL ANALYSIS

As the name implies, *predictive spatial analysis* aims to estimate a value or come up with a quality of an attribute that is non-existent in space (spatial prediction) or at present (temporal prediction). Spatial prediction is the estimation of the attribute value from its nearby observations within the defined neighbourhood. This is commonly known as spatial interpolation, in which spatial predictions are made about only one variable (e.g., rainfall) from itself. Only its value at different locations is predicted from that of the nearby observations, without involving other variables. The predicted value barely varies with time, so the temporal component is absent from this kind of predictive analysis.

Spatial predictive analysis can also refer to the estimation of a new attribute value at a given location from that of its contributing variables at the same location, such as the risk of landslides. The risk is related to topography, soil, surface cover, underlying lithology, and external events (e.g., rainfall and earthquakes), all of which exert an influence on the risk in one way or another. The exact influence may remain unknown. Also unknown is how the influence of individual variables should be combined in the estimation. This kind of predictive analysis is commonly known as modelling or simulation, and is able to predict future patterns and relationships. Spatial modelling and simulation are good at studying spatial processes by holding all variables unchanged except one. They are able to forecast what is going to happen in the future. As such, they are much more powerful and flexible than simplistic spatial prediction as they can predict the spatiotemporal behaviour of the dependent variable. They also allow the independent variables to be modified in order to run scenario analysis. Instead of producing one deterministic outcome as in simple predictive spatial analysis, spatial simulation can yield time-series results based on the changed input in the format of "what … if …". Nevertheless, it is not so easy to validate temporal predictions as what is going to happen in the future is still unknown. A common practice is to model what has happened in the past, and then compare the modelled result with the present reality.

Of the three types of spatial analysis, temporal predictive analysis is the most complex and difficult to undertake as it usually involves more variables, some of which are difficult or impossible to parameterise. It is also painstaking to construct a model, initialise model parameter values, and validate the modelled outcome. The topic is so complex that it will be covered in a separate chapter (Chapter 7).

The temporal relationship between the three types of spatial analysis is summarised in Figure 1.6. Of the five analyses indicated, descriptive analysis is the most fundamental and is always performed first. Only after spatial metrics have been derived from the observed data is it possible to analyse the relationship between one of the variables and others. Based on the calculated metrics, hypotheses can be formulated about their distribution. Thus, inferential spatial analysis is usually carried out after descriptive spatial analysis. Predictive spatial analysis, especially spatially modelling and simulation, must be based on what has happened in the past, thus it can be undertaken only after descriptive analysis. Single-variable spatial predictive analysis can still benefit from descriptive analysis. For instance, the value of spatial auto-correlation can shed light on the accuracy of prediction. After predictive analysis, the residuals of prediction can also be analysed descriptively to examine whether they are spatially correlated. Exploratory analyses may also be required to establish the relationship between the dependent variable and its explanatory variables. The established relationship may be used in predictive spatial analysis, even repeatedly, as in the simulation of urban growth based on the relationship between population size and urban area. Thus, predictive spatial analysis is always performed last, preceded by descriptive and even explanatory analyses.

1.7 SYSTEMS FOR SPATIAL ANALYSIS

Spatial analysis and modelling can be carried out in a number of computing environments or platforms. Some of them are commercially available packages with excellent

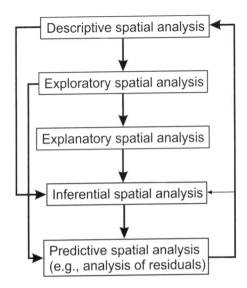

FIGURE 1.6 Relationship between different types of spatial analysis and the normal sequence of analysis.

technical support, while others are open source, developed by users themselves, with patchy and unreliable user support. In general, it is easier to seek assistance either online or from the user community with commercial packages than open-source platforms. Some of these computing systems can perform a comprehensive range of analysis, while others are designed for a niche field of application (Table 1.5). As far as capability is concerned, most systems allow fundamental spatial analyses to be undertaken. As a general rule, the more commonly performed spatial analyses can be realised in more platforms, and vice versa. Which system to choose is a matter of personal preference. However, the most sophisticated analysis or modelling has to rely on tailored systems. With the availability of so many mature systems, spatial analysis can be implemented with more ease and greater efficiency than ever before.

1.7.1 ARCGIS PRO

ArcGIS Pro is a desktop computer system designed to analyse and display spatially referenced data. It is a Windows-based, icon-driven GIS for analysing primarily vector data, with the functions of raster data analysis increasingly expanding and improving over the recent years. For many analysts, ArcGIS Pro is the default choice and preferred system for undertaking spatial analysis due to its longest history in existence, starting from the early 1990s. Primarily, as a vector-based system, ArcGIS Pro is good at analysing point, line, and polygon data. It contains a comprehensive suite of functions for processing spatially referenced data (Figure 1.7). Some standard functions are related to data compatibility transformation, such as raster to vector conversion and data projection transformation, changing enumeration units

TABLE 1.5
Comparison of Major Computing Systems (Environments) for Performing Spatial Analysis and Modelling

System	Main functions	Limitations
ArcGIS	A comprehensive range of functions for analysing vector data; easy integration of data from multiple sources	Limited but expanding raster data analysis capabilities; generic modelling only
ERDAS Image	A comprehensive range of functions for analysing raster data; rudimentary raster modelling possible	Difficult to incorporate non-spatial data; limited capability of analysing vector data
MATLAB®	Fulfilling niche fields of analysis; routine analysis possible with user-developed toolboxes	Unable to model; no user support; subject to availability of open-source code
NetLogo	Excellent for spatiotemporal dynamic modelling	Unable to incorporate external layers due to limited interface with other systems; scenario-based simulation only
QGIS	Open source; cross-platform; supports both vector and raster data; incorporates other analytical systems	Limited range of analysis and user support; inconsistent documentation
R(RStudio)	Versatile in implementing all kinds of analysis and modelling; flexible	Hard to learn; limited display functions; data preparation difficult

FIGURE 1.7   The ArcGIS Pro panel for vector data spatial analysis. Many more analytical functions can be searched in Geoprocessing. More raster data functions are available under Imagery.

from which statistical summary results may be derived and visualised in the tabular and graphic forms. Data may be prepared by running spatial operations that include split, merge (join), union, intersect, and buffer. Most of these spatial operations are characterised by the indiscriminate or identical treatment of all layers in the input, in which no spatial variation is taken into consideration. They are just manipulative processing of the spatial components of spatial data without generic and definitive objectives. For this reason, these spatial manipulations of the spatial database will not be further discussed hereafter.

Instead, the discussion will focus on the spatial modelling capability of ArcGIS Pro. In this regard, it supports efficient methods of data input, including import from other digital systems. In addition, it also supports alternative data models, particularly models of continuous spatial variations and conversions between them using effective methods of spatial interpolation. The other modelling capability of ArcGIS is spatial statistical analysis. It is able to transfer data to and from multiple analytical and modelling systems.

ArcGIS Pro is particularly strong at integrating data from different sources, a capability that is highly important in running spatial modelling that requires inputs of a large range of factors. ArcGIS Pro can perform a range of standard geometric operations, as well as spatial interpolation and surface analysis. It possesses an excellent and powerful suite of spatial analysis tools. However, when it comes to spatial modelling, the standard package may have a limited capacity. This capacity can be expanded by scripting the necessary operations using Python.

1.7.2   ERDAS IMAGINE

ERDAS Imagine is a computer system designed specifically for performing a wide range of analyses of imagery and topographic data (including LiDAR), usually in the multispectral domain. The most valuable contribution of ERDAS to spatial analysis is to supply current land cover information derived from analysis of satellite imagery and elevational data from LiDAR. Such data are essential in certain raster-based spatial analyses, especially in landscape ecology. ERDAS Imagine is excellent at handling raster data, but has a limited capability of processing vector data. Its spatial analysis functionality is confined to spatial filtering of raster data (e.g., majority filtering) and data transformation (e.g., principal component analysis). Another important function of ERDAS is its modelling environment, ModelBuilder, which allows multiple layers to be stacked or overlaid with each other in cartographic modelling. It has limited functions for descriptive spatial analysis.

1.7.3   RSTUDIO

RStudio is an open-source programming environment. This anarchic platform comprises hundreds of tools or functions developed by different users, some of which are relevant to spatial analysis, although scripting is essential in running the analysis. The type and variety of spatial analysis that can be performed using R or RStudio depend totally on how sophisticated the scripts are. Apart from writing the scripts oneself, R contains a comprehensive library of routines that can be used with minimal modifications. More importantly, many users have written a variety of scripts already, and these scripts are freely shared among the spatial analysis community around the world. Thus, it is not necessary to start scripting from scratch. The generic scripts can be downloaded and modified to suit the purposes and analytical needs of a given application perfectly, especially for rudimentary spatial analyses, such as calculation of spatial auto-correlation and geographically weighted regression analysis. However, more sophisticated, scenario-based analyses may require further investment of time and effort in producing a sophisticated and robust script.

So far, R has matured to such a degree that it is able to perform a number of spatial analyses with an ever-improving user-friendly interface (e.g., RStudio). In fact, a book has been dedicated to the topic (Bivand et al., 2013) which details how geostatistics can be performed using R scripts, even though the same analyses can be easily achieved in a commercial GIS, such as ArcGIS Pro, which offers better visualisation functions. Of particular note is R's capability in analysing spatial patterns of point data (see Chapter 4). R also contains functions for performing spatial interpolation and spatial-temporal modelling of disease propagation.

Of all the R functions, two are particularly pertinent to spatial analysis: **sp** (short for spatial) and **spatstat** (short for spatial statistics). The former is a special tool that provides classes and methods for handling points, lines, polygons, and grids. It enables easier interfaces and exchange of geo-referenced data with existing GIS packages, including support for coordinate (re)projection (some functions will not be available if **spatstat** is not initiated, as the functions have not been properly defined yet). **Spatstat** is an excellent tool for running exploratory and inferential analysis of point patterns, and for fitting models to the observed pattern to test their distribution.

A major drawback of R (or RStudio) is that it is command-driven. All spatial analyses have to be executed through codes, even though scripting has become easier with RStudio, in that prompts are provided to avoid typographic errors and incorrect expressions. While this scripting environment provides a high level of flexibility in running spatial analysis, users who have limited programming experience may find it rather daunting to grasp. Compared with commercial GIS systems such as ArcGIS Pro, R does not have a good display functionality. When the analytical results need to be visualised, the task is accomplished through a series of tedious commands to specify the display parameters, such as colour, size, shape, extent, and the attribute to be displayed. Thus, novice users will find spatial analysis using R to be rather slow, at least initially.

1.7.4 MATLAB®

Short for "matrix laboratory", MATLAB® is a multi-paradigm programming language and computing environment with symbolic computing capabilities developed by MathWorks. Apart from matrix manipulation, it is able to plot functions and data, implement algorithms, and create user interfaces. Although not a system designed specifically for performing spatial analysis, MATLAB has been explored widely in recent years for this purpose by various authors. They have produced a few toolboxes that can be used to perform certain spatial analyses (Table 1.6). Of these toolboxes, **spatialCopula** is a coupled spatial analysis toolbox that contains utilities for the analysis of spatially referenced data, such as parameter estimation, spatial interpolation, and visualisation (Kazianka, 2013). **TopoToolbox** contains a variety of MATLAB functions for both spatial and non-spatial numerical analysis (Schwanghart and Kuhn, 2010). This flexible and user-friendly software, whose codes are freely accessible (physiogeo.unibas.ch/topotoolbox), offers hydrologists and geomorphologists a tool for analysing digital elevation model (DEM) data with a focus on material fluxes and spatial variability of water, sediments, chemicals, and nutrients absent from commercial hydrological modelling packages. Its improved version, **TopoToolbox** 2, with a graphic user interface, allows visual exploration and interaction with DEMs (Schwanghart and Scherler, 2014). It runs much faster

TABLE 1.6

Comparison of Publicly Available MATLAB Toolboxes for Spatial Analysis and Modelling

Toolbox	Main use	Availability
Arc_Mat	Explanatory spatial analysis; visualisation	www.spatial-econometrics.com
spatialCopula	Spatial interpolation	fam.tuwien.ac.at/~hakazian/software .html
TecDEM	Topographic analysis; extraction of drainage networks and sub-basins	www.iamg.org/CGEditor/index.htm or www.rsg.tu-freiberg.de
TopoToolbox 2	Topographic and hydrological analysis and modelling	physiogeo.unibas.ch/topotoolbox

and more memory-efficiently than its predecessor, and lays the framework for building hydrological and geomorphological models in MATLAB.

Arc _ Mat is another MATLAB-based spatial data analysis toolbox for basic choropleth mapping and exploratory spatial data analysis (Liu and LeSage, 2010). It is able to generate exploratory views of spatial data graphically, such as histogram, Moran scatterplot, and three-dimensional plots, density distribution, and parallel coordinates. It is also able to perform spatial data modelling via the extensive Spatial Econometrics Toolbox functions. **TecDEM** is a similar MATLAB toolbox designed for hydrological analysis and modelling. It can extract tectonic and geomorphologic information from global DEMs, drainage networks, and sub-basins, generate morphometric maps, and calculate basin parameters such as isobase, incision, drainage density, and surface roughness, and draw basin hypsometry (Shahzad and Gloaguen, 2011a). It can also determine flow directions, vectorise streams, delineate watershed boundaries, label Strahler orders, generate stream profiles, select knickpoints, and calculate concavity, steepness, and Hack indices (Shahzad and Gloaguen, 2011b). Some of these functions are not available directly in the existing GIS analysis packages, so have to be run separately outside them.

1.7.5 NETLOGO

As an object-oriented programming environment, NetLogo is able to model and simulate a wide range of spatial phenomena that involve mobile agents. They can roam the modelling space under certain rules or instructions from the modeller. This system is particularly strong at running predictive spatiotemporal modelling and in exploring the connection between the micro-level behaviour of agents and the resultant macro-level spatial patterns emerging from the interactions among the agents, and between the agents and the environment. In the simulation, certain environmental parameters can be held constant, which enables the isolation of the effect of one variable from those of other variables. This powerful package can simulate not only the net outcome of a process, but also how the outcome is derived from an incremental stepwise simulation if needed.

This free and open-source system was initially authored by Wilensky (1999) and has been expanded, improved, and modified at the Centre for Connected Learning and Computer-based Modelling, Northwestern University, Evanston, Illinois. It can be run in a graphic environment in which the values of certain global variables can be entered using on-screen buttons. The values of other variables can also be changed by clicking and dragging the score bar. In this way, the analyst does not have to tamper with the script itself. Alternatively, values can be specified via text by right-clicking on the button. This ease of changing parameter values is crucial to running sensitivity analysis and scenario-based analysis. In addition, variables can be turned on or off via switches. This feature is rather handy in running scenario analyses in which some variables may be absent. The modelled results can be displayed graphically in real time on-screen if they are spatial, or as a line chart showing the statistical summary of the modelled outcome. If the results are displayed too quickly, ticks can be inserted into the script to force the display to update one at a time or for every increment.

NetLogo is a user-friendly simulation package that is easy to learn and use. It comes with a well-documented user manual in PDF format. Those users who have not had much experience with spatiotemporal modelling or simulation can become competent modellers after reading some chapters or by going through the given examples contained in the document.

REVIEW QUESTIONS

1. In your view, should spatial modelling (simulation) be included in the definition of spatial analysis? Why (not)?
2. Although geocomputation, spatial statistics, and spatial analysis are close to each other in their definitions, some analytical tasks fall uniquely to each of them. Use an example to illustrate each definition. Use another example to illustrate that spatial statistics and spatial analysis are identical.
3. What spatial relationship convinced Dr Snow that the Broad Street well was the culprit? How can it be derived using modern spatial analytical methods?
4. Is the ability to predict and explore alternatives confined exclusively to spatial simulation? How about spatial analysis and spatial modelling?
5. What are the common criteria used to separate the history of spatial analysis into four eras? Which of them is the most important?
6. Of the three targets of spatial analysis, which one is the easiest to quantify? Which one is the easiest to visualise?
7. Dissimilar to spatial modelling, spatial analysis always results in outcomes that are non-spatial. To what extent do you agree or disagree with this statement? Explain.
8. In your view, what is the relationship between descriptive spatial analysis and predictive spatial analysis?

REFERENCES

Bivand, R. S., Pebesma, E., and Gómez-Rubio, V. (2013) *Applied Spatial Data Analysis with R* (2nd edition). New York: Springer, p. 414.

Broome, F. R., and Meixler, D. B. (1990) The TIGER data base structure. *Cartography and Geographic Information Systems*, 17(1): 39–47, doi: 10.1559/152304090784005859.

Davis, J. C. (1986) *Statistics and Data Analysis in Geology* (2nd edition). New York: John Wiley & Sons, p. 646.

ESRI GIS Dictionary. https://support.esri.com/en/other-resources/gis-dictionary.

Goodchild, M. F. (1987) Towards an enumeration and classification of GIS functions. *Proceedings/CIS 87: The Research Agenda*, R. T. Aangeenbrug and Y. M. Schiffman (eds.), Vol. 11. Washington, DC: NASA, pp. 67–77.

Goodchild, M. F. (2003) Geographic information science and systems for environmental management. *Annual Review of Environment and Resources*, 28: 493–519.

Goodchild, M. F. (2018) Reimagining the history of GIS. *Annals of GIS*, 1–8, doi: 10.1080/19475683.2018.1424737.

Haining, R. (1990) *Spatial Data Analysis in the Social and Environmental Sciences*. Cambridge: Cambridge University Press, p. 409.

Haining, R. (2003). *Spatial Data Analysis: Theory and Practice*. Cambridge: Cambridge University Press, p. 432.

Haining, R., Wise, S., and Ma, J. (1998) Exploratory spatial data analysis in a geographic information system environment. *The Statistician*, 47(3): 457–69.

Hawkins, D. M., and Cressie, N. (1984) Robust kriging-a proposal. *Journal of the International Association for Mathematical Geology*, 16: 3–18.

Jiang, S., Ferreira, J., and Gonzalez, M. C. (2017) Activity-based human mobility patterns inferred from mobile phone data: A case study of Singapore. *IEEE Transactions on Big Data*, 3(2): 208–19, doi: 10.1109/TBDATA.2016.2631141.

Kazianka, H. (2013) SpatialCopula: A Matlab toolbox for copula-based spatial analysis. *Stochastic Environmental Research and Risk Assessment*, 27: 121–35, doi: 10.1007/s00477-012-0571-3.

Liu, X., and LeSage, J. (2010) Arc_Mat: A Matlab-based spatial data analysis toolbox. *Journal of Geographical Systems*, 12: 69–87, doi: 10.1007/s10109-009-0096-6.

Longley, P. A., Goodchild, M. F., Maguire, D. J., and Rhind, D. W. (2005) *Geographic Information Systems and Science* (2nd edition). Chichester, UK: John Wiley, p. 560.

Matheron, G. (1971) *The Theory of Regionalized Variables and its Applications*. École national supérieure des mines, p. 211.

Openshaw, S., and Abrahart, R. J. (1996) Geocomputation. *Proc. 1st International Conference on GeoComputation*, R. J. Abrahart (ed.) Leeds, UK: University of Leeds, UK, pp. 665–666.

Patterson, T., How to drape a satellite image onto a DEM in Bryce 5. http://www.shadedrelief.com/drape/, accessed on 26 May 2021.

Schwanghart, W., and Kuhn, N. J. (2010). TopoToolbox: A set of Matlab functions for topographic analysis. *Environmental Modelling & Software*, 25(6): 770–81, doi: 10.1016/j.envsoft.2009.12.002.

Schwanghart, W., and Scherler, D. (2014) TopoToolbox 2 – MATLAB-based software for topographic analysis and modeling in Earth surface sciences. *Earth Surface Dynamics*, 2: 1–7, doi: 10.5194/esurf-2-1-2014.

Shahzad, F., and Gloaguen, R. (2011a) TecDEM: A MATLAB based toolbox for tectonic geomorphology, Part 1: Drainage network preprocessing and stream profile analysis. *Computers & Geoscience*, 37(2): 250–60, doi: 10.1016/j.cageo.2010.06.008.

Shahzad, F., and Gloaguen, R. (2011b) TecDEM: A MATLAB based toolbox for tectonic geomorphology, Part 2: Surface dynamics and basin analysis. *Computers & Geosciences*, 37(2): 261–71, doi: 10.1016/j.cageo.2010.06.009.

Tobler, W. (1970) A computer movie simulating urban growth in the Detroit region. *Economic Geography* 46: 234–40.

Tomlinson, R. F. (1970) Environmental Information Systems. *Proceedings of the UNESCO/ IGU First Symposium on Geographical Information Systems, Ottawa, September 1970.* Ottawa: International Geographical Union, Commission on Geographical Data Sensing and Processing.

Visual Capitalist (2019) The History of the World in One Video. https://www.visualcapitalist.com/the-history-of-the-world-in-one-video/, accessed on 8 June 2021.

Wilensky, U. (1999) NetLogo User Manual. http://ccl.northwestern.edu/netlogo/.

2 Space in Spatial Analysis

In order to be analysed spatially, all data must have their geographic coordinates recorded. Coordinates themselves can show the location of spatial observations. A group of coordinates can reveal the shape of an entity, proximity between multiple entities, and dispersion for a group of spatial entities. This spatial component is so crucial to spatial analysis that it must be covered prior to any discussion of spatial analysis. In order to keep this chapter within a manageable length, only 2D space is presented here. The chapter will also discuss how different spatial entities are represented in 2D space, as the format of digital representation exerts a deterministic effect on the type of spatial analysis that can be performed on them. A number of important terms related to spatial analysis will also be introduced in this chapter, together with a few quantitative measures of space. This introduction is followed by spatial partitioning and spatial aggregation, two thorny issues that all spatial analysts have to grapple with frequently. The last section of this chapter is devoted to defining four rudimentary terms related to statistics and the accuracy of spatial data analysis. These definitions will lay a solid ground for the discussion in the subsequent chapters.

2.1 SPATIAL REFERENCE SYSTEMS

All spatial data encompass two essential components: thematic and geographic. The thematic component depicts the nature or properties of spatial entities. In most cases, it is the target of spatial analysis. The geographic or spatial component refers to the location of the observations. Only when these two components are properly linked to each other can some spatial analyses be undertaken on the attribute (thematic) data, and the derived analytical results are sensible. In some cases, it is possible for the spatial component to be the sole target of spatial analysis, such as shape, length, and spatial distribution. In the case of diverse data from multiple sources, the spatial component of all data must be standardised to a common reference system. There are two main systems for referencing the location of the collected spatial samples: local and global. Local reference systems are seldom used in spatial analysis as the collected data are routinely positioned using the Global Positioning System (GPS) nowadays, and GPS always makes use of the global system. Local systems are used only in spatial simulation, and will be covered after the global system in Section 2.1.3.

2.1.1 GEOID AND DATUM

For most spatial data, their horizontal coordinates are sufficient for spatial analysis and modelling. However, for topographic data, there must be a third dimension for representing height. Unlike horizontal distance, which can be referenced

DOI: 10.1201/9781003220527-2

from an arbitrarily selected origin, vertical relief must be referenced from a commonly accepted datum. The Earth's surface is highly irregular in shape, with the relief varying spatially in different parts of the world (Figure 2.1a). This irregularly shaped surface is difficult to depict and has to be modelled using a *geoid*. This model describes the shape of the Earth's surface under the influence of gravity and the Earth's rotation without considering the influences of winds and tides. When the surface is covered by water, it is an imaginary surface. For the land surface, it is irregular due to topographic relief. Because of the long-term effects of the Earth's rotation, the equatorial radius is longer than the polar one. Thus, such a surface cannot be precisely depicted mathematically as a sphere. This difficulty is overcome with the *ellipsoid* model. It is defined as a mathematically depictable surface that approximates the geoid, the truer, imperfect Earth's surface. Because of their relative simplicity, ellipsoids are commonly used as the reference surface for computing coordinates and converting them between different systems (Figure 2.1b), and for determining height. The vertical height at any terrestrial location, called *orthometric height*, is measured perpendicular to this ellipsoid if determined using a GPS receiver (Figure 2.2).

The value of the two uneven semi-axes (*a* and *b*) of the ellipsoid used to approximate the Earth's surface varies with the geodetic system (Figure 2.3). In the World Geodetic System 1984 (WGS84), one of the standard reference systems for altitude, the semi-major (*a*) and semi-minor (*b*) axes have a value of 6,378,137.0 m and 6,356,752.314 m, respectively. This commonly used vertical reference system is fully compatible with that of GPS as its origin is centred at the core of the Earth, the same as that used by GPS. Apart from this universal and popular system, a number of other systems are also in use by different countries, such as the New Zealand Geodetic Datum (NZGD). It was developed specifically for New Zealand, whose

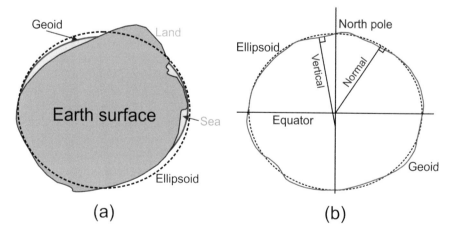

(a)   **(b)**

FIGURE 2.1   Approximation of the Earth's surface using a geoid (solid red line) and ellipsoid (dashed black): (a) relationship between geoid and ellipsoid for the land surface and sea; (b) reference of height from the geoid (solid red line) and ellipsoid (dashed black line).

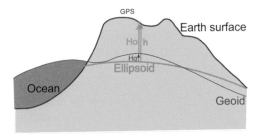

FIGURE 2.2   Relationship between geoid height (Hg), ellipsoidal height (h), and orthometric height Ho (h = Ho + Hg).

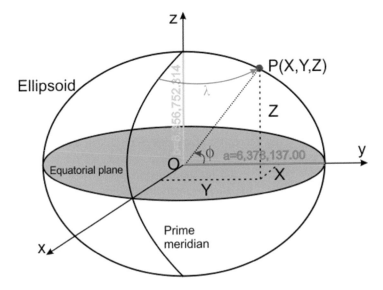

FIGURE 2.3   The angular coordinates (φ, λ) and metric coordinates (X, Y, Z) of point *P* in the reference frame of the World Geodetic System 1984.

uniquely shaped land mass is continuously shifting and deforming under the influence of the Australian and Pacific tectonic plates lying underneath, in contrast to WGS84, which is fixed to the Earth, and this tectonic influence causes the coordinates to continuously change. The NZGD has a few versions, one of which is called NZGD2000 and uses the 1980 geodetic reference system (GRS80) geoid. Its semimajor axis (*a*) has a value of 6,378,137 m, with the reciprocal of flattening (1/ε, ε = eccentricity) being 298.257222101.

When multiple layers of spatial data are used for analysis or modelling, it is essential to standardise their geodatum prior to analysis, in addition to their horizontal coordinates. This is especially critical in analysing elevational data, as the use of a different geodatum will lead to different heights directly.

2.1.2　Global Spherical System

The global spherical system is a 3D coordinate system for referencing geographical location that is inherently compatible with GPS. A pair of coordinates ((ϕ for latitude, λ for longitude) is needed to reference any points on the Earth's surface. In this system, the Earth is partitioned into two hemispheres of north and south from the equator to the pole, with latitude ranging from 0 to 90°N or S (Figure 2.3). The Earth is also divided into two hemispheres of east and west with longitude ranging from 0 to 180°E or W, starting from the prime meridian at the Greenwich Observatory. Although this angular, global system does not involve any distortions, it is cumbersome to use. A simple calculation of distance between any two points is complex as we need to take into consideration the hemisphere (north versus south) and curvature of the Earth surface. For this reason, this angular coordinate system is rarely used in spatial analysis. Instead, all coordinates in the curved 3D system are transformed into ones in a plane coordinate system mathematically. The use of plane coordinates expressed as (easting, northing, height) or (X, Y, Z) is the norm in spatial analysis.

2.1.3　Global Plane System

The global plane or 2D coordinate system references a location with a pair of horizontal coordinates, projected from the original 3D spherical surface. During the projection, inevitably some geometric properties of the Earth will be distorted, be they distance, direction (bearing), shape, or area. In spite of the huge number of projections that have been invented for the transformation, they are all subject to some sorts of distortions. Of the various methods available for coordinate projection, the most commonly used one is known as the Universal Transverse Mercator (UTM) projection, in which the globe is first intersected by a cylinder, a flattenable surface (Figure 2.4a). This cylinder is then rotated 90° (Figure 2.4b) and then spread out to form a 2D surface. As with all projections, the UTM projection also has its own distortions. Only the great circle intersecting with the globe, which is called the

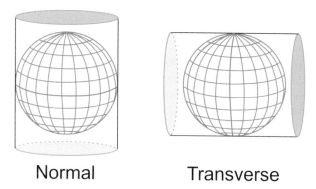

Normal　　　　　　　　　　Transverse

FIGURE 2.4　Transformation of the Earth's 3D surface into a 2D flattenable surface in the Universal Transverse Mercator projection.

central meridian, does not involve any distortion in length after being projected to the plane surface.

A number of measures can be implemented to suppress the geometric distortions to an acceptable magnitude that can be safely ignored in most cases. One effective way of suppressing the distortions is to divide the global 3D surface into longitudinal zones and project each zone individually and systematically (Figure 2.5). Naturally, the smaller the zones, the more accurately their curved 3D surface can be approximated by a flat, 2D surface. If the zones are sufficiently small, it is possible to confine the distortions to within an acceptable level at a given standard. In the UTM, the globe is partitioned into 60 zones of 6° each. During the projection to the global 2D system, the central meridian of each zone is located at 3°. Furthermore, geometric distortions are minimised by confining the projected surface to between 84°N and 80°S. As with all coordinate systems, all UTM coordinates are made positive by shifting the zonal origin to the west by 500,000 m (Figure 2.5). All zones are clearly identifiable by the zonal prefix preceding the easting coordinates. This zonal prefix can be easily determined from its longitude using Equation (2.1) after rounding up the ratio to the nearest integer. This zonal identifier needs to be removed from all easting coordinates when they cross multiple zones.

$$\text{Zone} = \text{longitude}/6 \tag{2.1}$$

For coordinates in the southern hemisphere, the northing coordinates are artificially increased by 10,000,000 m to render all coordinates positive (Figure 2.6). In undertaking global-scale spatial analysis, such conventions in rendering the coordinates in different hemispheres need to be taken into consideration to avoid inconsistent referencing of the coordinates. After being projected to the plane system, all UTM coordinates are metric and allow distance, area, and direction to be calculated easily,

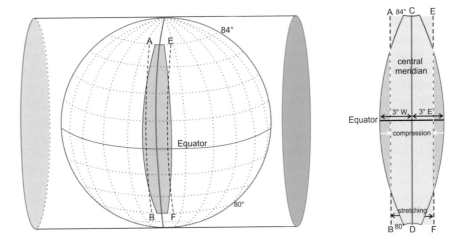

FIGURE 2.5 The projection of a 3D curved surface within a small zone (e.g., 6°) (left) to a cylinder surface of a uniform width (right) to minimise geometric distortions.

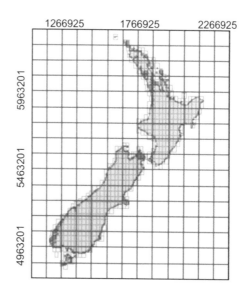

FIGURE 2.6 An example of projected plane coordinate system for referencing New Zealand. A pair of coordinates of easting and northing is sufficient to represent any points on the Earth's surface.

although the calculated results are subject to distortions caused during the 3D-to-2D transformation. Such distortions should not be a matter of concern as they are too small to exert any noticeable influence on the calculated results (e.g., the GPS-logged coordinates themselves may have a much higher uncertainty than these distortions).

The global system is commonly used in most spatial analyses, but is not suitable for local-scale spatial analyses and modelling. For this reason, it is worthwhile to introduce the local coordinate system briefly here. As with all plane coordinate systems, a local reference system is a modified version of the Cartesian system with two axes representing easting and northing (Figure 2.7a). It is local because the origin can be anywhere, usually arbitrarily set (Figure 2.7b). Like its global counterpart, the local system has a false origin located in the lower left corner so as to render all coordinates positive (e.g., only the first quadrant of the Cartesian system). This is because negative coordinates are inconvenient to handle in spatial analysis. For instance, in calculating the distance between two points from a pair of coordinates, it remains uncertain whether they should be subtracted from each other (all being positive) or added up (one coordinate is positive, and another negative), depending on the sign. This local coordinate system has a slightly varying version for raster data, such as scanned photos, satellite images, and digital elevation models (DEMs). In this case, the origin is located in the upper left corner, and the vertical axis increments downward instead of the usual upward. Local coordinate systems are not useful and are difficult to handle in most spatial analyses as it is impossible to transform all the input layers to a commonly adopted system. Thus, it is better to make use of the global system to reference all the geospatial data unless absolutely necessary.

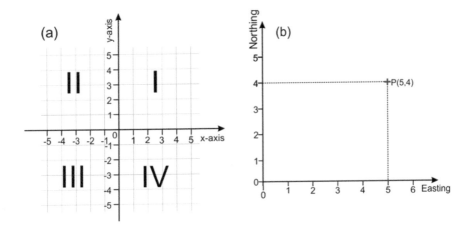

FIGURE 2.7 Comparison of the Cartesian coordinate system (a) and a local geographic coordinate system (b). In the Cartesian system, space is divided into four quadrants. The local geographic coordinate system has only one quadrant, and all coordinates are positive.

2.1.4 CONVERSION BETWEEN 3D AND 2D COORDINATES

When GPS-acquired spatial data are downloaded from a data logger, it is relatively easy to save the output in a plane coordinate system to conform to that used by most other geospatial data already saved. However, when only angular coordinates are available, they need to be converted into the metric format. The mathematical conversion of the angular geodetic coordinates expressed as (ϕ, λ) to metric coordinates of (X, Y, Z) (Figure 2.3) can be accomplished using the following formula:

$$\begin{cases} X = \left(\dfrac{a}{\sqrt{1-\left(\varepsilon\sin\phi\right)^2}} + h \right)\cos\phi\cos\lambda \\[4mm] Y = \left(\dfrac{a}{\sqrt{1-\left(\varepsilon\sin\phi\right)^2}} + h \right)\cos\phi\sin\lambda \\[4mm] Z = \left[\dfrac{a\left(1-e^2\right)}{\sqrt{1-\left(\varepsilon\sin\phi\right)^2}} + h \right]\sin\phi \end{cases} \qquad (2.2)$$

where ε refers to the eccentricity of the ellipsoid ($\varepsilon^2 = (a^2-b^2)/a^2$), and a and b are the major and minor semi-axis radiuses of the Earth, respectively.

Implementation of the above conversion equations is easily accomplished with many online systems. When only a few pairs of coordinates need to be converted, the conversion can be done manually at the website www.Earthpoint.us/BatchConvert. aspx. This online system allows the latitude and longitude coordinates expressed

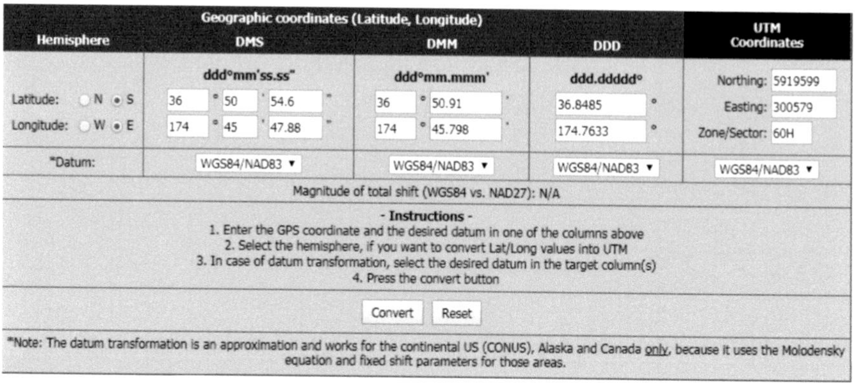

Hemisphere	Geographic coordinates (Latitude, Longitude)			UTM Coordinates
	DMS	DMM	DDD	
	ddd°mm'ss.ss"	ddd°mm.mmm'	ddd.ddddd°	
Latitude: ○ N ● S	36 ° 50 ' 54.6 "	36 ° 50.91 '	36.8485 °	Northing: 5919599
Longitude: ○ W ● E	174 ° 45 ' 47.88 "	174 ° 45.798 '	174.7633 °	Easting: 300579
				Zone/Sector: 60H
°Datum:	WGS84/NAD83 ▼	WGS84/NAD83 ▼	WGS84/NAD83 ▼	WGS84/NAD83 ▼
	Magnitude of total shift (WGS84 vs. NAD27): N/A			
	- Instructions - 1. Enter the GPS coordinate and the desired datum in one of the columns above 2. Select the hemisphere, if you want to convert Lat/Long values into UTM 3. In case of datum transformation, select the desired datum in the target column(s) 4. Press the convert button			
	Convert Reset			
°Note: The datum transformation is an approximation and works for the continental US (CONUS), Alaska and Canada <u>only</u>, because it uses the Molodensky equation and fixed shift parameters for those areas.				

FIGURE 2.8   The online coordinate conversion interface screen captured at www.Earthpoint. us/BatchConvert.aspx. The angular coordinates can be expressed as degrees with decimal points or as degrees–minutes–seconds.

in degrees-minutes-seconds (DMS) or decimal degrees to be converted to various coordinates (Figure 2.8). When a large number of coordinates need conversion, the batch mode can be used. Alternatively, the coordinates in degrees can be input into ArcGIS as points and saved as a point shapefile. The saved file can then be "projected" to any system supported by the software.

2.2   PROPERTIES OF SPATIAL ENTITIES

2.2.1   COMPONENTS OF SPATIAL ENTITIES

Spatial entities or objects exist in the 3D continuous expanse extending in all directions. The complete properties of an entity to be analysed may be expressed as:

$$(E, N, ht, Z, t) \tag{2.3}$$

where E (easting) and N (northing) refer to its horizontal position identical to X and Y, and ht (height) represents its elevation above the reference geodatum. These three parameters define the 3D position of point features or the centroid of areal (e.g., polygon) features. When only a flat 2D space is the target of analysis, ht automatically becomes zero and is excluded from consideration. Z is the attribute at the given location. In reality, it can be a function of (E, N). Namely, this attribute has a spatially varying value. In all spatial analyses, Z is always present, even though it may not be the target of analysis; t stands for the time at which the entity's attribute Z is observed or measured. In almost all spatial analyses, time is treated as invariable or sliced. In time slice analyses, time is treated as a special component, in that the same analysis is repeated numerous times, each time for a unique Z value. Multiple results are output at the specified temporal interval. Only in dynamic simulation that inevitably involves temporal variation is time explicitly treated as an extra dimension.

2.2.2 OBJECT VIEW OF SPATIAL ENTITIES

There are two fundamental means of representing spatial entities digitally, object view and field view (Figure 2.9). In the object view, space is regarded as continuous, and objects, be they points, lines, or areas, are clearly identifiable (Figure 2.9a). Point entities are represented as a single pair of coordinates without shape, area, or perimeter. Linear entities are represented by a string of coordinates in which any two adjacent pairs of coordinates are connected with a straight line. Linear entities have length, shape, and sinuosity, but not area. Line sinuosity can be used to characterise its complexity. Areal entities are identically represented as lines with a string of coordinate pairs, except that the first pair of coordinates is a replicate of the last

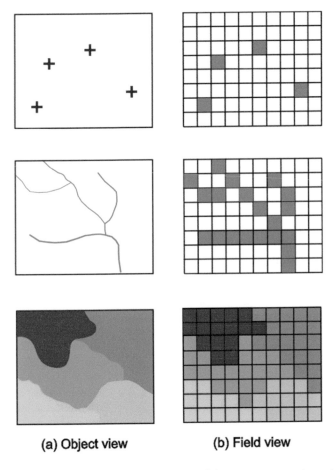

(a) Object view   **(b) Field view**

FIGURE 2.9 Comparison of the object view (a) of feature representation with the field view (b) of feature representation. In the object view of representation, space is discrete with objects distributed within it. In the field view of representation, space is partitioned into regular grids and the attribute of a grid is represented by its grey level or pixel value.

pair, indicating that an area is enclosed by the line. Areal entities are commonly called polygons in geographic information science (GIS) or patches in landscape ecology. They have the most comprehensive spatial properties among the three types of objects. Apart from perimeter length, area, and shape, areal entities also have compactness.

The object view of representation is compact as coordinates are needed only where entities are present. This view of representation is accurate. However, certain kinds of spatial analysis are difficult to implement with object (vector) data if the attribute of study varies continuously in space, such as elevation. In this case, extensive and intensive computations have to be carried out first to estimate the values at spots where no observations exist. This behind-the-screen interpolation can slow down the analysis considerably. Thus, it is better to convert vector data into raster data via rasterisation, during which only the cell size needs to be specified. Another disadvantage of the object view is the need to maintain a large overhead prescribing the topological relationship among all spatial entities and their compositional elements, such as belonging, adjacency, connectivity, and so on. However, the plus side of topological information is the ability to search and spatially query the data (Table 2.1).

2.2.3   Field View of Spatial Entities

According to the field view of representation, space is thought to comprise regular grids of a uniform size, shape, and orientation without any holes. A point is

---

TABLE 2.1

Comparison of the Object and Field Views of Spatial Entities and Their Best Uses in Spatial Analysis

Item	Object view	Field view
Philosophy of representation	Continuous 2D space, location explicit; representation of multiple attributes in one layer possible, with more columns in the tabular format	Discrete 2D space comprising regular grids of same size and shape; location implicit; one attribute per layer
Accuracy of representation	Precise, objects visible	Crude, accuracy subject to grid cell size; no objects, only pixels of different attribute values
Data volume	Compact, but overhead essential	Bulky, inefficient, and subject to grid cell size
Ease of updating	Complex (spatial component and attributes must be updated simultaneously)	Easy (only the concerned layer needs attention)
Best uses	Spatially discontinuous entities (e.g., roads); descriptive analysis; certain kinds of modelling requiring precise geometric properties (e.g., fire fronts)	Spatially varying attributes continuously; spatial modelling

represented as a single cell, a line as a string of cells, and an area as an array of cells (Figure 2.9b). This kind of representation is crude and inefficient, as data are still retained for places where no entities exist or where the attribute value does not change (e.g., a flat surface with little variation in elevation). In this kind of raster representation, entity location is implicit as it is not stored in the database. Instead, only the coordinates of the origin are stored, because the coordinates of all other cells can be inferred from their relative position to the origin, such as the number of rows and the number of columns from it (the cell size is known and uniform). The thematic attribute of a cell is represented by the cell value. The raster data can represent an imaginary or real (e.g., topographic) surface. This view of representation is the best for those variables that are spatially varying, such as soil moisture, temperature, and air pollutant concentration. The data can be easily updated as newly available layers can be added to the database without impacting any other layers already stored.

However, this field view of representation is utterly incapable of spatial search and query (e.g., finding the nearby observations) because the world is perceived to comprise an exhaustive array of cells without any objects. The accuracy of representation is affected by the cell size adopted. A smaller cell size allows fine details to be preserved, but may result in a bulky dataset. Since each cell is allowed to have only one value, multiple layers are needed to represent different aspects of the same spatial entity. For instance, a layer is needed to represent elevation, while another layer is needed to represent soil organic matter content. Despite these limitations, the field view of representation is still popular with certain kinds of spatial analyses, especially modelling. Besides, it is inherently compatible with satellite images that can serve as valuable inputs into some spatial analyses.

2.2.4 Dimension of Spatial Entities

In the object view of the world, the three fundamental types of objects have their unique topological dimensions. Point entities are 0D observations at particular spots, such as fire hydrants, schools, and bank branches. They can be easily located with a GPS receiver in the field. They are the simplest spatial entities to represent, requiring only a pair of coordinates. Linear entities such as rivers and roads are 1D features. They can be mapped with a GPS. 2D observations are made over an area. The most common 2D data are exemplified by socioeconomic attributes enumerated over a census track, such as population in a suburb. There is no specification as to what the minimum enumeration size should be. In certain kinds of spatial analysis, it is necessary to convert 2D entities to 0D entities, such as visualisation of population density over space. Population density is inherently areal in nature. In order to visualise it, the centroid of the enumeration unit is treated as the point from which the attribute value is sampled. In this way, a map showing the spatially continuous distribution of population density can be produced.

It is important to differentiate the topological dimension of different spatial entities as it has profound repercussions for spatial analysis. Data of different topological dimensions dictate the type of spatial analyses that can be performed on them sensibly (Table 2.2). For instance, point data are usually treated collectively in an analysis.

TABLE 2.2

Comparison of the Topological Dimension of Spatial Entities and the Types of Spatial Analysis That Can Be Performed on Them

Dimension	Entity	Examples	Analysis possible
0	Points	Schools, trees, crash spots, pollutant concentration, vehicles, soil pH and moisture	Pattern, density, nearest neighbour, dependency
1	Lines	Fault lines, wind directions, road networks	Length, sinuosity, fractal dimension, directional analysis
2	Polygons	Habitat, land cover, census units, surface represented as triangles, catchment area	Area, shape, compactness, density, spatial adjacency and proximity, diversity, evenness, juxtaposition, pattern analysis

It makes no sense to analyse individual points. In contrast, linear spatial entities can be analysed individually. For instance, analysis can be performed for a river channel to reveal its geometric properties, such as length, sinuosity, and fractal dimension. The attributes of linear entities can be stream runoff and discharge of a river channel, the traffic load of a road, and air pollution concentration along a route, to name just a few. Analysis of linear data, usually via network analysis, is far less common than analysis of areal data. Just like linear entities, 2D areal data can be analysed for individual polygons or a group of polygons collectively, and for more parameters, including shape, size, compactness, adjacency, and proximity. In particular, they are popular targets of analysis in landscape ecology, such as land use patches and habitats. Most of the analysis can be descriptive or inferential. In general, the higher the topological dimension of spatial entities, the more types of spatial analysis that can be performed on them.

If areal data are analysed collectively, descriptive spatial analysis metrics may include the mean and standard deviation of the geometric properties of spatial entities. Such analysis is especially significant in landscape ecology as a larger area (e.g., habitat size) may indicate more dominance in the landscape. For a group of areal entities, the two most important qualities are connectedness and barrier in studying the migratory behaviour of animals. Both refer to the physical traversability from one polygon (patch) to another. If two polygons are connected, there is no barrier between them. Animals can traverse them without any difficulty. Disconnection usually stems from barriers created by natural features (e.g., rivers and tall mountains) and infrastructure (e.g., highways) (Figure 2.10). Barriers are represented as polygons of different attributes in the digital land cover layer.

Finally, it merits mentioning that areas can overlap with each other, depending on their nature, especially those areas that share certain common constituents (e.g., a polygon of more than two land covers; overlapping habitats). A habitat (polygon) of kiwi can overlap with forest and shrub polygons if both types of vegetation have been inhabited by the bird. Certain kinds of rudimental spatial analysis, such as

FIGURE 2.10 Ecological barriers imposed by natural features (e.g., a river) or human-constructed infrastructure (e.g., a highway). They are commonly treated as elongated polygons in spatial analysis.

the overlay of multiple layers of different attributes (e.g., land cover maps and kiwi habitat maps), can yield rather insightful information on the preferred types of land covers inhabited by kiwi so that more efforts can be directed at protecting these land covers to ensure the long-term survival of the species.

2.2.5 SPATIAL ADJACENCY AND CONNECTIVITY

Spatial adjacency commonly applies to polygon entities. Its equivalency for point features is the shortest distance or nearest neighbour. When two polygons share a common boundary, they are spatially adjacent to each other, no matter how short it is. As shown in Figure 2.11a, polygons A and B and C are considered adjacent as they share one common boundary between any two of them. Furthermore, polygon C is adjacent to all the other four polygons. Polygons A and D are not adjacent as they do not share a common boundary. The concept of spatial adjacency is important to certain kinds of spatial analysis as it is synonymous with interactions and influences. When two polygons do not share a boundary, they do not interact with each other nor affect each other. In landscape ecology, two habitats represented as two polygons are traversable only when they are adjacent to each other without any ecological barriers between them. Spatial adjacency is also important to the analysis of spatial patterns. If polygons of the same identity are situated close to each other, they will form a spatially clustered pattern. Conversely, if they are spaced at a large distance from each other, they will form a dispersed pattern.

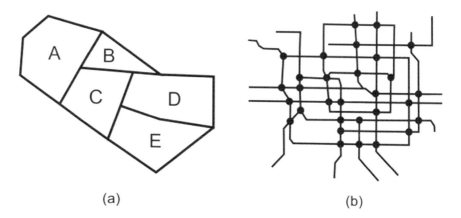

(a)   (b)

FIGURE 2.11   Spatial adjacency between polygons (a) and subway line connectivity (b). Two polygons are adjacent if they share a common boundary. Roads are connected if they are joined up via bridges or ramps.

It must be noted that spatial adjacency and even spatial intersection do not always equate to *spatial connectivity*, which means linkage between two adjacent entities, so one can traverse from one of them to the other. As illustrated in Figure 2.11a, polygons A and B may be adjacent, but they may not be connected if they are separated by a wired fence marking the section boundary in urban residential areas. In this case, it is not possible to traverse between them. Even if both polygons A and B are of the same type of habitat, they are not considered connected if separated by barbed wire. Connectivity is especially important to consider in analysing linear spatial entities, particularly road transport networks. As shown in Figure 2.11b, some of the subway routes may intersect with each other spatially, but they are not always connected. This means that passengers from one line cannot transfer to another. Connectivity can also occur between a highway and a local road. There must be an interchange between them for traffic to be connected, such as via an exit ramp from the highway to the local road or street. Connectivity plays an important role in routing and traffic modelling. The shortest route may prove untraversable if a road interchange is temporarily closed to traffic due to maintenance. Just like adjacency, connectivity can also signify interactions and influences. If two roads are not connected, their traffic volume does not affect each other. By default, river channels are always connected between the adjacent orders (e.g., first-order and second-order channels), but a channel itself can become disconnected by a dam, causing a hydrological disconnection between the upstream and downstream sections. Such disconnectivity must be factored into regional watershed–scale hydraulic modelling.

2.3   QUANTITATIVE MEASURES OF SPACE

In spatial analysis, it is frequently necessary to derive measures of space for a group of spatial entities. These measures serve as the evidence, based on which hypotheses may be tested and rejected. Commonly calculated and used measures of space include distance, proximity, neighbourhood, direction, and area.

2.3.1 DISTANCE

Distance is one of the most important and fundamental building blocks of other measures of space, from which more complex spatial measures can be derived. Distance can be defined in various ways, and one of the commonly used is the *Euclidean distance*. This refers to the shortest separation between two objects (points or even areas), represented by a straight line (Figure 2.12a). This distance is calculated from the coordinates of the two points, namely:

$$D = \sqrt{\Delta E^2 + \Delta N^2} = \sqrt{(E_B - E_A)^2 + (N_B - N_A)^2} \tag{2.4}$$

where E and N stand for the easting and northing coordinates, respectively. A and B in Figure 2.12 refer to the two points involved. This distance concept also applies to polygons (e.g., the distance between two cities), measured from their respective centroid.

It must be noted that the Euclidean distance, even though rather intuitive and easy to calculate, may not bear any relationship to the physical distance on the ground in the absence of a traversable road network between the two points. The alternative measure is the *block distance* or the *Mahalanobis distance* (Figure 2.12b). Variably known as the *Manhattan distance*, it is calculated by summing up the increment of coordinates in the easting and northing directions (Equation 2.5). The Mahalanobis distance measures the length along passable routes. It is longer and more complex to calculate than the Euclidean distance, and requires an additional layer of the road network for its determination, even though its calculation is much simpler mathematically. This measure of distance is more realistic than the Euclidean distance as one traverses the built environment in urban areas.

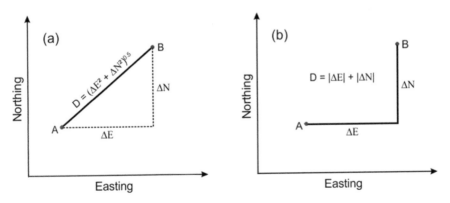

FIGURE 2.12 The concept of distance in spatial analysis: (a) Euclidean distance; (b) Mahalanobis distance.

$$D = |\Delta E| + |\Delta N| = |E_B - E_A| + |N_B - N_A| \qquad (2.5)$$

Distance plays a significant role in spatial analysis because of its three distinctive connotations:

(1) It is a measure of similarity, likelihood, or neighbourhood (nearest neighbour analysis), or even membership. As stated in Tobler's First Law of Geography, a shorter distance between two observations implies a higher similarity between their attribute values, while a longer distance suggests a lower likelihood of being a member of a group.

(2) It expresses a degree of interactions and influences (weight). Closely located observations have more chances of interacting with each other than distant ones. If two objects are too distant from each other, they may cease to interact with each other altogether. In this case, a distance beyond the threshold is not important any more. The threshold at which distance ceases to exert an influence on the neighbouring observations is unknown as it likely varies with the spatial variable under study. Nevertheless, the concept of threshold distance has seldom been studied except in geostatistics (which will be covered in Chapter 5). Associated with interactions is influence. Closer observations exert a stronger influence on the neighbouring observations than distant ones.

(3) It is indicative of the range of activities. A shorter distance can signify a narrower range of movement (e.g., migration), or a limited spatial extent centred around a dent. This is especially so when the target of study is mobile animals in ecologic analysis.

2.3.2 PROXIMITY

Proximity refers to the spacing among multiple spatial entities or objects in the Cartesian domain (Figure 2.13). It is measured by distance among the concerned entities. In general, those observations in close proximity to one another tend to have short distances between them. The lack of proximity is manifested by a large distance between neighbouring observations and is indicative of a dispersed distribution (Figure 2.13a). Nevertheless, no definite distance can be attached to proximity. Despite this ambiguity, proximity is the most useful in studying spatial interactions and influences. It is indicative of spatial adjacency and signifies the degree of influence by nearby observations in spatial correlation. Those observations in close proximity to one another have more chances of interacting (Figure 2.13b). Conversely, distant observations tend to interact less at a lower intensity. Similarly, those observations in closer proximity to one another exert a stronger influence on each other. Thus, in spatial interpolation, those neighbouring observations in close proximity to the point in question will count more in estimating its attribute value. However, the threshold of proximity is seldom defined. In other words, it remains unknown whether a neighbouring observation falls outside the proximity range, beyond which its influence on other observations ceases.

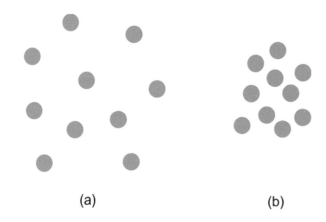

(a)   (b)

FIGURE 2.13 The concept of proximity in spatial analysis. It is measured by distance. The distribution is dispersed if observations are distant from each other (a). A group of observations are said to be in close proximity to each other if they are separated by a short distance from each other (b).

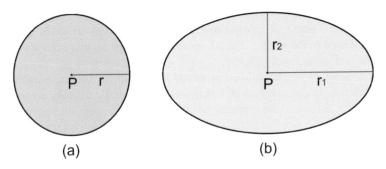

(a)   (b)

FIGURE 2.14 A neighbourhood centred at P, defined by a radius (r) is circular in shape (a), or oval in shape if defined by two radii (b) in the vector format.

2.3.3 NEIGHBOURHOOD

Certain types of spatial analysis require information on the neighbouring observations before they can proceed, such as in spatial interpolation. *Neighbourhood* refers to the spatial extent or range within which a spatial entity exists or should be considered. It can be defined in a number of ways, including continuity, distance, and general weights. An observation is considered a neighbour if it shares a common boundary with the observation in question. For instance, two measles cases within the same suburb can be considered as neighbours. In the object view of representation, the neighbourhood can be distance-based, defined by a radius, resulting in the neighbourhood always having a circular shape (Figure 2.14a). This shape can be made oval by defining the neighbourhood by two radii, one in the easting direction and another in the northing direction (Figure 2.14b). A circular neighbourhood means a uniform distance is applied to all qualifying neighbours, while an oval neighbourhood is defined by two distances. The neighbourhood centres at the point in question, and the radius length represents the neighbourhood size. The distance-based definition of neighbourhood is inclusive. All observations falling within the defined spatial extent (e.g., inside the circle or oval) are considered as neighbours, and as such, they will exert an influence on the observation in question.

However, this distance-based definition of neighbourhood is applied to the object view of representation only. In the field view of representation, neighbourhood is more complex to define as it involves both distance and adjacency. In terms of distance, neighbourhood can be defined by the window size surrounding the cell in question. This window is always square, and its size is usually an odd number, such as 3 by 3 or 5 by 5, with the central cell being the target of interest. Nevertheless, not all cells within this neighbourhood can be regarded as neighbours if additional criteria based on adjacency are imposed. Neighbourhood falls into two types of 4- and 8-connectivity in the raster format (Figure 2.15). Four-connectivity neighbours are those cells lying immediately to the left, right, above, and below the target cell, all located at exactly the same distance to the cell in question (Figure 2.15a). In addition to these cells, the 8-connectivity neighbours also include four diagonally adjacent ones that are slightly further away from the target cell than the four cardinal neighbours (Figure 2.15b). Thus, the distance between the neighbours and the target cell is variable. There is no rule governing which connectivity should be adopted for analysis. It is up to the analyst. The adoption of one connectivity versus another may affect the analytical results slightly.

The concept of neighbourhood is vitally important in running window-based spatial analysis in the raster form. For instance, in spatially filtering an image, the target pixel is subject to the influence of pixels within the specified neighbourhood, and neighbourhood size is critical to the analysis outcome. In certain types of spatial analysis, only those observations falling inside the defined neighbourhood are

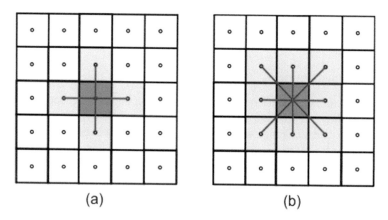

(a)   (b)

FIGURE 2.15   The concept of 4-connectivity (a) and 8-connectivity (b) neighbourhood (light shade) centred around the target cell (dark shade) in the field view of representation. Dots = cells; red lines = the distances to the neighbours. The neighbourhood is jointly determined by distance and adjacency.

considered because they are thought to exert an influence on the target cell, while those falling outside the neighbourhood are ignored. In this case, a neighbourhood can be used to demarcate the sphere of influences. Once a neighbourhood size is defined, statistical parameters may be derived based on all the neighbours, such as majority (for categorical data), mean, and standard deviation.

2.3.4 DISPERSION AND CLUSTERING

Dispersion refers to the wide scattering of observations over the area of study. A dispersed distribution is characterised by a lack of compactness for a group of observations of the same attribute (Figure 2.16). The observations spread out so widely that they occupy the entire area without obvious gaps. *Clustering*, in contrast, refers to the spatial grouping of certain observations that are much closer to each other than to the members of another group. Dispersion and clustering are analogous to the two sides of the same coin, in that the lack of clustering is manifested as dispersion, and vice versa. Thus, both can be defined by the distance among neighbouring observations. If all observations are widely scattered with a roughly equal distance between the nearest neighbours, they are said to be uniformly dispersed in their distribution (Figure 2.16a). If the distance is variable, then the pattern can be characterised as random dispersion (Figure 2.16b). In comparison, clustered dispersion is characterised by a huge variation in the distance between observations. If some of them tend to be in close proximity to one another, much further away than to the members of another spatial group, then they are said to be dispersed in clumps (Figure 2.16c). Clumps can also be dispersed if they spread out to fill most of the available space, just as with individual observations.

Both spatial dispersion and spatial clustering are rather subjective in reality. They are just the two extremes in the continuous spectrum of spatial distribution. There is no clear-cut threshold to distinguish one from another, as the criteria vary with the analyst. A conservative criterion will result in more clusters, while a generous criterion can cause all the observations to fall into one clump. Nevertheless, the

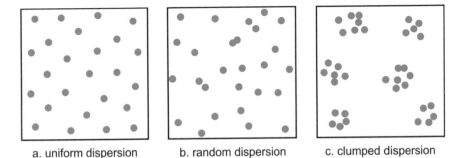

a. uniform dispersion   b. random dispersion   c. clumped dispersion

FIGURE 2.16 The concept of dispersion and clustering as measured by distance between neighbouring observations. The spatial patterns illustrated are: (a) uniform dispersion; (b) random dispersion; and (c) clumped dispersion or clustering.

degree of dispersion can be gauged by spatial density. Dispersed observations have low density, while clustered distributions tend to result in higher density locally. The formation of a dispersed or clustered distribution may be underlain by different physical processes. For instance, uniform dispersion of seeds in the landscape may result from a homogeneous environmental setting. A clumped dispersion may be associated with tree seeds in a forest shielded from the influences of winds and water, so they fall to the ground in close proximity to the parent trees.

2.3.5 Direction

Direction refers to the orientation or bearing in which an object moves or heads. It can also refer to the direction to which a motionless feature or phenomenon is facing, such as slope orientation. Thus, the direction is commonly associated with linear features or topographic surfaces. In the former case, it is indicative of the bearing to which an object is moving, such as water in a river channel flowing from upstream to downstream. Air pollutants are dispersed from the source to the nearby area of a lower concentration along the predominant wind direction. In both cases, direction is referenced from true magnitude north clockwise (Figure 2.17a). The direction of motion is determined by the relativity of the starting and ending positions in the local coordinate system:

$$\theta = \tan^{-1}\left(\frac{\Delta N}{\Delta E}\right) = \tan^{-1}\left[\frac{N_2 - N_1}{E_2 - E_1}\right] \tag{2.6}$$

where E_2 and N_2 denote the easting and northing coordinates of the ending point, identically defined as in Equation (2.1), and E_1 and N_1 are the easting and northing coordinates of the starting point. Dissimilar to Equation (2.4), the determination

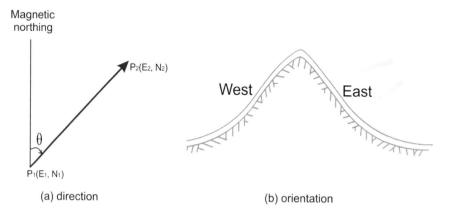

| (a) direction | (b) orientation |

FIGURE 2.17   Direction or bearing of linear features measured from magnetic north clockwise (a). It is commonly associated with movement. (b) Orientation to which a surface is facing. It is static and affects the energy received from the Sun, and hence surface moisture distribution.

of direction requires differentiation of the starting and the ending points as it can be totally opposite if the two are mixed in the calculation. It must be noted that the calculated direction applies to the object view of representation (e.g., vector coordinates) only.

In the field view of representation (e.g., raster), direction is determined from the observed height of grid cells in the defined neighbourhood. For a topographic surface, it refers to the orientation or the aspect to which a slope is facing (Figure 2.17b). The exact direction is subject to the relativity in height between all the cells within the defined neighbourhood. In the 8-connectivity neighbourhood, there are 9 primary, discrete directions to which a surface can potentially face: east, southeast, south, southwest, west, northwest, north, northeast, and 0 (flat) (Figure 2.18). This number drops to five in the 4-connectivity neighbourhood. This discrete expression of orientation can be made more precise by comparing the discrepancy in elevation among the four cardinal directions, with the two largest adjacent differences used in deriving the final direction or azimuth (Equation 2.7). This equation applies to a situation in which both ΔH_E and ΔH_N share the same sign. When they have the opposite signs, the azimuth needs to be incremented by either 90° or 270°, depending on which of them is positive, and which of them is negative:

$$\text{Azimuth} = \frac{180}{\pi} \tan^{-1} \frac{\Delta H_E}{\Delta H_N} \tag{2.7}$$

Irrespective of the nature of spatial features it is associated with, direction is unique in that it must have a pre-determined reference direction, and its range is confined to 0°–360°. This spatial measure is not as widely studied as distance. Its common fields of application in geography include transport (e.g., the direction of travel), hydrology (e.g., the flow direction of a channel and surface runoff on a slope), climatology (e.g., wind direction), geomorphology (e.g., slope orientation), hazard (e.g., volcanic lava flow direction), and oceanography (e.g., ocean current circulation direction). In ecology, it may include the direction of animal migration. Direction plays a decisively important role in spatial modelling of watershed hydrology. The direction of surface runoff flow dictates the watershed in which surface water converges and contributes to the discharge of a channel, and affects its peak flow.

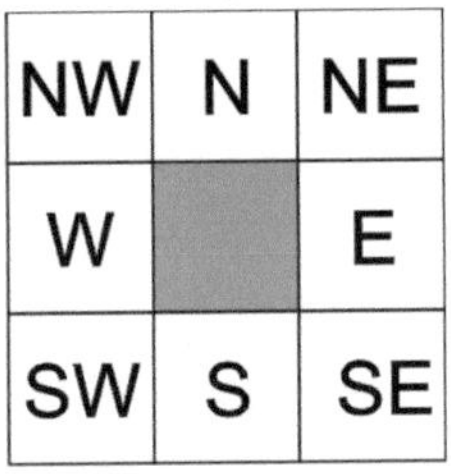

NW	N	NE
W		E
SW	S	SE

FIGURE 2.18   The potential discrete direction a cell (blue) in the defined neighbourhood can face in the 8-connectivity neighbourhood. The direction is the sky if all the cells in the neighbourhood have the same elevation.

2.3.6 AREA

Area refers to the physical size of a polygon encompassed by its perimeter, defined as the interior extent bounded by an enclosed border. The spatial extent of polygons is calculated by forming trapezoids using every edge of the polygon with the easting axis (Figure 2.19). For instance, the area of polygon ABCD is calculated by subtracting the two small trapezoids of BCcb and Abba from that of the two large trapezoids of DCcd and ADda. Areal data are commonly analysed in human geography and landscape ecology. All socioeconomic data are enumerated within an area, such as a census tract. In landscape ecology, area is associated with land cover or habitat type. Forest, grassland, and water are all areal-based features. The physical size or area of a polygon is indicative of the degree of landscape fragmentation. A landscape comprising many polygons of small area is synonymous with a high degree of fragmentation. A landscape comprising only a few large polygons tends to be intact. Area can also indicate the range of activity. A larger habitat suggests a longer range of foraging in the landscape.

Areal data are not as commonly analysed as point data in spatial analysis. However, areal data can be analysed in more versatile manners, such as for attributes thematically, as well as for polygon geometry spatially, both individually and collectively. The analysis of collective areal attribute data can reveal the spatial variation of distribution. The spatial relationship among polygons is analysed to identify their pattern (e.g., settlement pattern). Areal observations can be merged if they are spatially adjoining by dissolving their common boundaries to achieve a specific objective. For instance, in order to preserve anonymity and privacy, attribute data enumerated within a small census unit area are commonly aggregated to a more general level by combining adjoining polygons to form a larger polygon. How to combine these polygons to form reasonable and meaningful areas is a thorny issue in spatial analysis. The topic is so complex that it requires detailed discussion in Section 2.5.2.

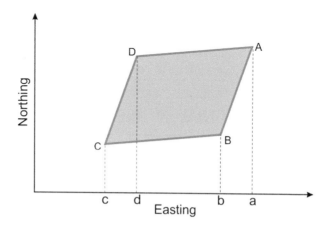

FIGURE 2.19   The concept of area enclosed by linked edges. The actual value of area is calculated through trapezoids formed by the dashed lines.

2.4   SPATIAL PARTITIONING AND AGGREGATION

In reality, it is rather common and necessary to divide a large geographic area into a number of smaller units or sub-regions so that each will become more manageable. This operation is commonly known as *spatial partitioning* in spatial analysis or *zoning* in urban planning. It is practised to fulfil diverse needs. For instance, a municipality may need to partition the area under its jurisdiction into a number of school zones so that schoolchildren living in a given street know which school they should attend. A city may be divided into zones to delegate administrative responsibilities (e.g., collection of household waste on different days), demarcate jurisdictional boundaries (e.g., territories of regional authorities), and facilitate provisions of services (e.g., delivery of mail and parcels within the same zip or postal code area). The same city may also be partitioned into different police precincts so that a given police station will know the spatial extent of its responsibility. Spatial partitioning is also crucial to the collection of socioeconomic data in general, and census data in particular, in which a large suburb is divided into a number of census units. Similarly, in geopolitics, a country may be divided into tens of electoral districts, from which a candidate is elected to represent his or her constituents in parliament. Spatial partitioning is important to spatial analysis as it dictates how spatial data are collected and enumerated, and thus the interpretation of the analytical results.

All of the aforementioned partitioning applications face the same issue of how to divide the space into small units properly. Opposite to spatial partitioning is *spatial aggregation* or merging of spatially adjacent observation units to enumerate the attribute over a larger polygon. The following discussion will elaborate on spatial partitioning first, followed by spatial aggregation.

2.4.1   SPATIAL PARTITIONING

Spatial partitioning is defined as the process of dividing the continuous, Euclidean space into non-overlapping regions or zones for the convenience of data enumeration and for the optimal allocation of resources. Spatial partitioning is precipitated by the fact that the space we live and work in is rather extensive and lacks discontinuity. It will be unmanageable and unimaginable if the entire space is treated as a single unit. Only through reasonable partitioning is it possible to achieve an equitable allocation of resources and delegation of responsibilities. In addition to fulfilling multiple needs, spatial partitioning also has its own requirements and principles to follow. Here three examples are provided to illustrate different purposes, including school zoning (Figure 2.20a), sub-division in urban residential development (Figure 2.20b), and in census data collection (Figure 2.20c). In school zoning, zones are formed to decide the schools that all schoolchildren residing in one zone should attend. In urban residential development, a large plot of land must be partitioned into smaller sections suitable for the construction of dwellings. A city may be zoned into areas of intensive housing and zones of low-density development. In a census, a municipal suburb may be divided into a number of small units, and samples can then be collected from each of them (Figure 2.20c). Such spatial partitioning facilitates data

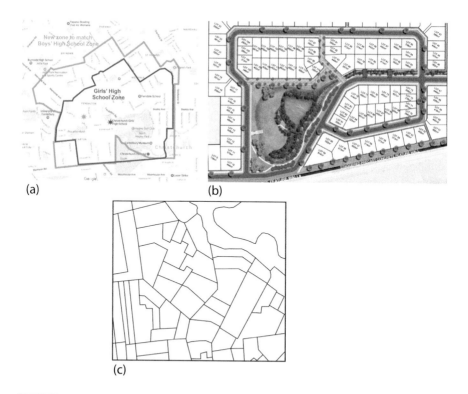

(a)   (b)

(c)

FIGURE 2.20   Three examples of spatial partitioning: (a) an adjustment of the zonal boundaries of the Christchurch Girls' High School to match those of the Boys' High School as a
result of population growth; (b) an example of urban subdivision in which the plane space is
divided into non-overlapping polygons of a roughly equal size (source: www.veros.co.nz/proj-
ects/tamakuku-terrace/, with permission); (c) a census area unit in which a suburb is divided
into a number of small meshblocks (tracks) for ease of administration and data enumeration.

collection, enumeration of the collected data, and subsequent data analysis, and can
sometimes ensure the spatial representativeness of the collected data.

In spite of the ubiquity and significance of spatial partitioning, surprisingly, there
are no prescribed rules governing how space should be partitioned properly. The
absence of concrete and repeatable formulas to follow means that the partitioned
outcome may be imperfect, unsatisfactory, and sometimes even controversial. This
is because each type of spatial partitioning has its own unique requirements and
special needs, such as the size and shape of the partitioned zones, as well as their
connectivity (Table 2.3). In the case of school zoning, it is the catchment area that
matters the most, as each school has its own enrolment capacity that can accommodate only a certain number of students. In the case of an urban subdivision, four
special considerations receive the highest priority of consideration: size, shape, connectivity, and accessibility of the sub-divided plots. Each sub-divided section should
be large enough to allow a dwelling to be constructed in it. The minimum section
size is stipulated by the local planning regulations. All sub-divided plots must also

TABLE 2.3
Comparison of Spatial Partitioning in Three Applications

Examples	School zoning	Urban subdivision	Census units
Purpose of partitioning	Allocation of resources	Creation of liveable space	Ease of data acquisition and enumeration
Requirements	Equitable student size (rolls)	Equitable physical size, uniform shape	Equitable population
Additional considerations	School capacity, commuting distance	Connectivity and accessibility	Preservation of common neighbourhoods
Principle of partitioning	Capacity matches catchment size, thematic uniformity	Geographic uniformity	Physical boundary or commonly accepted natural barriers
Temporal stability	Boundary may be periodically adjusted	Further sub-division possible if plot size is large enough	Certain units may be split or their boundaries adjusted periodically due to population growth

have a similar size, and a regular, and ideally close to square, shape suitable for constructing the dwelling. All the sub-divided sections must be connected to a road or a street for ease of access and should be accessible by vehicles (including firefighting vehicles and medical ambulances) for mail delivery, household waste collection, and emergency evacuation and response (Figure 2.20b). They must be able to navigate to all the constructed dwellings from the street. Therefore, roughly squared sections must be juxtaposed with elongated skinny zones of roads that can connect as many lots as possible to minimise the cost of service provision to the residents.

Such requirements, however, do not apply to census units, as the population is not uniformly distributed in space, so the physical size of a census unit can vary widely with population density. The shape and size of the partitioned census units are not a major concern as long as each unit contains roughly the same number of residents (e.g., thematic uniformity). Common neighbourhoods must be preserved in the name of the census units, though. Efforts should be made to avoid partitioning a common suburb into census tracks that belong to different suburbs.

How to partition the space optimally is a delicate issue that has profound consequences. For instance, how a country is partitioned into electoral districts may have profound ramifications for the candidate who is going to win an election. The lack of rules to follow and the absence of agreeable zonal boundaries can cause the partitioning to be highly contentious, such as partitioning of ethnically and socioeconomically diverse neighbourhoods in the US elections. This is because the partitioned zonal limit can be artificial (e.g., highways and roads) or even invisible (e.g., telephone area code) in human geography. In contrast, in physical geography, the partitioned zones can be bordered by concrete physical boundaries (e.g., watershed divide). In this case, sensible partitioning can be achieved with greater ease devoid of controversy. Apart from watershed divides, the boundaries of the partitioned space can be rivers or mountains.

If there is a universal rule that must be adhered to in all spatial partitioning in human geography, it is the maximisation of uniformity, either thematic or spatial. *Thematic uniformity* means that all the partitioned areas encompass a roughly equal number of subjects. Thematic uniformity is important to consider when the partitioned space affects capacity (e.g., school rolls in school zoning) or workload equity, as in police precinct zoning. The downside of thematic uniformity is that the partitioned zonal boundaries must be adjusted periodically with the growth of the population (Figure 2.20a). Sometimes an existing zone has to be split to create more meshblocks or electorates as the population grows. In comparison, *spatial uniformity* is much easier to achieve as the space to be partitioned is stagnant and static. Once a minimum size is specified, general considerations are given to the shape of the partitioned polygons. Spatial uniformity is the primary concern in the urban subdivision. All the sub-divided plots must have a similar size in accordance with the zoning regulations.

2.4.2　Spatial Aggregation and the Modifiable Area Unit Problem

Opposite to spatial partitioning, *spatial aggregation* refers to the process of amalgamating spatially adjoining observation units (e.g., polygons sharing a common boundary) to form a large unit of observation by combining the attribute value of all the merged polygons and by dissolving their shared boundaries. During amalgamation, the elimination of common boundaries between adjoining polygons results in fewer but larger polygons in the aggregated spatial data. It is virtually a process of data generalisation. Spatial aggregation is commonly carried out on socioeconomic data, prompted by the need to preserve anonymity and privacy of the census subjects before they are released to the public. Identical to spatial partitioning, there are no definite rules regarding the proper level or degree of aggregation. A commonly adopted criterion is the minimum size of the aggregated unit, such as a population of 50,000 (thematic uniformity), or a minimum physical size of 450 m² as in an urban subdivision. What is certain is that as more adjoining, small polygons are merged to form larger ones, the less detailed the aggregated data become, and the attribute of interest is generalised on an ever-broader scale.

Of special note is that not all polygons destined for spatial aggregation share the same trait or are of the same type. In fact, they can have drastically different and even contrasting socioeconomic traits, which creates a serious issue in geopolitics, namely how to form polygons of an ethnically and socioeconomically homogeneous population from diverse census tracks that serve as electorates so as not to favour certain kinds of candidate. This is not an easy feat, and it is a rather contentious issue facing contemporary US geopolitics. Unfortunately, there is no satisfactory solution for many aspiring candidates.

Spatial aggregation is commonly performed for attribute data enumerated over an area. As mentioned in Chapter 1, one of the objectives of spatial analysis is to study spatial patterns. Some kinds of patterns emerge or become apparent only after the observations have been spatially aggregated. Since there are no definite rules dictating how adjoining spatial entities should be merged, the same attribute can be made to have different patterns, depending upon how neighbouring polygons are amalgamated. In many analyses of census data, the areal units (zonal objects) used are arbitrary, modifiable, and subject to the whims and fancies of the data aggregator. This brings out the *modifiable area unit (MAU) problem*, initially observed by Openshaw (1984). Later, Fotheringham and Wong (1991) presented strong evidence of the unreliability of multivariate analysis with data of area units. Implicit in this problem are two related concepts: *scaling* and *zoning*. Scaling refers to the physical size of the merged polygons, or the influence of the size of the mapping units on the attribute being depicted (Figure 2.21). As more small polygons are merged, fewer polygons are preserved in the amalgamated result, causing the representation of the depicted attribute to shift from a local scale to a broad scale. Scale increases as the level of aggregation rises. Scale is so important to spatial analysis that the issue will be revisited in Chapter 5 on geostatistics. Thus, the same phenomenon can have drastically different patterns, dependent upon the level of spatial aggregation, and the degree of generalisation to various scales.

Zoning refers to the manner in which adjacent polygons are aggregated to a similar physical size, hence all amalgamated polygons have a similar scale

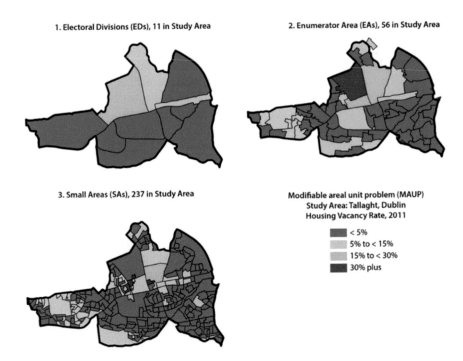

FIGURE 2.21 The spatial pattern of the same variable, Dublin housing vacancy rate in 2011, mapped at three scales. The pattern becomes more detailed as the size of the mapping units decreases (source: Kitchin et al. 2015, open access, permission not required).

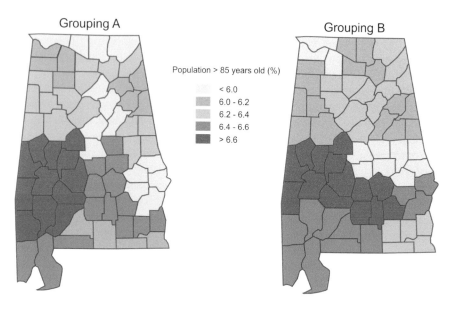

FIGURE 2.22   Comparison of the zoning effect on the resultant spatial pattern in the modifiable area unit problem. All the amalgamated polygons have a similar size, and the number of amalgamated polygons is equal in both amalgamations.

(Openshaw, 1977). The scaling issue, as discussed in the preceding section, differs from the zoning effect in that the finally generalised pattern is based on the same number of polygons of an approximately equal size (Figure 2.22). If the spatial partitioning has been carried out flawlessly, then the between-polygon variation should be maximised while within-polygon variation is homogenised. This will also reduce the aggregation bias, and avoid pseudo patterns if only polygons of a similar value are merged. Naturally, the same attribute will have different spatial patterns formed by different ways of merging adjacent polygons.

The joint effect of scaling and zoning can cause the same geographic phenomenon to exhibit variable spatial patterns, some of which are not genuine. They can be misleading if polygons of drastically differing attribute values are combined (Figure 2.23). The visualised phenomenon is so variable that it prompted Monmonier (1996) to write a book on the topic. Although the MAU problem has been widely acknowledged in the literature, it has been conveniently ignored in practice, probably because there is no easy solution to it. The best way of handling it is to run the same analysis at multiple scales to identify the most appropriate one, bearing in mind that the studied attribute may even vary at this scale within the same study area. In this case, the MAU can serve as an analytical tool to enable a better understanding of spatial heterogeneity and spatial auto-correlation. If high values of two variables are close together (i.e., positive spatial auto-correlation), their correlation is likely to be intensified through aggregation. Conversely, if either high or low values are spatially scattered (i.e., low spatial auto-correlation), the grouping of such observations will diminish the effect of the MAU.

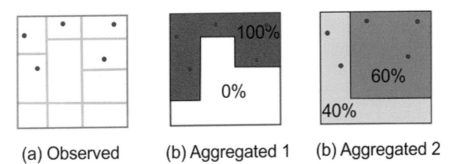

(a) Observed   (b) Aggregated 1   (b) Aggregated 2

FIGURE 2.23   Spatial patterns formed by aggregating spatially adjacent polygons of COVID-19 cases: (a) the observed data; (b) aggregation of polygons with the same value; and (c) aggregation of polygons of different values. Both aggregations contain two classes of five polygons each. Percentage refers to the proportion of polygons affected.

2.5   SPATIAL TESSELLATION

The spatial partitioning mentioned in the preceding section takes place at a broad scale mostly for resources allocation and easing collection and enumeration of socioeconomic data. However, spatial partitioning can take place at a local scale, known as tessellation. *Spatial tessellation* is a process of partitioning the 2D space into a mosaic of very small blocks, such as regular grid cells or lattices (Laurini and Thompson, 1992). The systematically partitioned space is composed of small areas or units of uniform shape (and uniform size in certain cases), within which the observed value is enumerated. Spatial tessellation is commonly undertaken as a preparatory step to represent a surface, either physical or imaginary. This representation aims for maximum accuracy, but at a reduced data volume. Spatial tessellations differ from spatial partitioning in that they always abide by rules, so the tessellated results are replicable no matter who is performing the partitioning. The only difference between different tessellations is whether the partitioned space comprises regular lattices or irregular tiles, an outcome governed by the data format (e.g., vector or raster).

2.5.1   REGULAR TESSELLATIONS

In regular tessellations, space is partitioned into regular lattices of squares, triangles, or hexagons, all having an equal size (Figure 2.24), so the depicted attribute value is available at a uniform spatial interval. Of these possible lattices, squares are the most common building blocks owing to their rotation-invariant shape. They also enable the area under study to have neat and straight borders. In addition, hexagonal lattices are also possible. However, they cannot always guarantee clear-cut boundaries in some cases. For this reason, they are far less common than square lattices. In all regular tessellations, observations are made at a constant interval both horizontally and vertically, as with grids, DEMs, and images. Virtually, regular tessellation is a process of converting point data (e.g., elevation at a specific location) to areal data

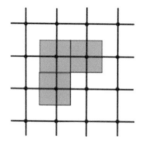

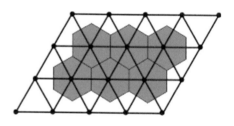

(a) Proximal areas for regular
square lattice

(b) Proximal areas for isoceles
triangle lattice

FIGURE 2.24 Examples of partitioning space into lattices in regular tessellations: (a) squares; (b) hexagonal lattices.

(e.g., cell value over an extent). As with all sorts of raster data, regular tessellations are best used for spatial modelling, even though they are plagued with the issue of representation inefficiency as data redundancy is rife and the norm with them.

2.5.2 IRREGULAR TESSELLATIONS

In irregular tessellations, the tessellated units are of different sizes, but have a similar shape. The commonest shape is a triangle, owing to its simplicity. Triangle size varies with the variability of the attribute being represented. A special irregular tessellation is exemplified by the Triangulated Irregular Network (TIN), a type of vector data devised to represent the surface. In this tessellation, space is partitioned into a network of irregularly shaped triangles of varying sizes (Figure 2.25). The surface is represented by a series of continuous, non-overlapping triangles, with adjoining triangles sharing common boundaries and vertices. This tessellation is suitable for representing angular topography by locating the vertices at critical topographic positions, such as peaks and pits along ridges. The density of vertices varies with the surface regularity. It can be reduced for a flatter topography, but increased for highly irregular surfaces. TIN is able to achieve a similar representation accuracy to that of regular tessellations, but with a reduced number of vertices. Higher efficiency is achieved owing to the adoption of a variable distance between vertices.

In representing topography using a TIN, only points critical to the surface morphology are captured, and any two adjacent points are linked with a straight line segment. And three adjacent points are connected to form a triangle. Therefore, the surface is represented by points or nodes capturing elevations at special spots, lines (edges), and polygons (triangles). Within each triangle, the surface is approximated as a plane, as it is assumed to be flat (e.g., the surface is smooth and all variations within it are ignored), but the transition from one triangle facet to the adjoining one is angular. This model is thus unable to capture local variations within the triangle. As with all vector data, TIN is a compact and efficient means of representing surface.

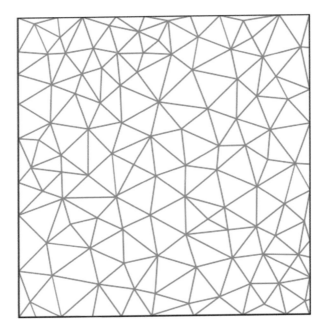

FIGURE 2.25 A network of triangles of varying size and orientation. This irregular form of tessellation is a vector form of representing topography.

Nevertheless, the high representation efficiency is undermined by the necessity of maintaining a large per-point overhead, as with all vector data. This overhead spells out the topological relationship among different triangles. The accuracy of representation depends on the nature of the surface, and the size and shape of triangles. High accuracy can be expected if all the crucial points of the surface are captured. In addition, the accuracy of representation can be improved by preserving special topographic features, such as ridges, streams, and transitional points on the surface (Figure 2.26). Other lines may include drainage divides, shorelines, and even fault lines. With these additional lines as triangle edges, it is possible to preserve critical topographic features in a TIN. The TIN model is a significant alternative to the regular grid tessellation of terrain, even though it is usually constructed from regular lattices by eliminating insignificant grid cells from a DEM.

Despite its compactness, TIN is severely handicapped in spatial modelling in which the attribute value under study must be available at a regular interval. On the other hand, it is time-consuming to estimate this value at regularly spaced points from a TIN, such as height and slope that have to be interpolated from the triangle facets individually. Thus, the speed of modelling is slowed down considerably in complex spatial modelling. For this reason, TINs are not used as commonly as grid DEMs. One reason for this lesser popularity is the scarcity of data available in the TIN format, as they must be constructed from other sources, such as elevational data sampled along contours (see Section 3.2.2), spot heights from GPS data, and even light detection and ranging (LiDAR) point cloud data, all of which can be

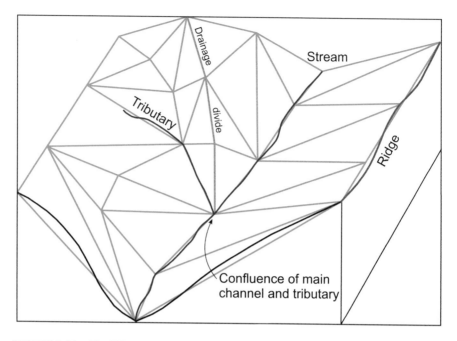

FIGURE 2.26   The TIN method of representing terrain, in which a surface is approximated as a network of triangles whose vertices are located at critical spots.

converted to regular grid DEMs via spatial interpolation (which will be covered in Chapter 5).

The generation of a TIN from a grid DEM involves three key components: (a) what points to retain; (2) how to connect the retained points to form triangles; and (3) how to model the surface within each triangle. The last question is the simplest to answer, and has already been addressed. Since the surface is treated as a plane within each triangle, the retained points must be rather critical to capturing the surface. Whether a point should be retained in a TIN depends on its importance to the accuracy of representation, as judged by its elevation in relation to that predicted from its immediate neighbouring cells. The operation is implemented within a window of 3 by 3 cells iteratively. After the average height of a pair of opposite elevations or the average of the four heights (e.g., the elevation interpolated from its two immediate neighbours) is calculated under the assumption that the surface between them is smooth, it is compared with the observed height H (Figure 2.27). The discrepancy between the observed and predicted height (half of $H_1 + H_2$) is compared to the pre-defined tolerance threshold. The cell is retained only if it is "significant" (e.g., the discrepancy exceeds the allowable threshold) (Figure 2.27), otherwise it is deleted . This process is iterated until the number of points or the significance reaches the pre-determined limit. All the retained points make a significant contribution to the surface to be represented by the TIN. How to connect the retained vertices to form triangles is a complex topic that will be discussed in Section 2.5.3.

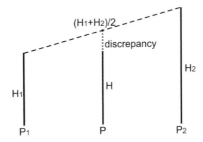

FIGURE 2.27   The discrepancy (red dotted line) in height caused by the removal of point *P*. It is the difference between its elevation (*H*) and the average of the two adjoining cells' elevations. This discrepancy in relation to the threshold is critical to deciding whether point *P* should be removed from the constructed TIN.

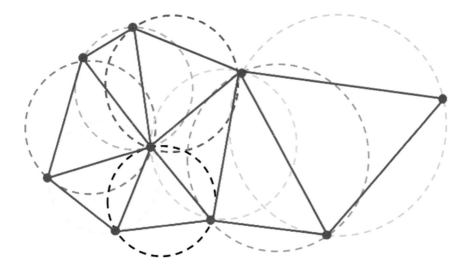

FIGURE 2.28   A network of triangles formed in Delaunay triangulation. All the formed triangles are as equilateral as possible owing to the maximum-minimum angle criterion adopted.

2.5.3   DELAUNAY TRIANGULATION

In the constructed TIN, triangles are formed from the retained vertices using the *Delaunay triangulation* method, in which the formed network of triangles must meet two criteria: (a) the empty circle (Delaunay) criterion; and (b) the maximum-minimum angle criterion. The first criterion ensures that each circle is formed out of three vertices and the circumcircle of every such formed circle does not encompass any other retained vertices in its interior (Figure 2.28). The second criterion guarantees that the minimum measure of angles of all the constructed triangles in the triangulation is the largest (Tsai, 1993). This criterion effectively ensures that the constructed triangles are as equiangular or equilateral as possible. This is desirable as the intersection of triangle edges always involves a degree of uncertainty caused by inaccurate coordinates. This uncertainty is minimised if the two edges intersect at an angle close to 90°, but is magnified disproportionately if they intersect at a small acute (e.g., <30°) or a large obtuse (e.g., >120°) angle.

To meet the second criterion, a number of trial connections must be carried out, and all the resultant connections are compared with each other (Figure 2.29). Only the one achieving the maximal total angle is retained (e.g., the last connection in Figure 2.29).

Delaunay triangulation can be implemented using several methods. Some of the commonly used ones are divide-and-conquer, incremental insertion, and triangulation growth. In divide-and-conquer, the entire planar dataset is recursively divided into two disjoint subsets until each subset contains no more than four points, and they are triangulated (Lewis and Robinson, 1978). These sub-triangulations are merged to form the final triangulation recursively in a bottom-up breadth-first manner. In incremental insertion, an initial Delaunay triangulation network encompassing all the retained data points is established as a super triangle or a polygon (Lawson, 1977), to which new triangles are added. The constructed triangular network can be optimised with the constrained local optimisation procedure (Dwyer, 1987). In triangulation growth, one of the points is arbitrarily selected, and then its

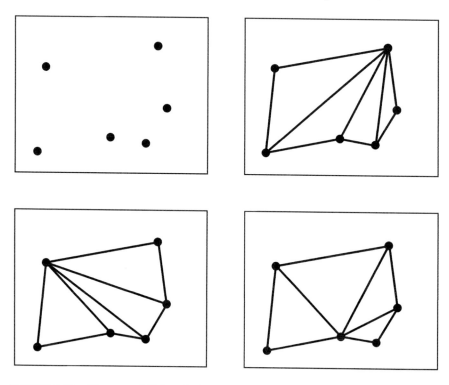

FIGURE 2.29 Three possibilities of triangulating a group of six vertices to form four triangles, the last being the best.

two nearest neighbours are searched. Once located, these two points are linked to form a Delaunay triangle edge. The criteria of constructing Delaunay triangulation are applied to search for the third vertex. This process is iterated until all the data points have been processed.

After the retained points are successfully triangulated, the resultant triangle network can be used to construct the Voronoi (also known as Dirichlet) polygons or Thiessen polygons, in which each vertex serves as the centroid of a polygon and every polygon encompasses only one vertex (Figure 2.30).

2.5.4 THIESSEN (VORONOI) POLYGONS

Dissimilar to all the aforementioned spatial partitioning that does not involve any external inputs, *Thiessen polygons* are a way of partitioning the space into regions from a number of existing point observations. Virtually, this partitioning identifies the sphere or domain of influence of each observation, also known as its catchment area. Thiessen polygons extend point-based observations to an area, opposite to the conversion of polygon-based observations to point ones using centroids. The 2D space is partitioned into non-overlapping, irregularly shaped polygons of various sizes and shapes known as the *Voronoi polygons* (Boots, 1986) or the Thiessen diagram. After partitioning, space is composed of unique, contiguous, and space-exhaustive polygons, all being convex in shape, though some are not bounded by other polygons, but by the border of the study area (Figure 2.31). The proximal regions are formed by subdividing lines joining two nearest observations, or the Delaunay triangle edges. These bisectors form the edges of the Voronoi polygons, and the intersection of the three bisectors of each triangle forms a polygon vertex. The sequential connection of adjacent polygon vertices forms a Thiessen polygon. It encloses only one point, whose distance to all of the polygon boundaries is the shortest. Consider a set, S, of n labelled points in the plane:

$$S = \{p_1, p_2, \ldots p_n\} \tag{2.8}$$

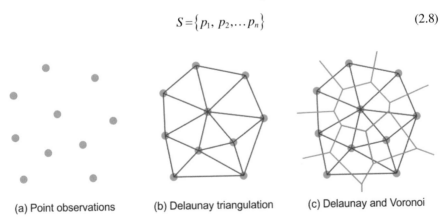

(a) Point observations   (b) Delaunay triangulation   (c) Delaunay and Voronoi

FIGURE 2.30 Relationship between a collection of point observations (a): Delaunay triangulation to form a TIN (b) and Voronoi diagram (c). Each Voronoi polygon encloses only one triangle vertex.

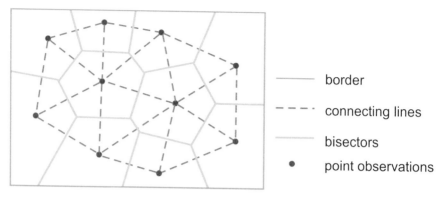

border

connecting lines

bisectors

point observations

FIGURE 2.31   Formation of Thiessen (Voronoi) polygons from known points to identify their sphere of influence.

p_i is a point in S associated with all locations x in the plane which are closer to p_i than to any other point, p_j, in S (i≠j). The creation of a Thiessen polygon, p_i, is based on the Euclidean distance from x to p_i, or d(x, i), then:

$$p_i = \{x \mid d(x,i) \ni d(x, j); \; i \neq j\} \tag{2.9}$$

Once constructed, Thiessen polygons can find a number of applications. The commonest application in spatial analysis is to identify the relationship between points and areas, such as the school that children residing in a suburb should attend. In this application, the known points are schools, and they are used to construct the Thiessen polygons to identify their catchment areas. All the students living in a Thiessen polygon should attend the school enclosed in the same polygon, as the distance of travel for all of them is the shortest. Another example of application is the hospitalisation of traffic accident victims. With the aid of Thiessen polygons, it is possible to decide the nearest hospital to which a patient should be admitted under the assumption that all other conditions are the same. In both examples, the spatial partitioning problem evolves to a spatial allocation problem, namely the identification of the points belonging to a given polygon. The allocation of observations to a given polygon should minimise the cost of travel or maximise capacity to reduce operation cost. Thus, all students residing in one Voronoi polygon should attend the school it encompasses if its capacity has not been exceeded.

Another very important application of Thiessen polygons is the calculation of rainfall over an area. In climatology, rain gauges are positioned in a spatially dispersed manner from each other. It is not accurate to average the readings of all gauges to derive the mean precipitation for a city as each weather station may have a varying sphere of influence. The area under the influence of each rain gauge needs to be taken into consideration. This area can be used to weigh the observed rainfall reading at each station using the following formula:

$$\text{rainfall } P = \frac{\sum_{i=1}^{n} p_i \cdot a_i}{\sum_{i=1}^{n} a_i} \tag{2.10}$$

where a_i stands for the area of the ith Thiessen polygon with a rainfall reading of p_i. As shown in Figure 2.32, the rainfall of the same Voronoi polygon is treated as spatially uniform. The rainfall reading changes abruptly across the polygon boundaries, so its attribute value does not vary continuously across space. This discrete distribution of rainfall reading can be made continuous using spatial interpolation if there is a sufficient number of rain gauges within the area of study.

2.6 RUDIMENTARY SPATIAL MEASURES

A few statistical terms will be used frequently throughout the subsequent chapters. It is thus necessary and useful to introduce them, especially their meaning and significance in spatial analysis, in this section. Included in the following discussions are mean, variance, standard deviation, correlation, and root-mean-square error (RMSE).

2.6.1 MEAN

Statistically, mean is synonymous with the average of all observations. It is calculated by averaging all observations, which can be realised by firstly summing

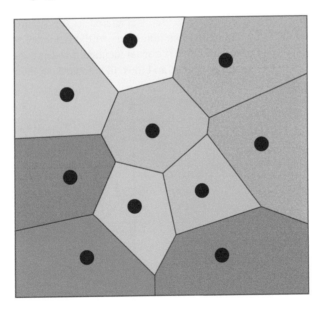

FIGURE 2.32 Use of Voronoi polygons to extend point observations (dots) to areal observations. In each polygon, the attribute is assumed to be spatially uniform and is represented as a single value identical to that at the point of observation.

up all the observed values, and then dividing the sum by the number of observations (n) (Equation 2.11). This average is also called the arithmetic mean or mathematical expectation for a collection of observed values. In the context of spatial analysis, mean refers to a measure of the central tendency for a group of point observations (Figure 2.33a). For a polygon perimeter represented by a string of pairs of coordinates, the mean of all coordinates represents the centroid of the polygon (Figure 2.33b), namely:

$$\bar{X} = \frac{1}{n}\sum_{i=1}^{n} x_i \tag{2.11}$$

Mean values can be calculated for distance and area. In spatial analysis, mean distance is a useful parameter that can indicate the density of point data and proximity between polygons (e.g., the distance from the centroid of one polygon to that of another).

2.6.2 Variance and Standard Deviation

Similar to mean, variance is also a measure of central tendency. It measures the deviation of all observed values from their mean (Equation 2.12). A small variance suggests that these observations have values highly resembling each other and they are rather compact in their spatial distribution. Conversely, a large variance means that the value of all the observations varies widely, indicating that they are widely scattered or dispersed in space (Figure 2.33a). Calculated as the square root of mean variance (Equation 2.13), standard deviation refers to the tendency to deviate from the mean for a group of observations. It is almost identical to variance except that it is adjusted by the number of observations, and thus all standard deviations are directly

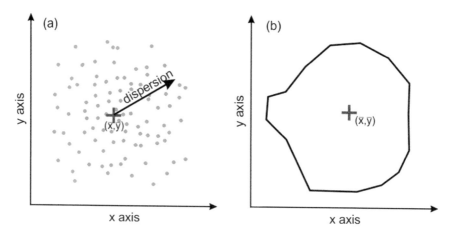

FIGURE 2.33 Meaning of mean $(\bar{x},\bar{y})$ in spatial analysis: (a) the central tendency of a group of observations; (b) centroid of a polygon.

comparable to each other. A large standard deviation can be caused by dissimilarity in the observed values. A small standard deviation in geographic coordinates indicates that all the observations are very close to each other in space and their distribution leans towards clustering. Conversely, a large standard deviation suggests that observations are distant from each other spatially or have a large degree of spatial dispersion.

$$\sigma = \Sigma \left(Z_i - \bar{Z} \right)^2 \tag{2.12}$$

$$\text{Standard deviations}\, \sigma = \sqrt{\frac{1}{n} \sum_{i=1}^{n} \left(Z_i - \bar{Z} \right)^2} \tag{2.13}$$

2.6.3 CORRELATION

Correlation refers to the predictability of the attribute value of one spatial variable in relation to that of another. If one variable behaves predictably with regard to the behaviour of another, then they are said to be correlated with each other, or one of them is dependent on the other. Thus, correlation indicates a degree of predictability and/or dependency. The variable that can cause the other variable to change is termed the *independent variable*. The variable whose value is subject to the change in the independent variable is called the *dependent variable*. The correlation between these two types of variables can be calculated using different methods, such as Pearson's and Spearman's. Pearson product-movement correlation is expressed mathematically as:

$$r(\rho_{xy}) = \frac{\text{cov}(x,y)}{\sigma_x \sigma_y} = \frac{\Sigma x_i y_i - \dfrac{\Sigma x_i \Sigma y_i}{n}}{\sqrt{\Sigma x_i^2 - \dfrac{(\Sigma x_i)^2}{n}} \sqrt{\Sigma y_i^2 - \dfrac{(\Sigma y_i)^2}{n}}} \tag{2.14}$$

where x_i and y_i refer to the ith observed value of variable x and variable y, respectively. The Pearson correlation coefficient has a value between -1.0 and 1.0. It must be noted that this coefficient measures only the linear correlation between the two variables (Figure 2.34a). Linear correlation means that all the observations follow a clear trend or lie in close proximity to a straight trend line. If the nature of the variables involved is known, the portion of variation "explained" by the independent variable is also known, and commonly expressed as the R^2 value (variable 2). Sometimes the two variables may be correlated with each other exponentially or logarithmically (Figure 2.34b). In this case, it is still possible to calculate their Pearson correlation using Equation (2.14) only after the data have been properly transformed. Prior to calculating the correlation between two variables, it is advisable to plot their scatter diagram (Figure 2.34) to visually assess the nature of the correlation between them so as to avoid mistakes.

Equation (2.14) makes use of the actually observed values in calculating the correlation. In reality, it is also possible to use the rank of the observed values instead of

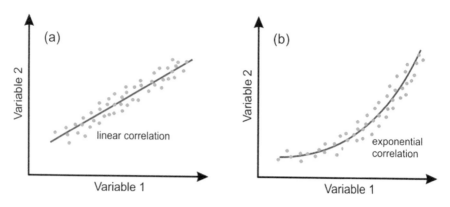

FIGURE 2.34   Correlation between two variables: (a) linear; (b) non-linear (exponential). If calculated using the Pearson correlation equation, the non-linear data must be transformed accordingly prior to the calculation to avoid mistaken results.

their value. In this case, the correlation is termed the *Spearman's rank correlation*, a non-parametric version of the Pearson product-moment correlation. The use of the ranked order (e.g., first, second, third, etc.) instead of the actual values is advantageous when the values themselves involve a high degree of uncertainty. Spearman correlation can show the strength and direction of association between two ranked variables.

Finally, it must be emphasised that the correlation between two variables simply shows that they are associated with each other. There is no guarantee that the relationship is always cause–effect. Even if this is indeed the case, then the relationship needs to be scrutinised closely to determine which variable is the cause and which is the effect. It is relatively easy to calculate correlation, but the proper interpretation of the calculated correlation requires exercise of caution.

2.6.4 RMSE

In spatial analysis, RMSE is a statistical measure of the agreement between the predicted and observed values at the same location. It is commonly used to assess the quality of spatially interpolated and modelled outcome based on a number of evaluation points (n), and provides an indication of accuracy about the derived results. It is calculated as:

$$RMSE = \sqrt{\frac{\sum_{1}^{n}(Z_i - \overline{Z}_i)^2}{n}} \qquad (2.15)$$

where Z_i refers to the attribute value of the ith evaluation point at a given location, $\overline{Z}$ is the estimated value at the same location, and n denotes the total number of evaluation points used. In order for the calculated RMSE to be statistically viable, n should not drop below 30. Mathematically, this equation closely resembles Equation (2.13). The difference lies in the comparison of the observed value with the estimated value instead of the group mean. RMSE is a critical non-spatial indicator of the accuracy of spatial operations, such as spatial interpolation. It statistically and quantitatively indicates the deviation of the interpolated values from the observed ones. A small RMSE suggests that the predicted values closely resemble the observed ones. Conversely, a large RMSE error is synonymous with a large degree of uncertainty in the interpolated results. However, there exist no standards regarding the acceptable RMSE. What RMSE is considered acceptable depends highly on the spatial analyst (e.g., how much risk to take) and the purpose of spatial analysis. Since RMSE is just an absolute value, it does not provide a perspective to the magnitude or relativity of disparity, which may be remedied by expressing RMSE relative to the mean attribute value in percentage. Apart from RMSE itself, attention should also be paid to the spatial distribution of *residuals*, the discrepancy between the observed and the estimated value at the same location, and their spatial pattern in assessing estimation accuracy. A good estimator should produce residuals that are spatially uncorrelated. If large residuals are clustered in certain areas, additional processing may be undertaken to remedy the situation.

REVIEW QUESTIONS

1. In theory, all spatial data should be referenced to the global system using a GPS to facilitate their analysis. Why is this not practical or not necessary in certain cases where the local system has to be used?
2. In spatial analysis, it is essential that all the spatial layers be georeferenced to the same coordinate system. To what extent do you agree or disagree with this statement?
3. How the spatial data are referenced spatially has no bearing on the analytical results. True or false?
4. Although the field view of spatial entities is much inferior to the object view, why is it still widely used in spatial analysis?
5. Of the five components of any geospatial data, which one is analysed most commonly in which kind of analysis?
6. What is the significance of spatial adjacency in spatial analysis?
7. What is the relationship of distance with proximity, neighbourhood, and clustering?
8. In which ways does spatial adjacency differ from proximity?
9. How can area be studied in landscape ecology?
10. In your view, how space should be partitioned properly to fulfil different application needs?
11. What is going to happen to the modifiable area unit problem if the spatial entities are evenly distributed over the area of study?
12. What are the main differences between spatial partitioning and spatial tessellations in terms of the partitioned space?

13. What is the relationship between Delaunay triangulation and Thiessen polygons? Compare and contrast the main characteristics of the same surface represented by Thiessen polygons and by an irregular tessellation.

14. What are the main differences between standard deviation and RMSE in spatial analysis?

REFERENCES

Boots, B. N. (1986) *Voronoi (Thiessen) Polygons.* Norwich, UK: Geo Books, p. 51.

Dwyer, R. A. (1987) A fast divide-and-conquer algorithm for constructing Delaunay triangulations. *Algorithmica,* 2: 137–51.

Fotheringham, A. S., and Wong, D. W. S. (1991) The modifiable areal unit problem in multivariate statistical analysis. *Environment and Planning A,* 23: 1025–1044.

Kitchin, R., Lauriault, T., and McArdle, G. (2015) Knowing and governing cities through urban indicators, city benchmarking and real-time dashboards. *Regional studies, Regional Science,* 2(1): 6–28, doi: 10.1080/21681376.2014.983149.

Laurini, R., and Thompson, D. (1992) *Fundamentals of Spatial Information System.* London: Academic Press, p. 680.

Lawson, C. L. (1977) Software for C surface interpolation. In *Mathematical Software III,* J. Rice (ed.) New York: Academic Press, pp. 161–94.

Lewis, B. A., and Robinson, J. S. (1978) Triangulation of planar regions with applications. *Computers Journal,* 21(4): 324–32.

Monmonier, M. (1996) *How to Lie with Maps* (2nd edition). Chicago: The University of Chicago Press, p. 222.

Openshaw, S. (1977) Optimal zoning systems for spatial interaction models. *Environment and Planning A: Economy and Space,* 9(2): 169–84, doi: 10.1068/a090169.

Openshaw, S. (1984) *Concepts and Techniques in Modern Geography Number 38: The Modifiable Areal Unit Problem.* Norwich: Geo Books, p. 41.

Tsai, V. J. D. (1993) Delaunay triangulations in TIN creation: An overview and a linear-time algorithm. *International Journal of Geographical Information Systems,* 7(6): 501–24.

3 Spatial Data and Association

3.1 DATA AND THEIR SOURCES

3.1.1 SPATIAL DATA

Spatial data have two vital components: geographic and thematic. The geographic or *spatial component* refers to the location at which thematic data are collected. It can also refer to the spatial extent over which the attribute value is enumerated. These data may pertain to the boundaries of the enumeration unit in the case of polygon data, and the actual coordinates for point and linear features. Spatial data may be acquired by several means, depending upon their topological dimension (Table 3.1). Point data are usually collected with a positioning device such as a Global Positioning System (GPS) data logger if the number of points is small, or a light detection and ranging (LiDAR) scanner if a huge quantity of points is collected. The latter is particularly suited to the acquisition of dense points over an area. All digital personal assistants enabled with GPS are also able to acquire point data in a manner strikingly similar to that of GPS. Although digital topographic maps are also a source of point data, they are not commonly used as they are secondary in nature and subject to mapping inaccuracy. Apart from point data, GPS is also a fast means of collecting linear data by mounting a GPS receiver on top of a moving vehicle driven along the linear feature to be captured. Areal data are normally collected from satellite images and thematic maps, even though it is still possible to acquire such data using GPS, albeit less effectively than point and linear data, especially in mountainous regions and at inaccessible locations. Besides, some governmental organisations, such as the Statistics Department, may release boundary maps about census data and administrative boundary maps. Irrespective of their sources, all spatial data must be geo-referenced properly to the desired projection at the appropriate scale and be reasonably current. More importantly, they must be associated with the thematic component to be useful.

3.1.2 THEMATIC DATA

Also known as attribute data, *thematic data* are pertinent to the nature or property of geographic entities. They can be socioeconomic, physical, and environmental. Most socioeconomic data are second-hand, collected mostly by relevant government agencies, such as the Department of Labour (Table 3.2). Another common source of socioeconomic data is the census, which collects the most comprehensive socioeconomic data dating back decades, if not centuries. Census data are updated

DOI: 10.1201/9781003220527-3

TABLE 3.1

Comparison of Methods for Acquiring Spatial Data of Different Topological Complexity

Type of data	Examples	Main methods of collection/sources of data
Point	Location of schools, sampling sites, elevation	GPS (limited number), LiDAR (a large collection), Digital personal assistant
Line	Roads, streams, boundaries	GPS
Areal data	Land covers, census units, electoral precincts	Satellite imagery, thematic maps, census maps

TABLE 3.2

Comparison of Main Sources of Thematic Data for Spatial Analysis

Type of data	Sources	Pros	Cons
Census (socioeconomic)	Government statistics department	Long-term data for longitude studies, comprehensive, detailed	Aggregated, census boundaries may have changed over the years
Crime data	Local police department	Can be current, detailed, and accurate	May be aggregated by suburbs
Traffic	Local transport authority	Traffic levels along particular roads are known, traffic accident spots are known	May be aggregated by the time of travel, or by roads
Health	The health department or private healthcare providers	Communicable disease data are always collected, available nationwide	May be aggregated to the suburb level, patient information may be removed.
Pollution	Local council or environmental protection agencies	Locations are precisely known, can be long-term and continuous	Collection devices could have different precisions, spatial distribution may not be balanced, data may be aggregated temporally
Questionnaire	Participants in surveys	Data about individuals is known	May not be representative, expensive, data aggregation essential
GPS	Field surveys	Fast and accurate positions only	Error-prone, attributes acquired manually, subject to site accessibility

periodically at an interval of five years or longer. Socioeconomic data in general, and census data in particular, are enumerated in various units, the commonest being meshblocks or census tracks. Meshblock-level data are not normally released to the public. Instead, they are spatially aggregated to a more general level to preserve anonymity and privacy prior to releasing to the public. Since the relevant government agencies are responsible for their collection and ensuring their quality, there is not much the spatial analyst can do about these data. In most cases, they are taken at face value, with little regard for data accuracy and completeness. The biggest advantage of using such second-hand data is that they are free or mostly free and have been collected for decades, so they are ideal for longitudinal studies. In using such data, nevertheless, attention must be paid to data currency and scale, as some data, especially population data, can become obsolete very quickly, even within years. Currency does not mean just up-to-datedness. It also implies that all the data used in the same analysis must have a roughly similar currency. Any discrepancy in currency among the thematic data being used can lead to potentially erroneous results of analysis or modelling.

Apart from currency, census data also need to be scrutinised closely before being used in longitudinal studies because boundaries of census tracks may have been altered, so minor adjustments are needed to standardise some of the observations to a consistent spatial extent. In addition to census data, many special statistical data will have been collected already by various government agencies, from which the data may be acquired. For instance, migration and international tourist data can be obtained from the border control authorities, and crime statistics from the local police stations. If the data are collected at the address level, some kind of geocoding known as *address matching* is essential. Through this processing, an address is converted to a pair of coordinates that can be displayed and visualised graphically. However, data concerning sensitive information such as addresses of crimes are normally aggregated to the suburb level, even though they are initially reported at specific places and addresses. Another important type of data for spatial analysis is traffic accidents. These data highly resemble crime data except that they are collected and maintained by different government agencies (e.g., Department of Transport). Most traffic data are also point data that may require geocoding before they can be analysed properly.

Apart from the aforementioned socioeconomic data, special statistical data such as housing price data have to be collected from real estate institutions. Similarly, any data related to epidemiological infections and health care must be collected from hospitals, private healthcare providers, and the local health authorities. Since such data relate to individuals, patient confidentiality or data privacy is a major concern with their use. Data aggregation and anonymisation of the patients involved are the primary ethics that must be observed in analysing such data. In contrast, environmental quality data, such as air pollutants collected at points, may be obtainable from regional or central governments. Prior to making use of such data, it is always a good practice to check that the data come with coordinates, preferably with the time of data collection as well.

Physical data, such as soil pH and nutrient content, are not normally collected by government agencies systematically, as they are user-driven. Spatial analysts have to collect such data themselves. For special types of statistical data, such as people's perception of natural hazards, special questionnaires have to be designed to solicit responses from targeted participants in a survey. Questions in the questionnaire should be concise, clearly stated, and easily understood, without jargon. If the answers are in multiple-choice format, then the choices provided should be non-overlapping. Scaling, if necessary, must be explained and consistent. Some research organisations may require prior ethics approval before a questionnaire is allowed to be distributed. The questionnaire samples may be distributed as widely as possible, but should always be representative. It is also possible to upload the questionnaire online and solicit responses from the wider community. However, this kind of data collection may be biased, as only those with access to computers can participate in the survey. The collected data should be aggregated to such a degree that they do not allow the identification of individuals. Whenever individuals become the target of a study (e.g., case analysis), pseudonyms should be used to protect the identity of the subject. A group of observations can be described according to a common criterion, such as area, time, and socioeconomic status. Finally, it is unethical or even illegal to use the collected data for a purpose other than that explicitly stated beforehand and that has been made known to the survey participants.

3.1.3 ENUMERATION SCALES OF THEMATIC DATA

All thematic data must be enumerated at one of the four scales: nominal (categorical), ordinal, interval, and ratio. *Nominal data* are numbers (codes) that merely establish identity or difference. Common examples of nominal data include land cover codes, student IDs, telephone area codes, and telephone numbers. Of the four scales, nominal data are the most general and imprecise. They can be categorical or even binary, such as different kinds of land cover (e.g., forest, shrubland, and grassland) or the presence/absence of kiwis in a forest. Regardless of their exact form, all nominal data are unable to show importance, order, or relative rank, so it is meaningless to compare nominal data unless the purpose is to identify differences, such as land cover change from one type to another, or make comparisons at the categorical level, such as the quantity of change for a given type of land cover. In contrast, ordinal data are more prescriptive as they are indicative of relative importance or establish order, such as high, medium, and low income or severe, moderate, and slight contamination, without indication of differences between orders, such as best students, top achievers, the second-highest score, etc. Therefore, the actual value is not important, only the order or rank matters. Ordinal data are more informative than nominal data. Both scales are qualitative in nature. Because of this, it is impossible to carry out further processing or analysis on them.

Interval data refer to numbers expressed over a certain range, such as suspended sediment concentrations at multiple levels (80–90, 90–100, 100–110 mg·l⁻¹). Thus, the difference (interval) between numbers is meaningful, even though the numbering scale may not necessarily start from zero or arbitrarily set zero (e.g., temperature). Another good example of such data is contour interval in representing topography, such as 90–100 m. In analysing interval data such as elevation, attention must be paid to the height reference system or the geodetic datum used, as the exact height of the same spot can vary with it, so it is essential that a common datum is adopted for topographic data from different sources in the same analysis. Certain arithmetic operations, such as subtraction and addition, can be performed on interval data meaningfully, but not division. For instance, it is allowable to subtract the elevation before an earthwork project from the elevation after the project to determine the volume of earth materials that have been excavated.

Ratio data are the numerical or quantitative manifestation of an attribute, such as population density (e.g., 55 people per km²), usually calculated from two variables (e.g., population and area in this case). Ratio data have an absolute zero, and the difference between ratios is meaningful. Owing to the existence of a common benchmark (e.g., 0), all ratio data are directly comparable with each other meaningfully. However, it must be noted that the difference between interval and ratio data may not be as clear-cut as has been discussed in some cases.

It is important to understand the differences between these four scales of enumeration as they directly dictate the kind of spatial analysis that can be sensibly performed on them. Only certain kinds of analysis are permissible with data enumerated at a given scale (Table 3.3). Although attribute data enumerated at either of the four scales can be analysed, they cannot be analysed indiscriminately. In fact, the more general the enumeration scale, the more restrictive the types of spatial analysis that can be performed to them. For instance, nominal data (e.g., thematic land cover maps classified from satellite imagery) can be filtered spatially in a neighbourhood-based analysis. In post-classification processing of an image, a tree pixel may be generalised as urban if it is surrounded by urban pixels. In addition, nominal data can also be used in spatial modelling. For instance, a grid cell may represent a pasture covered by healthy vegetation, or it can become degraded to barren if overgrazed over an extended period. This bare ground pixel can transition to grass if sufficient time is allowed for the vegetation to regenerate in the absence of external disturbances. Of the four scales, ratio/interval data are the least restrictive and the most analysed in spatial analysis, especially in descriptive analysis. However, inferential analysis can still be performed on spatial data enumerated at all four scales. In this case, the target of analysis is the spatial component, not the attribute value, such as whether pine trees in a forest are spatially clustered in their distribution. Similar analysis can be undertaken for house prices to explore whether wealthy suburbs are spatially clustered.

TABLE 3.3

Relationship between the Enumeration Scale of Attribute Data and the Type of Spatial Analysis Permissible

Enumeration scale	Examples	Descriptive analysis	Explanatory analysis	Predictive/ Inferential analysis
Nominal	Land cover, degraded versus intact	No	No	Yes, as in simulation, but not value
Ordinal	Severe degradation, moderate burn, slight pollution	No	Yes	Yes (e.g., whether socioeconomically deprived neighbourhoods are clustered)
Interval	Contour interval of 90–100 m	Yes	Yes	Yes
Ratio	Population density, pollutant concentration	Yes	Yes	Yes

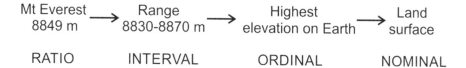

RATIO INTERVAL ORDINAL NOMINAL

FIGURE 3.1 Four main scales of thematic (attribute) data and their relationship. The conversion from one scale to the next is only one-directional, from specific to general.

Under normal circumstances, there are no relationships between the four scales of enumeration. However, on some special occasions, it may be necessary to convert the data from one scale to another before any kind of spatial analysis is permissible. For instance, whether a city is racially segregated spatially can be studied by converting the percentage of the black population in a suburb into the ordinal scale first. Conversion between the four scales is one-directional, in that all data can be grouped more generally from specific ones (Figure 3.1). For instance, the highest peak of Mt Everest is known to have an elevation of 8,849 m (ratio) above sea level. Because of survey inaccuracy, its actual elevation can range from 8,830 to 8,870 m (interval). Regardless of the exact elevation, it is still the highest elevation on Earth (ordinal). The peak is located in a land area covered by perennial snow (nominal). As the level of generalisation rises from left to right in Figure 3.1, the detail about the spatial entity is increasingly lost. But such a conversion in the opposite direction is not feasible, as it is from general to specific.

3.2 SPATIAL SAMPLING

In spatial analysis, it can be difficult to study the entire population because of the enormously large extent of a study area or the sheer volume of data that have to be collected. This difficulty necessities *spatial sampling*. It is a way of determining the attribute value at certain selective and representative locations *in situ* directly (e.g., measurement of soil pH). Quantitative measurement is produced about the attribute to be studied. Through analysis of such collected data, it is possible to yield unbiased estimates of the population if the sample size is sufficiently large. What remains unresolved is how to design the sampling scheme and how many samples should be collected. Spatial sampling differs from conventional non-spatial sampling in that the sampling scheme to be adopted must take into account the location and the distribution of the points to be sampled. This section will cover all matters related to spatial sampling, including sampling strategies, the quantity of samples to be collected and their distribution, and the spatial dimension of sampling units.

3.2.1 SAMPLING CONSIDERATIONS

In a number of fields of study, ground samples must be collected to facilitate data analysis and modelling. For example, ground control points must be selected for geo-referencing remotely sensed images and aerial photographs that do not conform to any global coordinate systems. Samples are needed to assess the accuracy of land cover maps produced from satellite images. Samples are also essential to establish the empirical relationship between their ground properties and those on satellite imagery at the corresponding locations so that point-observed properties can be extrapolated to spatial distributions. In summary, spatial sampling is indispensable in collecting data in the field. Apart from fulfilling the usual requirement of yielding an unbiased sample, spatial sampling has its own methods and requirements to fulfil. Namely, it must enable a sufficient number of spatially representative samples to be collected at the lowest cost possible. A delicate balance between the minimum number of samples to be collected and sufficient representativeness of these samples must be struck. A high level of geographical representativeness can be achieved by collecting samples all over the study area. Such a strategy can guarantee that all the possible values are captured by and represented in the collected samples. Consideration should also be given to site accessibility beforehand. It is inconvenient or impossible to collect samples that are located at inaccessible sites, such as in the middle of a swamp. If relevant, the vertical distribution of samples also warrants consideration, for instance in studying the distribution of vegetation over a mountain range. Equally, the slope aspect at which vegetation is distributed may also need to be taken into consideration in sampling design. Depth of samples is another factor to consider in studying the spatial distribution of soil nutrients and moisture, as both vary with depth.

3.2.2 STRATEGIES OF SPATIAL SAMPLING

Conventionally, the two main methods of random and systematic sampling are both applicable to spatial sampling. In addition, spatial sampling can also be clustered or stratified (Figure 3.2). *Random sampling* means that samples are collected without any bias or preference over the entire study area (Figure 3.2a). This strategy of sampling is important in ensuring the scientific objectiveness of the collected samples. Namely, there is no bias towards the collected samples, and they are geographically representative. However, this may not result in a sufficient number of samples being collected if the attribute concerned is not randomly distributed in space. For instance, forests are not randomly distributed. Settlements are usually located in flat areas close to transport infrastructure. Bare patches in an alpine meadow are not randomly distributed spatially (Figure 3.3). In this case, there is a high chance that

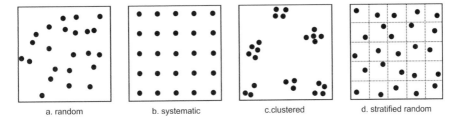

a. random   b. systematic   c.clustered   d. stratified random

FIGURE 3.2   Four common methods of spatial sampling: (a) random, (b) systematic, (c) clustered, and (d) stratified random.

FIGURE 3.3   The spatial distribution of bare patches and rodent burrows in a degraded meadow. A sufficient number of burrow samples can be collected using only the stratified random strategy.

random sampling will lead to an insufficient number of samples being collected. The alternative is *systematic sampling*, which means that samples are collected at a constant interval both horizontally and vertically (Figure 3.2b). Thus, the number of sampling spots is known once the sampling interval is fixed for a given study area. Similar to random sampling, systematic sampling cannot guarantee that a sufficient number of samples will be selected in the end (e.g., at the pre-determined sampling site, the meadow may be rather healthy instead of being denudated). The number of sampling spots can be increased by decreasing the sample interval. However, there is still no guarantee that the pre-designated spots of sampling will be accessible or suitable in the field. For instance, trees of the same species are likely to be located on certain slopes in a mountain range, and settlements are likely to be located along a road. In both cases, they are better sampled with the *clustered sampling* strategy, in which certain samples are collected in closer proximity to each other than to members in other groups. Once a target of sampling is hit, then more samples are collected in its vicinity or the partitioned quadrat (Figure 3.2c). This method can improve sampling efficiency. However, it still remains uncertain how many clusters should be adopted, nor the number of samples to be collected in each cluster.

All of these three sampling methods face the same dilemma of whether an adequate number of samples can be collected. Therefore, the same sampling process has to be repeated to add more samples in case of inadequacy, slowing down the sampling process and potentially increasing sampling cost. One strategy to overcome this limit is *stratified sampling* (Figure 3.2d). Actually, stratifying the area to be sampled has nothing to do with sampling. It is only a preparatory step for sampling by dividing the entire study area into a number of sub-areas of a uniform shape and size, and a specific number of samples are then collected from each sub-area randomly (Figure 3.2d). In other words, it works only if it is combined with one of the three sampling strategies discussed earlier. The distribution of samples within each sub-area can be random or clustered, depending upon with which sampling method it is combined. If combined with random sampling, stratified random sampling guarantees that the collected samples are widely distributed over the entire area of study, and hence are geographically representative. Similarly, stratified sampling can be combined with clustered sampling, and a large number of samples are collected from a sub-area where the target is hit, but may not be collected from other sub-areas if the target is absent or the required number of samples have already been collected.

Apart from such *geographic stratification*, samples can also be collected via *thematic stratification*, in which the number of samples having a pre-defined attribute value will be selected, usually randomly. Such thematic stratification can guarantee the selection of samples with pre-determined values. Which sampling strategy is the best to use depends on the purpose of sampling. For instance, thematically stratified random sampling is the best for selecting samples in assessing the thematic accuracy of land cover maps produced from digital analysis of satellite imagery in remote sensing. In this example, an adequate number of sample points must be selected randomly for each type of mapped land cover, which can hardly be satisfactory if they are selected purely via random sampling.

The above-discussed sampling strategies are generic, in that they do not fulfil any prescribed objectives, so they may not be applicable to certain specific instances, such as the sampling of elevations from digitalised contours for the construction of digital elevation models (DEMs). In this case, the position and distribution of contours (i.e., elevation) are known already. Points are best sampled along contours whose elevations are known and of the same value along the same contour (Figure 3.4). This sampling strategy is termed *purposive sampling*, which refers to the acquisition of samples along pre-existing lines. In purposive sampling, sample points are collected by tracing contours on topographic maps. It is virtually identical to sampling along transects, except that the lines are already in existence and they are seldom straight. Since all contours are traced to achieve an accurate representation, sampling density is automatically linked to the complexity of the terrain to be represented. More points are sampled for a terrain with more vertical variations and from contours of higher complexity (e.g., a higher curvature), and vice versa, as few contours are needed to represent the vertical relief of flat terrain. The sampling interval can be further lengthened along contours with a low curvature (Figure 3.4). Purposive sampling has another added advantage in that all the points sampled along the same contour have a known and the same height, thus there is no need to estimate the elevation at all the sampled locations.

Irrespective of the exact strategy adopted, all sampling methods should minimise biases and maximise objectivity. They all face a common issue of sampling accuracy that can be assessed by comparing the samples or a surface constructed from them with

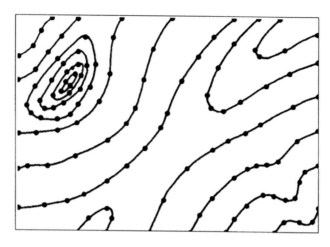

FIGURE 3.4   Purposive sampling in collecting elevations along contours whose elevation is known in constructing a DEM (source: Gao, 1995).

TABLE 3.4
Comparison of Accuracy of DEMs (50 by 50) Constructed with Purposively Sampled Points Followed by Spatial Interpolation with Systematic Sampling without Interpolation (source: Gao, 1995)

		Purposive sampling			Systematic sampling	
Resolution (m)	Terrain type	RMSE (kriging)	RMSE (moving averaging)	No. of sample points	RMSE	No. of sample points
48	Rugged	4.46	5.48	905	9.40	625
	Gentle	3.21	4.73	622	7.62	625
24	Rugged	2.04	2.99	1,809	3.35	2,500
	Gentle	1.62	2.26	1,866	3.33	2,500

Source: Gao (1995).

the ground truth at a certain number of points. The differences between the two sets of elevations are analysed statistically and expressed in root-mean-square error (RMSE). Unsurprisingly, the most important factor influencing sampling accuracy is the sampling interval (Table 3.4), followed by the complexity of the topographic surface (e.g., gentle versus rugged) in elevation sampling. If the samples are converted to a regular grid, then the method of conversion also affects the accuracy. According to Table 3.4, purposive sampling is able to lead to a more accurate representation of the terrain than systematic sampling, especially at a large grid size of the DEM. This is true regardless of the complexity of the terrain to be represented, even with far fewer sample points. For instance, only 1,866 sampling points were retained for the gentle topography, and they caused the representation to have a RMSE of 1.62 m, only half of the 3.33 m achieved using 2,500 points sampled using the systematic method. Admittedly, purposively sampled points can only be converted into regular grids with the assistance of a spatial interpolator, which inevitably introduces uncertainties to the interpolated elevations. Even so, the final DEM generated using moving averaging (RMSE = 2.26 m) is still about one-third more accurate than its counterpart constructed from systematically sampled points (RMSE = 3.33 m). With the wide availability of computer software for spatial interpolation, the purposive method of spatial sampling should be widely adopted to construct DEMs from contours. Purposive sampling is more accurate than systematic sampling as the collected samples all have a definite elevation associated with them. In other words, they have a high level of certainty. Systematic sampling, on the other hand, likely involves the collection of samples at which no definite height is indicated. So a degree of uncertainty is introduced in estimating the height from the nearby contour values via spatial interpolation.

In addition to the aforementioned classic methods of sampling, other more innovative strategies such as *adaptive* or *progressive sampling* have been proposed in the literature (Thompson and Seber, 1996). In this relatively recent sampling strategy, information about the sampled data is generated and used to continuously guide subsequent sampling (Stein and Ettema, 2003). Thus, the sampling scheme is not fixed during sampling, but evolves constantly based on what has been sampled already. It comprises two steps:

(1) The sampling area is divided into a number of contiguous blocks. Each block is further neatly partitioned into an equal number of non-overlapping plots. Thus, each plot is surrounded by four neighbouring plots. During sampling, a numerical criterion is formulated for a plot.
(2) If the criterion is met for a randomly selected plot, then its four neighbouring plots will be sampled.

This strategy differs from all the aforementioned schemes, in that the procedure is adaptive to the population values observed in the field. A very common application of adaptive sampling is digital photogrammetry, in which the collected height is examined. If it does not vary much spatially, then a larger sample interval is adopted. In this way, it is still possible to capture the spatial variability of elevation with a smaller number of samples as the chosen data points minimise the mean squared error of the estimate of the already collected samples.

3.2.3   SAMPLING DIMENSIONALITY

The *dimension of sampling* refers to the topological dimension of the sampling site or sampling unit. Basically, there are two fundamental types of sampling units: points and plots. Of the two, point samples are much more common than plot samples (Figure 3.5). In the field, whether samples should be collected at a point or over a plot depends solely on the nature of the attribute to be sampled. *Point sampling* is the default choice if the attribute is observable at every given point, such as soil

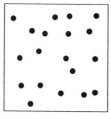

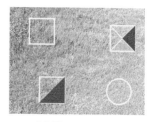

| (a) Point sampling | (b) Transect sampling | (c) Plot sampling |

FIGURE 3.5 Common dimensions of sampling units: (a) point sampling, (b) transect sampling to capture the dune profile at strategic positions (the distribution of sampling points along a transect is proportional to surface variability), and (c) plot sampling. Plots can be square or circular in shape. Square sampling plots allow sub-area sampling (e.g., within half or a quarter of a plot).

pH, organic matter content, elevation, and so on. Point samples can be distributed randomly in the space of study (Figure 3.5a). In studying the morphology of coastal sand dunes, the point sampling strategy is not satisfactory as the number of collected samples may not necessarily capture the true variation along with certain directions even with a sufficiently large sample size. In this case, *transect sampling* is commonly adopted. In order to generate a reliable representation of dune morphology, it is essential to sample elevations along a number of strategically located, parallel transects, all perpendicular to the shoreline (Figure 3.5b). These one-dimensional samples can reveal the change in beach morphology economically. Along each transect, samples are collected at variable intervals. As expected, the number of samples along each transect should be proportional to the surface complexity, with more samples collected in highly variable sections of a transect.

However, when it comes to sampling attributes whose value is available only at spatially aggregated units, such as plant biomass, vegetation cover, and carbon stock, point sampling is utterly unsuited. The sampling unit has to be expanded to an area or plot. *Plot sampling* is the norm in sampling attributes that do not exist at points (Figure 3.5c). In addition, plot sampling is essential if the *in situ* sampled attributes are to be linked to their spectral properties on a satellite image that is composed of pixels covering a ground area. In this case, the spectral properties of the sampled sites are always recorded over a square area. In fact, this kind of spatial sampling is applicable to all area-aggregated variables, such as all sorts of socioeconomic data. Population, income, and even crime rate must be collected at the meshblock or the district level. Compared with point sampling, area sampling takes much longer to complete and is more subjective, as the results vary with the size of the sampling plot, especially when the attribute to be sampled is highly variable in space. Invariably, the sampling plot has a square shape that can be partitioned into sub-sections conveniently with the assistance of the two diagonal axes (Figure 3.5c). This partitioning is quite useful in studying vegetation biomass change with time. For instance, biomass can be collected in one of the four quarters of the sampling plot in the first year, and any one of the remaining three quarters in the subsequent year.

In addition to being arduous and time-consuming, plot sampling can also be destructive (e.g., clipping grasses to determine their fresh above-ground biomass). To minimise the destruction and expedite sampling, it is critically important to select an appropriate plot size. A small plot size means less fieldwork as the grasses can be clipped very quickly, and the results can be more reliable. However, a too-small plot may not be representative. On the other hand, a large plot can be more representative, but the amount of fieldwork required is going to quadruple. Thus, a delicate balance must be struck between sample plot size and representativeness. There is no definite answer to what is the optimal plot size as it depends on what is going to be done with the collected samples. If they are to be associated with image spectral properties, then plot size should be commensurate with the pixel size or the spatial resolution of the images to be analysed. However, the actual plot size may not be crucial if the attribute under study is spatially uniform (e.g., it is scale-independent).

There is no stated requirement of the orientation of the sampling plots, but the positioning of the samples should be the plot centroid. Besides, samples must be collected randomly, which may prove impossible with socioeconomic data, as data at every enumeration unit must be collected. Thus, certain spatial analysis results must be treated with caution as data are not independent (e.g., the selection of one sample in one round does not exclude its selection in the next round of sampling). In plot sampling, a circle or a ring is tossed into the air randomly, and wherever the ring lands is then sampled. Thus, both randomness and independence of spatial sampling are guaranteed. Such sampled data allow inferential spatial analysis to be carried out without any restrictions.

Once samples are located, the attribute data (e.g., biomass) are collected first, followed by the recording of the sample location with GPS. In the case of plot sampling, the plot's centroid should be logged as the position of the collected sample that can also be analysed spatially. GPS receivers come with vastly differing functionality and accuracy levels. Irrespective of the GPS model used, in general the positioning accuracy at a fixed sampling site should be very high after averaging, especially if the coordinates are differentially corrected. Samples should be positioned as accurately as possible by logging the locations repeatedly and averaging them. This is especially critical if the sampled attributes are going to be linked to their spectral properties in remote sensing imagery, and the attribute value varies drastically even over a short spatial range. Inaccurate positions of the collected samples can cause the *in situ* sampled attribute values to be linked to the corresponding image property of the neighbouring pixels. Inevitably, the mismatch between *in situ* samples and their corresponding spectral values on the satellite image will lead to unreliable results. Typical positional accuracy used to be around 3–5 m, but can now be as accurate as sub-metres after differential correction. More accurate positioning is possible with Ramon filtering of the logged data, but may not bring much benefit if other geospatial data used in spatial analysis have a coarse spatial resolution (i.e., around 3 m). In the worst case, the positioning inaccuracy of *in situ* samples should not exceed the spatial resolution of the imagery to be used.

3.2.4 SAMPLE SIZE AND SPACING

Sample size refers to the quantity or volume of samples to be collected. It is an important sampling factor to consider, as a small size does not allow the population to be estimated at a sufficiently high confidence. Inferences drawn from a small sample about the population face high risks. On the other hand, a larger than necessary sample size will incur excessive and unnecessary expenses, effort, and time in the field and in the lab to process the collected samples. Therefore, it is crucial to select an optimal number of samples. There is no universal rule regarding the optimal sample size, which varies according to the observations and the field of study. In order to be statistically viable, ideally, at least 30 samples should be collected. However, this bare minimum does not leave any room for validation, so the desirable actual number is larger than this. The drawbacks associated with a larger sample size are prolonged sampling, a higher cost, and more destruction to the target of sampling, as in the determination of vegetation biomass through clipping of grasses to ground level. In this case, a smaller sample size means less destruction to the target of study. Whenever permissible, as large a sample size as possible should be adopted, as it offers the flexibility of discarding those samples with incomplete records or of dubious quality in subsequent data analysis.

The appropriate size of samples is related to two statistical parameters, population size (N) and the desired level of precision (the margin of error, e). The former refers to the total quantity of observations, which is usually unknown in spatial sampling. The latter is a percentage that is indicative of the degree to which the collected samples can reflect the overall population. A smaller margin of error means that the samples are closer to the whole population at a given confidence level. The theoretical sample size (n) can be calculated using the Cochran formula:

$$n = \frac{Z^2 p(1-p)}{e^2}$$

(3.1)

where Z value indicates the confidence level and can be found in a Z table, p represents the (estimated) portion of the population with the desired attribute, and e is expressed in decimal form (e.g., 3% = 0.03). For instance, at a confidence level of 90%, $Z = 1.65$, if p = 0.6, then

$$n = 1.65^2 \times 0.6 \times (1-0.6)\big/(1-0.90)^2 = 65.$$

If the population size (N) is taken into account, then Equation (3.1) needs to be modified as follows:

$$n = \frac{\dfrac{z^2 \cdot p(1-p)}{e^2}}{1 + \dfrac{z^2 \cdot p(1-p)}{e^2 N}}$$

(3.2)

Thus, a larger sample size is helpful to reduce the margin of error. The achievement of a higher confidence means a collection of more samples in the field.

Another consideration related to sample size is sample spacing. In systematic sampling, a uniform spacing is adopted for the entire area of sampling, and sampling spacing is strictly proportional to sample size. However, in all other methods of spatial sampling, spacing is spatially variable. Theoretically, sample spacing should vary with the variability of the attribute under study. Thus, samples should be spaced at close proximity to each other in highly variable sections of the area, and vice versa (Figure 3.6). Of the various sampling strategies presented in Section 3.2.2, systematic sampling has a fixed spacing, and is inferior to purposive sampling as the attribute to be sampled may not have an indicated value at the sampled spot. In comparison, purposive sampling allows the sampling spacing to be variable, while all sampled points along the same contour have an indicated value. This sampling strategy conforms to the requirement that in sampling elevations from contours, sampling spacing or interval should vary with the complexity of the terrain. It can be measured by the standard deviation of elevations per unit area. Ultimately, sampling interval implicates the accuracy of topographic representation (Gao, 1998). If modified by the density of the sampled points, RMSE bears a linear relationship with sampling spacing that does not vary with the complexity of terrain noticeably (Figure 3.6). The accuracy of the constructed DEM is a function of sampling interval and the complexity of the terrain to be represented. The sampling interval also affects the topographic variables derived from the sampled elevations.

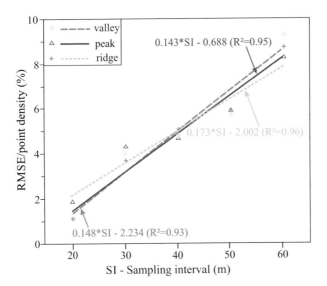

FIGURE 3.6 Relationship between sampling interval and the accuracy of the constructed DEM for three types of geomorphic units (source: modified from Gao, 1998).

3.3 SPATIAL ASSOCIATION AND PATTERN

3.3.1 Spatial Continuity versus Spatial Pattern

One of the main targets of spatial analysis is the spatial variation of the attribute of a variable under study. Spatial variability can be treated as either spatially continuous or spatially discrete. *Spatial continuity* refers to the gradual and continuous change in the attribute value in space, usually two-dimensional (Figure 3.7a). Continuity is a term commonly reserved for describing point-based attributes expressed at the ratio scale, such as topography. The depicted attribute may be tangible in reality, such as a topographic surface. It can also be invisible (e.g., aerosol concentration and population density), or even imaginary (e.g., risk of landslides). In either case, the attribute is spatially continuous and varying. The behaviour of the attribute is said to follow some kind of continuity if its value enumerated at points can be predicted from that of observations in close proximity to it. When the attribute value behaves predictably (e.g., without any sudden, abrupt variations), it is said to be continuous locally. The continuous surface can be expressed as a function of its location or its distance to a reference point numerically. The attribute value varies only gradually with a high degree of predictability within a certain range or distance. Thus, observations that are further apart vary more widely in their attribute values than those that are closer. In other words, continuity refers to change in the attribute value, not where the samples are collected. A continuous attribute can be predicted more reliably than a discontinuous one.

In contrast to spatial continuity, *spatial pattern* refers to the overall spatial arrangement or distribution of observations that share the same or a similar attribute value (Figure 3.7b). It is normally reserved to describe area-based nominal (e.g., categorical) and ordinal data (Table 3.5). Common spatial patterns may refer to land use patterns and settlement patterns. If *in situ* samples are collected at specific sites, such

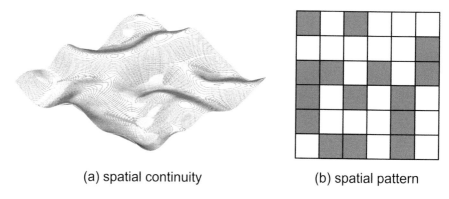

(a) spatial continuity   (b) spatial pattern

FIGURE 3.7 Comparison between spatial continuity and spatial pattern: (a) spatial continuity is commonly reserved to describe point data of a continuous attribute value, such as elevation (source: Joseph Claghorn, 2014, with permission); (b) spatial pattern is reserved to describe attributes enumerated over an area expressed at the nominal scale (e.g., presence or absence of bushfires).

TABLE 3.5
Comparison between Spatial Continuity and Spatial Pattern in Spatial Analysis

Items	Spatial continuity	Spatial pattern
Dimension of sampling	Point-based	Area-based
Scale of attribute	Interval/ratio	Nominal/ordinal
Nature of analysis	Predictive, quantitative, less restrictive	Descriptive, qualitative, restrictive
Inferential analysis?	No	Yes
Accuracy of analysis	Known	Irrelevant
Reference pattern	Not needed	Essential
Visualisation	Perspective surface	Prism/choropleth shading

as air pollutant concentration, they cannot be used to study spatial patterns. In order to qualify, the attribute value has to be converted to the ordinal or nominal scale. This is because in order to be studied for its spatial pattern, an attribute must have an invariable or constant value within a polygon, not at particular spots. Attribute values between adjoining polygons are jumpy and discontinuous. They appear as prisms of various heights if the attribute is visualised in a perspective prism map. However, the disparity between spatial continuity and spatial pattern diminishes with some point-based observations, such as aerosols and particulate matter suspended in the atmosphere. They can be analysed for their spatial pattern, in which the location of points, not their attribute value, becomes the target of analysis.

It is important to differentiate spatial continuity from spatial pattern as they are analysed differentially. Spatial continuity is exploited in predictive spatial analysis using geo-statistics. The analysis is quantitative, and the accuracy of the analysis can be quantified as well. In contrast, spatial pattern is normally studied via either descriptive or inferential analysis. For instance, spatial patterns can be quantified using special statistics such as spatial auto-correlation based on the popular Moran's I statistic. Commonly used descriptive analysis methods, such as nearest neighbour and k-function, can all be used to study spatial patterns. In addition, spatial patterns can also be tested to see whether they obey some kind of distribution (see Section 3.5.1). All of these methods are applicable to the global spatial auto-correlation using all data. The Moran statistic may not be completely valid when the data show some local clustering, though (Ord and Getis, 1995).

3.3.2 SCATTERPLOT AND H-SCATTERPLOT

The association of one variable with another can be revealed via their *scatterplot* (Figure 3.8). A scatterplot is a diagram illustrating the distribution of the observed values, usually along two axes. It is able to reveal graphically the association that may exist between two variables. They are said to be highly associated or correlated with each other if their scatterplot follows a neat trend line. A scatterplot is good at

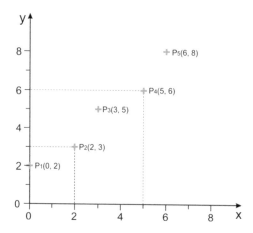

FIGURE 3.8   A scatterplot illustrating the relationship of one variable x to another y. It shows all possible pairs of data values whose locations are separated by a certain distance in a particular direction.

revealing the nature of the relationship, whether linear or non-linear. The definition of a scatterplot must be modified in spatial analysis in which the target of study is a sole variable or attribute. It refers to the relationship of this variable's attribute value at one location with those of its neighbours. In the spatial domain, a scatterplot becomes an *h-scatterplot*, or a diagram illustrating the distribution of attribute values between a pair of observations separated by a spacing of *h*, namely a scatter diagram of vector(t) versus v(t+h) (Figure 3.9). *h* is a vector with direction and magnitude. It simply means separation in certain directions, but does not stipulate the actual direction. Therefore, observations can be paired up horizontally, vertically, and even diagonally for raster grid cells. This pairing of observed values can take place in more than one direction, resulting in multiple h-scatterplots from the same dataset. Since an h-scatterplot results from pairing the attribute values of all possible observations separated by a certain distance in a particular direction, it varies with two spatial parameters, distance and direction. It is particularly useful in studying whether the attribute behaves predictably in space.

As illustrated in Figure 3.9, the number of paired observations decreases as the separation *h* increases. Moreover, as *h* rises, the observed attribute value of the paired observations deviates increasingly further from the dashed 1:1 trend line, suggesting that the association between neighbouring observation values is loosened. As with all scatterplots, an h-scatterplot is able to reveal the correlation of the same variable at different separations or scales.

3.3.3   CORRELOGRAM

Correlogram is a portmanteau contraction of "correlation and diagram". A correlogram is a function of |*h*|, and shows the variation of the correlation coefficient with separation. Its horizontal axis indicates the absolute value of separation |*h*|

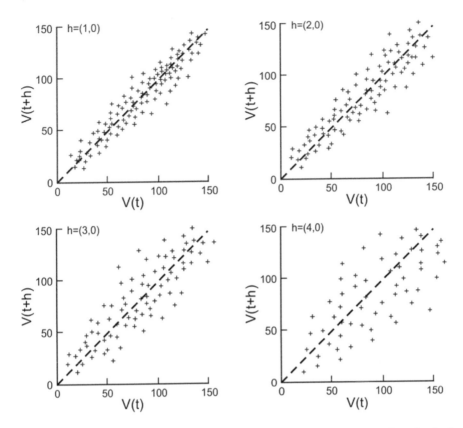

FIGURE 3.9 The concept of the h-scatterplot, which illustrates the distribution of paired attributes in different directions. In this case, observations are paired only horizontally ($x = 1, 2, 3,$ and 4, $y = 0$).

irrespective of its direction, and its vertical axis illustrates the correlation coefficient ranging from 0 to 1 (Figure 3.10). A correlogram graphically illustrates how the correlation of an attribute value with that of its neighbouring observations' values decreases as the separation rises, which just confirms Tobler's First Law of Geography. Observations in closer proximity resemble each other more closely in their attribute value than those distant from each other. A correlogram represents another means of illustrating the relationship between the correlation coefficient and *h* without involving direction.

It must be noted that Figure 3.10 just illustrates the simple linear correlation coefficient or Pearson's product-moment correlation coefficient calculated using Equation (2.14). This linear correlation coefficient has a theoretical value ranging from –1.0 to 1.0. A value of 1.0 means perfect correlation (e.g., with itself), while –1.0 means just the opposite (e.g., negation). To some degree, this correlation is indicative of spatial association of the same attribute at different locations only. It does not imply dependency between them (e.g., elevation at location A does not depend on the elevation

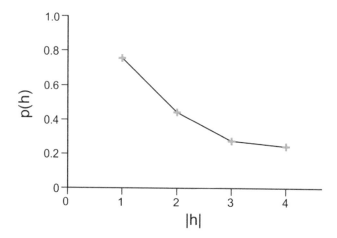

FIGURE 3.10   A correlogram showing the variation of correlation coefficient with separation |h|.

at location B), nor a cause–effect relationship (e.g., the high elevation at location A does not cause the elevation at location B to be high), even though it is possible for the relationship to be cause–effect in some instances.

3.3.4   Spatial Auto-correlation

Also known as serial correlation, *auto-correlation* refers to the correlation of adjacent observations of the same variable that are separated by a lag either spatially or temporally, such as the pairwise correlation of univariate observations at the same location, but at different times (e.g., traffic flow and river discharge). In spatial analysis, it is called *spatial auto-correlation* or the correlation of the same attribute at different locations within a neighbourhood (Figure 3.11). It is much more complex than simple auto-correlation as it involves direction and is multi-dimensional. As mentioned earlier, spatial auto-correlation likely varies with spacing and direction. In other words, it varies with scale.

Spatial auto-correlation has three unique connotations in spatial analysis:

(1) It is a quantitative measure or index of spatial distribution. A high spatial auto-correlation of a variable means that it enjoys a high degree of spatial continuity and is highly predictable.

(2) It can be used to explore potential factors responsible for the formation of the observed pattern. Through correlation analysis of the observed spatial auto-correlation with other variables of the same area, it is possible to identify their spatial linkage. A closer correlation suggests a higher possibility of association with this factor.

(3) It can be used to predict the occurrence (probability) of the attribute value in new locations, as in spatial interpolation. The value at a new location can be reliably predicted from its neighbouring observations if the attribute is highly auto-correlated across space.

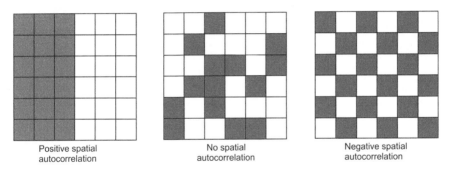

| Positive spatial autocorrelation | No spatial autocorrelation | Negative spatial autocorrelation |

FIGURE 3.11   The concept of spatial auto-correlation for binary observations at regular grids.

Since spatial auto-correlation is directional, the correlation may be stronger in one direction than in another, a phenomenon known as *anisotropy*. It refers to the variation of spatial auto-correlation with direction. Anisotropy is commonly caused by the influence of the environment, such as topography that affects airflow. Thus, those temperatures at a higher elevation are less subject to the influence of terrain-induced atmospheric turbulence, hence there is a higher spatial auto-correlation than at a lower elevation closer to the Earth's surface. Anisotropic auto-correlation is exemplified by the latitudinal distribution of air temperature of the conterminous US that has a strong latitudinal gradient, and water temperature at different depths. Air temperatures at the same latitude are more closely correlated with each other than with those along the latitudinal gradient. Similarly, water temperatures at the same depth in a stagnant lake are correlated spatially more closely to each other than those at different depths. If the correlation does not vary with direction, it is termed *isotropic*. For instance, at a local scale, surface elevation of a residual hill behaves isotropically.

3.3.5 Cross-correlation

Cross-correlation refers to the correlation of two series of observations at successive lags, either temporal or spatial. *Temporal cross-correlation* measures the similarity or variation of the same attribute at different times for time-series observations. Time-series data analysis has found useful applications in real life. For instance, the temporal cross-correlation coefficient can be used to match hydrographic diagrams observed at multiple stations (Figure 3.12a). A match of multiple hydrographic diagrams can reveal the time it takes for floodwater to flow from one station to the next. Similarly, temporal cross-correlation analysis is commonly carried out to match stratigraphic profiles at multiple locations. It can reveal the time lag of sedimentation among them. In both cases, the match is based on a series of calculated cross-correlation coefficients, in which the time-series observations in the two datasets are shifted by the pre-determined temporal interval each time. The best match is achieved when the cross-correlation coefficient reaches the maximum. The magnitude of shift to achieve this match is the temporal lag between the two time-series observations. Temporal cross-correlation is thus one-dimensional, as the coefficient varies with time only. The diagram of the cross-correlation coefficient and the

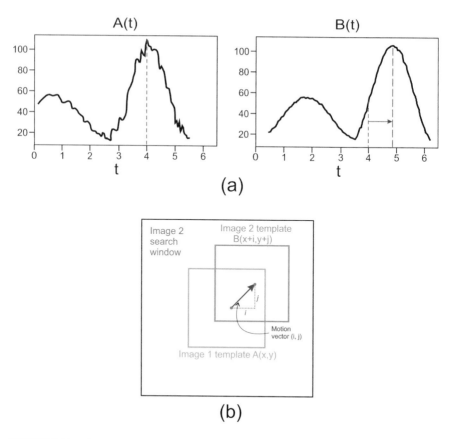

FIGURE 3.12 Comparison of temporal (a) and spatial (b) cross-correlation. The two time-series observations can reach an almost perfect match by shifting one of them slightly. The same feature is searched in image 2 in all possible positions and directions within the specified window via calculating spatial cross-correlation.

magnitude of the temporal shift is termed the *cross-correlogram*. It illustrates how the correlation coefficient varies with the temporal lag.

Spatial cross-correlation involves two spatial variables of the same area, such as satellite images that are separated by a certain time, so their cross-correlation is location-dependent. It measures the correlation of one variable at location A with the attribute value of another variable at location B. Both images may not cover the same spatial extent, but must encompass a common sub-area. It is within this search window that the cross-correlation coefficient between the two sub-images is calculated (Figure 3.12b). It is possible for one of the images to cover a larger spatial extent than the other, which enables the search for the feature of interest in the same, but shifted window in the second image. Whenever either of the images is shifted slightly by any magnitude in any direction, their spatial cross-correlation changes accordingly. A match is considered to occur at the shifted location where the cross-correlation coefficient is maximal. It is calculated using Equation (3.3):

$$\rho(i,j) = \frac{\text{cov}[A(x,y), B(x+i, y+j)]}{\sqrt{\text{var}[A(x,y)] \cdot \text{var}[B(x+i, y+j)]}} \tag{3.3}$$

where

$$\text{var}[A(x,y)] = \sum_{x=1}^{I}\sum_{y=1}^{J}\left[A(x,y) - \overline{A}\right]^2 \tag{3.4}$$

$$\text{cov}\left[A(x,y), B(x+i, y+j)\right] = \sum_{i=1}^{I}\sum_{j=1}^{J}[A(x,y) - \overline{A})(B(x+i, y+j) - \overline{B}(i,j)] \tag{3.5}$$

where I and J refer to the number of rows and the number of columns of the search window, respectively, i and j are the counters, and $\overline{A}$ and $\overline{B}(i,j)$ stand for the mean attribute value of the first and second images (variables) within the search window that has shifted by (i, j), respectively.

Spatial cross-correlation is much more complex than temporal cross-correlation because both variables must be spatial. Their cross-correlation coefficient varies not only with the magnitude of shift, but also its direction. As illustrated in Figure 3.12b, a search window must be specified beforehand. There is no restriction on its size. Naturally, it must be small enough to ensure that a search area common to both layers can always be found. The actual area must be commensurate with the physical size of the feature to be searched (e.g., an eddy). In comparison, the actual range of search is much easier to decide as it is governed by the magnitude of motion between the two variables or layers, or the speed at which the feature of interest moves. If the layers are images, it depends on their temporal separation as well. This search window is usually shifted in all possible directions and to all possible positions. At each position, the spatial cross-correlation coefficient between the two variables is calculated based on the search window. A spatial (lag) correlogram can be produced to show the correlation coefficient at different lags.

As in temporal cross-correlation calculation, the two variables used in calculating spatial cross-correlation are also implicitly time-series in certain cases. They can cover the same or roughly the same ground area, but at different times, for instance multi-temporal satellite images of an area suffering from landslides (before and after) in detecting the landslide distance. The main applications of spatial cross-correlation lie in the detection of motion from multi-temporal satellite images. It can also be used to identify the vector of movement of a feature, such as the motion of landslides following an earthquake (Peppa et al., 2017). Other common applications include detection of ocean circulation patterns (speed and direction) from time-series satellite images that are separated by 12 hours (Gao and Lythe, 1998) or less. By matching all the possible rows and columns within a search window, it is possible to find a position where the two images have a maximal spatial cross-correlation. The direction and magnitude of the second image relative to the first one indicate the vector of motion. In this application, the two input layers depict sea surface temperature derived from thermal images. The location at which the maximal cross-correlation coefficient is found from the initial search position represents the distance of motion, and their bearing, the direction of motion.

Another field of application is digital photogrammetry, in which multiple overlapping stereoscopic aerial photographs need to be positioned relatively to achieve 3D viewing, known as image interior orientation. In this application, the photographs are automatically matched through their maximal cross-correlation. There is no guesswork involved, as the two photographs are taken within seconds of each other and encompass the same ground area, albeit from two slightly different perspectives. It is permissible for the two photographs to rotate slightly or have differing scales. All of these applications are underpinned by one assumption: namely, the same feature to be searched in all the images acquired at a short separation from each other moves only translaterally without any deformation within any field of it, even though the two images can have slightly different scales. Whenever this assumption is violated, the quality of the detected motion will be degraded.

3.3.6 SCALE IN SPATIAL ANALYSIS

Scale can have vastly differing connotations in different contexts. In cartography, map scale simply refers to the ratio of a length or distance on a map to its actual distance or length on the ground. Scale in this section refers to the areal extent to which the observation is enumerated, not the enumeration scale of its attribute value. Scale is so inherently linked to spatial data that without mentioning it, all the data collected and the results derived from them via spatial analysis are not meaningful. Scale is important in three senses:

(1) The data to be analysed have a scale component. Certain data, especially those collected in the field, are point- or plot-based. Attention must be given to their spatial distribution if the generated results are generalised to the entire study area.
(2) The scale can be regional or even global if generated from satellite imagery of a coarse spatial resolution.
(3) Scale is critical to understanding the processes underlying some geographic phenomena, and exerts a profound impact on the analytical results.

Scale affects the spatial behaviour of the same variable, and hence its spatial auto-correlation, which is scale- and perhaps even direction-dependent. Certainly, the outcome of spatial analysis is scale-dependent if the variable of analysis is spatially variant. At a local scale, it may be predictable, but its predictability declines at a broader scale. If area-enumerated data are aggregated, spatial aggregation affects the scale of the data and the attribute's auto-correlation, as testified by the modifiable area unit (MAU) problem. The increasing scale at a higher level of spatial aggregation will generalise local variations of the attribute values, homogenise the landscape, and lead to a closer auto-correlation. The auto-correlation observed at one scale is hence likely to be different from that at another scale.

Since a process may not take place at the same scale as that of the data collected, it is important to acknowledge the scale at which the data are collected in reporting the results of spatial analysis. They likely vary with the scale of data analysis. The findings of data analysis apply mostly to the scale at which data are collected and analysed. Any generalisation of the findings from one scale to another should be treated with caution as it can lead to misleading conclusions. If extrapolated to another scale, there is a chance of falling victim to the *ecological (inference) fallacy* (Openshaw, 1984), because the same process operating at one scale may be totally different from that at another. The issue is so complex that a book has been written on the topic (Zhang et al., 2014).

3.4 SPATIAL AUTO-CORRELATION

Spatial auto-correlation can be measured quantitatively using Geary's ratio and Moran's I, both applicable to attributes that are enumerated either categorically or nominally over an area, such as socioeconomic data collected via the census.

3.4.1 GEARY'S RATIO

Also called Geary's contiguity ratio (Geary, 1954), *Geary's ratio* measures the auto-correlation of area-based observations. Although applicable to both one- and two-dimensional data, it is invariably two-dimensional data that are analysed to derive Geary's ratio using the following formula:

$$GR = \frac{(n-1)\sum_{i=1}^{n}\sum_{j=1}^{n} w_{ij}\left(z_i - z_j\right)^2}{2\sum_{i=1}^{n}\sum_{j=1}^{n} w_{ij}\sum_{i=1}^{n}\left(z_i - \bar{z}\right)^2} \tag{3.6}$$

where w_{ij} refers to the similarity weight of attribute value between area unit i and area unit j, and $w_{ij} = 1$ if they are adjacent, or 0 otherwise. Thus, this ratio is calculated from those observation units that share a common boundary; z_i and z_j denote the attribute value of area units i and j, respectively; n stands for the total number of area units; $\bar{z}$ is their mean attribute value; $\sum_{i=1}^{n}\sum_{j=1}^{n}$ calculates the total number of boundaries between adjacent area units that share a boundary. For a pair of polygons, it equals 2 (A versus B, and B versus A).

Geary's ratio is always positive, with the maximum value permissible to be larger than 1. A lower ratio is synonymous with a more spatially homogeneous pattern. A spatial auto-correlation coefficient <1 or z < 0 implies spatial clustering of high and/or low values. There is no distinction between high or low values. Coefficients significantly smaller than 1 suggest increasingly positive spatial auto-correlation, while coefficients significantly higher than 1 demonstrate increasingly negating spatial auto-correlation. A negating spatial auto-correlation (e.g., c > 1 or z > 0) manifests a chequerboard pattern, or competition.

3.4.2 Moran's I

Devised by Moran (1950), *Moran's I* is another way of measuring spatial auto-correlation. Similar to Geary's ratio, it also takes into account the interactions between neighbouring observations through weights. It is calculated as:

$$\text{Moran's } I = \frac{n\sum_{i=1}^{n}\sum_{j=1}^{n} w_{ij}\left(z_i - \overline{z}\right)\left(z_j - \overline{z}\right)}{\sum_{i=1}^{n}\sum_{j=1}^{n} w_{ij}\sum_{i=1}^{n}\left(z_i - \overline{z}\right)^2} \tag{3.7}$$

where n refers to the total number of spatial entities indexed by i and j, z_i represents the value of the ith observation unit, $\overline{z}$ denotes the mean of all the observation units, w_{ij} stands for the matrix of binary spatial weights or the strength of interactions between observations i and j, and $w_{ij} = 1$ if they share a common border, or 0 otherwise, and it is the sum of all the observations.

Both Equations (3.6) and (3.7) share the same denominator. Geary's ratio and Moran's I differ from each other in calculating the difference of the attribute value, or the value to which the observed value is compared. In Geary's ratio, it is the pairwise observations whose difference is calculated. The difference in the attribute value of two adjacent polygons is calculated if they share a common boundary (only 4-connectivity neighbours are considered for raster lattice data). In Moran's I, the difference is calculated by subtracting the attribute value of the respective observation unit from the mean of all the observation units. In both calculations, the weight is adjacency-related, not distance-based as stated in Tobler's First Law of Geography. A value of 1 is assigned to any two observations that share a common boundary, or a value of 0 otherwise. There is nothing in between, as adjacency is synonymous with interaction in this case. It can be potentially modified by assigning a weight of 1 to all observations that lie within a certain distance from the observation in question, or a value of 0 to all observations lying beyond this distance threshold.

Figure 3.13 compares Moran's I and Geary's ratio for three spatial patterns. A Geary's ratio of 0.5 corresponds to a Moran's I coefficient of 0.375, or a ratio of 1.2 is equivalent to −0.3625 of Moran's coefficient (Figure 3.13). Moran's I has different meanings, subject to its value, which ranges from −1.0 to 1.0. All Moran's I coefficient values can be grouped into two categories, >0 and <0 in terms of the spatial pattern revealed (Table 3.6). A positive spatial auto-correlation value ($I > -1/(n-1)$ or $z > 0$) suggests spatial clustering of high and/or low values without distinction between high or low, or a similarly regionalised smooth pattern. A small coefficient value indicates a spatially independent, uncorrelated pattern. A small, negative spatial auto-correlation ($I < -1/(n-1)$ or $z < 0$) shows a dissimilar, contrasting and uncorrelated random pattern. A value of −1.00 indicates a chequerboard pattern or a pattern of competition. Thus, its meaning is less precise than Geary's ratio, which is inversely related to Moran's I, but not strictly. The distinction between high or low Geary's ratios is ambiguous since it has an opposite sign from Moran's I, which measures global spatial auto-correlation, whereas Geary's ratio is more sensitive to local-scale spatial auto-correlation.

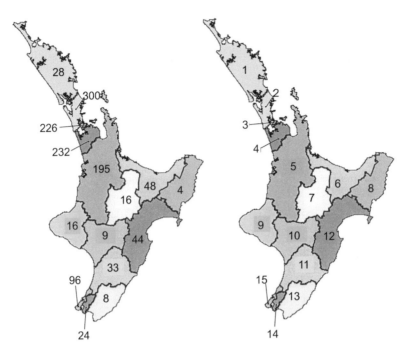

FIGURE 3.13 The number of COVID-19 cases in April 2021 in the North Island of New Zealand (left) by district health boards that have been numbered for the convenience of calculation (right)(data source: Ministry of Health, New Zealand, www.moh.govt.nz).

TABLE 3.6

Comparison of Geary's Ratio and Moran's I Values and the Spatial Pattern Revealed

Geary's ratio	Moran's I	Spatial pattern revealed
0<R<1	>0	Similar regionalised, smooth, clustered
1	<0	Independent, uncorrelated, random
>1	<0	Dissimilar, contrasting, chequerboard

3.4.3 An example

In this example, the spatial auto-correlation of community-transmitted COVID-19 cases (e.g., those arriving from overseas in managed isolation and quarantine were excluded) in the North Island of New Zealand in April 2021 will be analysed for both Geary's ratio and Moran's I . The information was obtained from the New Zealand Ministry of Health website (www.moh.govt.nz), and is broken down by district health boards (n = 15) in Figure 3.13 (left). In order to facilitate manual calculation, all polygons of the district health boards must be numbered (Figure 3.13, right).

The first step to prepare for the calculation of Geary's ratio is to square the difference between a health board's number of cases and the mean of all district health boards $(\bar{z} = 85.17)$ after it has been calculated (Table 3.7, column 2). The squared differences are then summed for all the observation units. The next step is to calculate the disparity between a board's value and the mean of all the boards, multiplied by its neighbouring boards' difference, then all the difference products are summed up by board (e.g., row sum in Table 3.7).

The procedure for calculating Moran's I is illustrated in Table 3.8, in which the square of the difference between a health board's number of cases and the mean of all district health boards has been omitted as it is identical to the second column in Table 3.7. After the difference in the number of cases between adjacent health boards is calculated and squared, it is summed by board (e.g., row sum):

$$\text{Moran s } I = \frac{15 \times (70,986.30 \times 2)}{2 \times 23 \times 140,446.93} = 0.33 \tag{3.8}$$

Both Geary's ratio and Moran's I suggest that the number of infected cases is spatially clustered and correlated. This correlation is explained by the fact that infectious people are interacting more intensively with other adjacent unaffected members of the population than with distant people. Neither Geary's ratio nor Moran's I take into account the physical size of the observation units, even though weight appears in the calculation equations. This weight simply indicates adjacency. It has nothing to do with weighting the observation units by their size. So one potential improvement to both calculations is to weight v(v–1) by the ratio of the size of the polygons concerned to the total area of all polygons. This weighting reflects the fact that not all observation units (polygons) are of the same size, and they have a varying influence on the auto-correlation. This weight is irrelevant if all the observations have the same area, as shown in Figure 3.13.

3.4.4 LOCAL-SCALE SPATIAL ASSOCIATION

The spatial association between observations at a local scale can be studied using two methods. The first one is the G_i and $G_i{}^*$ statistic proposed by Getis and Ord (1992). It enables the evaluation of whether there exist pockets of spatial association of variable x. G_i is a function of distance calculated as:

$$G_i(d) = \frac{\sum_{j=1}^{n} w_{ij}(d) x_j}{\sum_{j=1}^{n} x_i} (i \neq j) \tag{3.9}$$

where w_{ij} is a symmetric weight matrix of a binary value of 0 and 1, with 1 for all observations lying within a distance of d from the given observation i, and 0 for all other observations beyond d. The numerator calculates the sum of all x_j within d of i exclusive of x_i itself. Its variation for i=j is termed $G_i{}^*(d)$. The denominator calculates the sum of all x_j except x_i. $G_i(d)$ is able to measure the degree of clustering or concentration

TABLE 3.7
Difference Square of Neighbouring District Boards' Cases in Preparation for Geary's Ratio Calculation ($\bar{z}$ = 85.27, n = 15)*

No.	$(z_i - \bar{z})^2$	2	3	4	5	6	7	8	9	10	11	12	13	14	15	Row sum
1 (28)	3,279.47	73,984†														73,984
2 (300)	4,6110.40		5,476													5,476
3 (226)	19,805.87			36												36
4 (232)	21,530.67				1,369											1,369
5 (195)	12,041.40					21,609	32,041		32,041	34,596						120,287
6 (48)	1,388.80						1,024	1936				16				2,976
7 (16)	4,797.87									49		784				833
8 (4)	6,604.27											1,600				1,600
9 (16)	4,797.87									784						784
10 (9)	5,816.61										576	1,225				1,801
11 (33)	2,731.80											121	625	81	3,969	4,796
12 (44)	1,702.94															0
13 (8)	5,970.14													256		256
14 (24)	3,753.60														5,184	5,184
15 (96)	115.20															
Total	**14,0446.93**															**219,382**

*The matrix is symmetrical, and only half of it is shown in the table. The table is symmetrical and lists only the upper half of the matrix, so the total of the row sum needs to be multiplied by 2.

†(28 – 300) × (28 – 300):

$$GR = \frac{(n-1)\displaystyle\sum_{i=1}^{n}\sum_{j=1}^{n} w_{ij}\left(z_i - z_j\right)^2}{2\displaystyle\sum_{i=1}^{n}\sum_{j=1}^{n} w_{ij}\sum_{i=1}^{n}\left(z_i - \bar{z}\right)^2} = \frac{(15-1)\cdot 219382 * 2}{2\cdot(23*2)\cdot 140446.93} = 0.48$$

23 – number of edges shared between adjacent polygons shown in Table 3.7; both neighbouring polygons and their variance must be multiplied by 2 because the table just shows half of all the neighbouring districts (e.g., the table is symmetrical).

TABLE 3.8
Difference Square of a Board's Number of Cases Minus the Mean Number between Neighbouring Boards in Preparation for Moran's I Calculation*

No.	2 (300)	3 (226)	4 (232)	5 (195)	6 (48)	7 (16)	8 (4)	9 (16)	10 (9)	11 (33)	12 (44)	13 (8)	14 (24)	15 (96)	Row total
+	-12,297.06†														-1229.06
2		30,220.14													30,220.14
3			2,0650.27												2,0650.27
4				16,101.54											16,101.54
5					-4,089.40	-7,600.86		-7,600.86	-8,369.00						-27,660.12
6						2,581.34	3028.54				1,537.87				7,147.75
7									5,282.74		2,858.40				8,141.14
8											3,353.60				3,353.60
9									5,282.74						5,282.74
10										3986.20	3,147.27				7,133.48
11											2,156.87	4,038.47	3,202.20	-561.00	8,836.55
12															
13													4,733.87		4,733.87
14														-657.60	-657.60
															70,986.30

* The table is symmetrical and lists only the upper half of the matrix, so the total of the row sum needs to be multiplied by 2.

† (28 − 85.27) × (300 − 85.27).

of observations via the sum of values associated with variable X. Whether observations are concentrated spatially can be determined by comparing the observed $G_i(d)$ with a set of values within d of location i randomly. Under the assumption of normal distribution, the Z-score that measures spatial auto-correlation can be calculated as:

$$Z_i = \frac{G_i(d) - E\left[G_i(d)\right]}{\sqrt{VarG_i(d)}} \tag{3.10}$$

where $E\left[G_i(d)\right]$ is the expected value, calculated as:

$$E(G_i) = \frac{W_i}{n-1} = \frac{\sum_j w_{ij}(d)}{n-1} \tag{3.11}$$

$VarG_i(d)$ is the variance of $G_i(d)$, calculated as:

$$Var(G_i) = E(G_i^2) - E^2(G_i) = \frac{1}{\left(\sum_j x_j\right)^2} \left[\frac{W_i(n-1-W_i)\sum_j x_j^2}{(n-1)(n-2)} \right] + \frac{W_i(W_i-1)}{(n-1)(n-2)} - \frac{W_i^2}{(n-1)^2} \tag{3.12}$$

Equations (3.9)–(3.12) assume a binary weight of 0 or 1. They will become much more complex in format for non-binary weights. Although vector data are commonly used to calculate d, it can also be derived from raster lattice data. In this case, d is based on the grid dimension. The sign of Z_i can indicate the direction of correlation (i.e., positive or negative). A large positive Z_i value suggests that the observations within d of point i are above the mean value, and vice versa. After the local association has been calculated, it is possible to test the existence of local dependency. In the test, the null hypothesis is no local spatial association or dependency, and the alternative hypothesis is otherwise.

The second method is the *Local Indicators of Spatial Association* (LISA) statistic proposed by Anselin (1995). It serves several purposes, such as measuring spatial heterogeneity, detecting local clusters, and assessing the influence of individual observations. Influential observations are determined by decomposing the global measure of spatial association into observation-specific components. The LISA statistic must meet the following two requirements:

(1) The LISA of a given observation is indicative of the extent of significant spatial clustering of observations with a similar value around the observation.
(2) The total of LISAs for all observations is proportional to the global spatial association.

This family of indicators includes local instability, local gamma, local Moran's I, and local Geary's ratio. The only difference between the global and local versions of these indicators is distance (d). In the local version, only those observations within d of observation i are used in the derivation. The LISA statistics may be interpreted as indicators of local pockets of non-stationarity, or *hotspots*, rather reminiscent of the G_i and $G_i{}^*$ statistics of Getis and Ord (1992). The influence of individual observations on the global statistic is then used to identify "outliers", as in Anselin's Moran scatterplot (Anselin, 1995):

$$I_i = \frac{N\left(X_i - \bar{X}\right)\sum_j^N W_{ij}\left(X_j - \bar{X}\right)}{\sum_i^N \left(X_i - \bar{X}\right)^2} \tag{3.13}$$

where I_i stands for the local Moran's I index of sub-region i. It indicates the degree of similarity between observations in sub-region i and its neighbours. There are four critical permutations of I_i: high-high, low-low, high-low, and low-high. The last two combinations indicate a hot spot and a cold spot, respectively.

3.4.5 GLOBAL BIVARIATE SPATIAL AUTO-CORRELATION

Both Moran's I and LISA are applicable to studying the correlation of the same variable at different locations either globally or locally. In practice, it is highly useful and necessary to identify the spatial auto-correlation of two variables, such as crash sites and road layout, to explore how road design affects traffic accidents. This can be fulfilled using global bivariate spatial auto-correlation, which can quantify the spatial dependency between two variables (Matkan et al., 2013). This analysis can help us identify the crash risky locations in urban areas (Figure 3.14).

The spatial dependency between two global variables x_l and x_k can be quantified using the global bivariate Moran's I statistic, or I_{kl}:

$$I_{kl} = \frac{z_k w z_l}{n} \tag{3.14}$$

where n refers to the number of observations and w is the row-standardised spatial weight matrix. It has a value of 0 for non-neighbouring observations, and 1 for neighbouring ones; Z_k and Z_l are calculated as $z_k = \frac{x_k - \bar{x}_k}{\sigma_k}$ and $z_l = \frac{x_l - \bar{x}_l}{\sigma_l}$. In essence, both are standardised with a mean of 0 and a standard deviation of 1.

Apart from identifying the influence of road layout on crashes, I_{kl} can find other applications, such as accounting for the spatial variation of rainfall caused by topography within a local area. It is similar to spatial cross-correlation in that two variables are involved in the calculation. They must have exactly the same spatial extent. Both variables must be standardised before the calculation. Local bivariate LISA can be expressed as:

$$I_{kl}^i = z_k^i \sum_j w_{ij} z_l^j \tag{3.15}$$

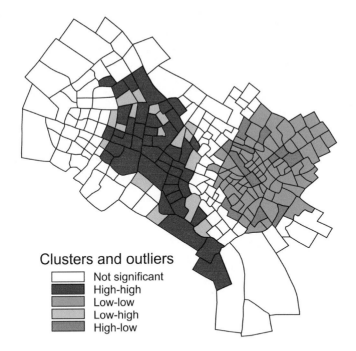

Clusters and outliers

☐ Not significant
■ High-high
▨ Low-low
▨ Low-high
■ High-low

FIGURE 3.14 The LISA map of crashes in the urban area of Mashhad, Iran in 2007–2008(source: modified from Matkan et al., 2013).

It measures the linear association between variable l at location i and the average value of variable k at neighbouring location j. A variation of global bivariate spatial auto-correlation is time–space auto-correlation, which applies to the same variable, but at different times.

3.5   AREA PATTERN AND JOINT COUNT STATISTIC

3.5.1   AREA PATTERNS AND JOINT COUNTS

A number of questions can be asked about the spatial pattern of two spatial variables that have a continuous distribution. For instance, are the accident-prone spots related to road layout? The answer to this question can be derived through a comparison of the joint count statistic of the observed pattern with some standard (reference) patterns, such as spatial even distribution and random distribution (Figure 3.11). The joint count statistic is able to detect spatial dependency between two variables whose attribute value is enumerated over area units at the nominal (categorical) scale. It addresses the pattern of their spatial distribution, not the spatial pattern of the attributes themselves (Figure 3.15). Furthermore, the nominal attribute value of study can be binary, such as the presence/absence of weeds in a grassland patch, or whether the ground covered by healthy meadow has been denuded. Other examples of binary attributes suitable for joint count analysis pertain to the nature of burglaries and the

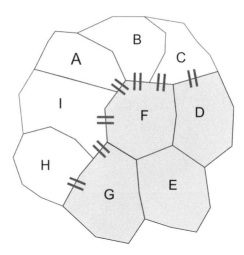

FIGURE 3.15 Spatial distribution of a polygon-based attribute of a binary nature. The distribution can be studied using spatial auto-correlation.

prevalence of a disease (affected versus intact). It could also be categorical, involving several attributes, such as kobresia, forb, and weeds for vegetation types. In the case of binary data, the joint count statistic relates to the number of observed connections between the attribute values of "present" cells and those of "absent" cells to the theoretical number of connections in a random distribution. A *joint count* refers to the number of adjoining observation units having the same attribute value.

There are two approaches to addressing the question posed at the start of this section of how to count the occurrences of a nominal variable and undertake a χ^2 test, or to compare pairwise polygons that share a common boundary. The latter is known as joint count statistic. Compared with χ^2 tests, joint count statistics produce much more informative outputs that include three types of statistical parameters, a Z-score indicating the closeness of auto-correlation for each of J_{ab}, J_{aa}, and J_{bb}; the last two are scores measuring the positive spatial auto-correlation for variable a and variable b joints, respectively. The J_{ab} Z-scores measure the spatial auto-correlation between the two variables. They are calculated as:

$$Z_{obs} = \frac{O_{PA} - E_{PA}}{\sigma_{PA}} \tag{3.16}$$

where O_{PA} represents the observed joint count and E_{PA} is the expected joint count or the number of connections under a theoretically random distribution, calculated as:

$$E_{PA} = \frac{2CPA}{n(n-1)} \tag{3.17}$$

where C stands for the total number of connections between the observation units or polygons, P denotes the number of "presence" polygons, A refers to the number of "absence" polygons, and n stands for the total number of polygons ($n = P + A$).

σ_{PA} in Equation (3.16) stands for the standard deviation of the expected joint count. It is calculated as:

$$\sigma_{PA} = \sqrt{E_{PA} + \frac{\Sigma V(V-1)PA}{n(n-1)} + \frac{4\left[C(C-1) - \Sigma V(V-1)\right]P(P-1)A(A-1)}{n(n-1)(n-2)(n-3)} - E_{PA}^2} \qquad (3.18)$$

where V refers to the number of neighbouring polygons and $\sum V$ denotes the summation of all the neighbouring polygons.

3.5.2 JOINT COUNT TEST

The procedure for undertaking a joint count test is rather complex, consisting of a number of steps. A preliminary step is to count the number of neighbouring polygons and represent the outcome in a binary matrix of adjacency, with 0 denoting non-adjacency and 1 adjacency (Table 3.9), very similar to Tables 3.7 and 3.8 except that the actual values should all be 1. If the study area is partitioned into an array of grid cells, then the task of finding the neighbouring observations is much easier as each cell is surrounded by four neighbours. In the case of irregularly shaped polygons, all of them need to be numbered for ease of identification (see Figure 3.15). These polygons are grouped into two types: those that share a common boundary (1) and those that do not (0). They are differentiated by their unique identity, and the number of

---

TABLE 3.9

The Adjacency Matrix Showing Polygons Sharing a Common Boundary (1) Based on Figure 3.16

	A	B	C	D	E	F	G	H	I	Sum
A	0	1	0	0	0	1	0	0	1	
B	1	0	1	0	0	1	0	0	0	
C	0	1	0	1	0	1	0	0	0	
D	0	0	1	0	1	1	0	0	0	
E	0	0	0	1	0	1	1	0	0	
F	1	1	1	1	1	0	1	0	1	
G	0	0	0	0	1	1	0	1	1	
H	0	0	0	0	0	0	1	0	1	
I	1	0	0	0	0	1	1	1	0	
V	3	3	3	3	3	7	4	2	4	**32 (2C)**
V–1	2	2	2	2	2	6	3	1	3	
V(V–1)	6	6	6	6	6	42	12	2	12	**98**

each type of polygons is counted. A spatial matrix of 1 and 0 (1 = adjacent; 0 = non-adjacent) is then constructed (Table 3.9). Its size is a square equalling the number of polygons (n). In Table 3.9, V stands for the number of neighbouring observations of each polygon, and the total number of connections (C = 16) is half of the total number of neighbours (32). The actual implementation of the joint count test comprises three steps:

Step 1: Calculate the theoretical number of connections (C) between "presence" polygons P and "absence" polygons A, or E_{PA}, according to the null hypothesis of dependency using Equation (3.17) for a random spatial distribution:

$$E_{PA} = \frac{2CPA}{n(n-1)} = \frac{2 \cdot 16 \cdot 5 \cdot 4}{9(9-1)} = 8.889 \tag{3.19}$$

Step 2: Calculate the standard deviation of the number of E_{PA} for random spatial distribution (theoretical), σ_{PA}, using Equation (3.18):

$$\sigma_{PA} = \sqrt{8.889 + \frac{98 \cdot 5 \cdot 4}{9 \cdot 8} + \frac{4(16 \cdot 15 - 98) \cdot 5 \cdot 4 \cdot 4 \cdot 3}{9 \cdot 8 \cdot 7 \cdot 6} - 8.889^2} \tag{3.20}$$

$$= \sqrt{8.889 + 27.222 + 45.079 - 79.014} = 1.475$$

Step 3: Calculate the observed joint count O_{PA} ($\sum C_{PA}$), and test the significance of the difference between the random and the observed distributions using the Z statistic.

The similarity between the two observed distributions or the dependency of one of them on another in a general or specific way can be determined using two types of test, the bilateral test or the unilateral test, respectively. The former asks the question of whether the spatial distribution of "presence" polygons significantly differs from a random one. The null hypothesis is H_0: $O_{PA} = E_{PA}$ (area pattern is random), and the alternative hypothesis is H_1: $O_{PA} \neq E_{PA}$ (spatial pattern is not auto-correlated or dependent).

The unilateral test goes one step further by checking whether the observed "presence" polygons are significantly "grouped" or "dispersed" in their spatial distribution. In the test, the alternative hypothesis is expressed as H_1: $O_{PA} < E_{PA}$ or $O_{PA} > E_{PA}$. All the steps of calculation and test are shown in Table 3.10 using two examples: a clustered pattern first, then a dispersed pattern.

As shown in Table 3.10, in the first example the distribution is not significantly different from a random distribution in the bilateral test. In the unilateral test, the observed distribution is significantly different from randomness, and its pattern leans towards clustering (grouped). In the second example, the observed distribution is not significantly different from being random in the bilateral test, but is

TABLE 3.10
Z Test of a Group of Polygons for Theoretical Dependent Distribution

Observed distribution   **Theoretical dependent distribution**

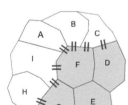

$$Z_{obs} = \frac{O_{PA} - E_{PA}}{\sigma_{PA}} = \left(7 - 8.889\right)/1.475 = -1.281$$

Bilateral test with $\alpha = 5\%$ (two-tailed distribution)

$H_1: O_{PA} \neq E_{PA}$

$Z_{test-0.025} = -1.960$

$|Z_{obs}| < |Z_{test-0.025}|$

Thus, H_0 is rejected.

The observed spatial pattern is not significantly different from a random distribution.

$O_{PA} = 7$

Unilateral test with $\alpha = 2.5\%$

$H_1: O_{PA} < E_{PA}$

$Z_{test-0.025} = -1.96$

$Z_{obs} (-1.281) > Z_{test-0.025} (-1.96)$

Thus, H_0 cannot be rejected. The observed distribution is significantly different from a random distribution, and the distribution leans towards clustering (grouped).

$E_{PA} = 2CPA/n(n-1) = 8.889$

$$Z_{obs} = \frac{O_{PA} - E_{PA}}{\sigma_{PA}} = \left(11 - 8.889\right)/1.475 = 1.431 \text{ Bilateral test with } \alpha = 5\%$$

(two-tailed distribution)

$H_1: O_{PA} \neq E_{PA}$

$Z_{test-0.025} = -1.960$

$|Z_{obs}| < |Z_{test-0.025}|$

Thus, H_0 cannot be rejected. The observed spatial pattern is not significantly different from a random distribution.

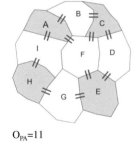

$O_{PA} = 11$

Unilateral test with $\alpha = 2.5\%$

$H_1: O_{PA} > E_{PA}$

$Z_{test-0.025} = -1.96$

$Z_{obs} (1.43) > Z_{test-0.025} (-1.96)$

Thus, H_0 cannot be rejected, and the observed spatial pattern is significantly different from a random distribution. Its distribution leans towards dispersed.

significantly different from a random one in the unilateral test. The pattern leans towards dispersed.

REVIEW QUESTIONS

1. In your view, which type of data is much more challenging to keep up-to-date, spatial or thematic? Explain.
2. Discuss the general role of GPS in the acquisition of thematic data.

3. Use an example each to illustrate the type of spatial analysis that is permissible for thematic data recorded at each of the four enumeration scales.
4. Spatial sampling can be carried out in 2D or in 3D. Discuss why 3D spatial sampling is more difficult to be spatially representative than 2D sampling.
5. Although point sampling is very easy to implement, it has to be replaced with plot (quadrat) sampling in certain applications. Why? Discuss how plot size should be determined in sampling.
6. What are the main factors that affect the density of samples in constructing a DEM?
7. Compare and contrast the pros and cons of random sampling, transect sampling, and plot sampling. What is the relationship between point sampling, transect sampling, and plot sampling?
8. What are the advantages and disadvantages of collecting samples as many as possible?
9. In your view, which one is much easier to analyse, spatial pattern or spatial continuity? Explain.
10. Define h-scatterplot. What spatial properties can it reveal about an attribute?
11. Compare and contrast spatial autocorrelation with spatial cross-correlation. What specific objectives can each fulfil?
12. What is ecological fallacy and how it should be avoided?
13. Compare Geary's ratio and Moran's I. What is the critical limitation to both of them?
14. What are the main differences between cross-correlation and global bivariate spatial autocorrelation?
15. Compare and contrast spatial autocorrelation and joint count statistic in analysing spatial pattern.

REFERENCES

Anselin, L. (1995) Local indicators of spatial association – LISA. *Geographical Analysis*, 27(2): 93–115.

Claghorn, J. (2014). Generative Landscapes – Random Topography and Surface Analysis – Example 4.1. https://generativelandscapes.wordpress.com/2014/08/15/random-topography-and-surface-analysis-example-4-1 (accessed on 8 June 2021).

Gao, J. (1995) Comparison of sampling schemes in constructing DTMs from topographic maps. *ITC Journal* 1995–1: 18–22.

Gao, J. (1998) Impact of sampling intervals on the accuracy of topographic variables mapped from grid DEMs at a micro-scale. *International Journal of Geographic Information Science*, 12(8): 875–90.

Gao, J., and Lythe, M. (1998). Evaluation of the MCC method in detecting oceanic circulation patterns at a local scale. *Photogrammetric Engineering and Remote Sensing*, 64(4): 301–8.

Geary, R. C. (1954) The contiguity ratio and statistical mapping. *The Incorporated Statistician*, 5(3): 115–27, 129–46.

Getis, A., and Ord, J. K. (1992) The analysis of spatial association by use of distance statistics. *Geographic Analysis*, 24 (July), 189–206.

Matkan, A. A., Shahri, M., and Mirzaie, M. (2013) Bivariate Moran's I and LISA to explore the crash risky locations in urban areas. *N-AERUS Enschede 12–14 September 2013*, p. 12.

Moran, P. A. P. (1950) Notes on continuous stochastic phenomena. *Biometrika*, 37(1): 17–23, doi: 10.2307/2332142.

Openshaw, S. (1984) Ecological fallacies and the analysis of areal census data. *Environment and Planning A*, 16: 17–31.

Ord, J. K., and Getis, A. (1995) Local spatial autocorrelation statistics: Distributional issues and an application. *Geographical Analysis*, 27(4): 286–306.

Peppa, M. V., Mills, J. P., Moore, P., Miller, P. E., and Chambers, J. E. (2017) Brief communication: Landslide motion from cross correlation of UAV-derived morphological attributes. *Natural Hazards and Earth System Sciences*, 17: 2143–50, doi: 10.5194/nhess-17-2143-2017.

Stein, A., and Ettema, C. (2003) An overview of spatial sampling procedures and experimental design of spatial studies for ecosystem comparisons. *Agriculture, Ecosystem and Environment*, 94: 31–47.

Thompson, S. K., and Seber, G. A. F. (1996) *Adaptive Sampling*. New York: Wiley, p. 288.

Zhang, J., Atkinson, P., Goodchild, M. F. (2014) *Scale in Spatial Information and Analysis*, CRC Press, p. 367.

4 Descriptive and Inferential Spatial Analysis

In practice, it is frequently necessary to come up with a quantitative measure of the spatial distribution of points and polygons so that they can be compared against each other quantitatively, even though this quantitative measure itself may not be perfect, such as the shape of tropical cyclones and newly urbanised suburbs. Such quantitative measures can shed light on the importance of certain factors in shaping the observed pattern. The derivation of such quantitative measures is routinely performed in *descriptive spatial analysis*, which aims to attach a quantitative value to an individual observation or a group of observed entities and generate descriptive information about them (including their spatial patterns), identifying unusual or interesting features, distinguishing accidental from important observations, and formulating hypotheses about them (Haining et al., 1998). Also known as *exploratory data analysis*, descriptive data analysis can be extended to spatial data whose analysis can fulfil the same objectives as non-spatial exploratory data analysis except that hypotheses can be formulated about the spatial distribution of data and assessing spatial models. The targets of analysis can be points, lines, polygons, or even directions. Apart from points and directions, which must be analysed collectively, the number of observations in an analysis can range from one to multiple. This chapter presents the analysis of such data, both descriptively and inferentially.

4.1 ANALYSIS OF POINT DATA

Point pattern analysis aims to quantify the spatial relationship between point features, and test whether they obey a certain kind of spatial distribution as a whole. Point pattern analysis started as early as the late 1950s and early 1960s, but did not gain much popularity until the mid-1980s due to the lack of suitable computer software (Gartall et al., 1996). It bloomed with the advent of geographic information systems (GISs) in the early 1990s, when geo-referenced point data became widely available in the digital format and the results of analysis could be visualised almost instantly in graphic format.

4.1.1 POINT DATA

Point data are the most common and simplest type of spatial data that are collected at specific locations. In addition to the thematic attribute at these locations, their

DOI: 10.1201/9781003220527-4

coordinates are also recorded. With the ubiquitous use of Global Positioning System (GPS)–enabled devices, point data can be collected with unprecedented ease. Point data are routinely collected in a wide range of fields of study, ranging from law and order (e.g., burglary and homicide addresses), transport (e.g., car crash spots), natural hazard (e.g., earthquake epicentres), and ecology (e.g., location of trees and forest fires), even to epidemiology (e.g., cases of infection by a communicable virus such as COVID-19). These points share one commonality in that they are confined to the 2D plane space referenced by a pair of horizontal coordinates, even though they may be located on a slope, such as trees in a mountainous forest. The elevation or height of tree locations is not normally considered in point pattern analysis. All point features are distributed within a physical boundary known as the study region. Each point observation can be treated as an event. There is no limit to how many points can be analysed at a time. These points may be of the same nature, such as incidents of traffic (e.g., univariate), or of different nature, such as trees of different species (e.g., multivariate).

All point observations are considered dimensionless topologically. However, some areal features can also be represented in the point format, depending on the scale. For instance, on the global scale, individual cities can be represented as points even though they are areal entities at a local scale. In order to produce a diagram to illustrate the spatial distribution of per capita income that must be enumerated over an area (e.g., census unit), the observations can be considered to be made at a particular point within each census unit. In the conversion from area-based observations to point data, the centroid of the area is commonly taken as the point where observations are made, even though it is generally understood and accepted that the attribute value is enumerated over the entire city limit. The size of the study area is the same as the spatial extent within which the points are distributed. This polygon encloses all points to be analysed. It can be of any shape, but in general, has no holes in it.

The targets of study in point pattern analysis are collective points and their spatial relationship to one another. The spatial properties of point data can be studied in two orders, *first order* and *second order*. The former simply describes the spatial variation of the expected mean or average or the distance between adjacent points such as nearest neighbours. The *second-order analysis* aims to describe the co-variance (or correlation) of points in different regions and seeks to understand the spatial pattern and detect region-wide "trends".

4.1.2 Fundamental Types of Point Patterns

Points of the same type can be spatially distributed in a number of ways. The three distinctive patterns against which an observed pattern can be compared are random (complete spatial randomness), uniform, and clustered (Figure 4.1). Similar to random spatial sampling, a *random point pattern* implies haphazardness in the distribution (Figure 4.1a). A higher concentration of points at some locality may lie right next to a lower concentration in its vicinity, all being totally unpredictable. The formation of a random point distribution is a manifestation of lack of external control or a completely homogeneous space. Random point distribution is perfectly exemplified

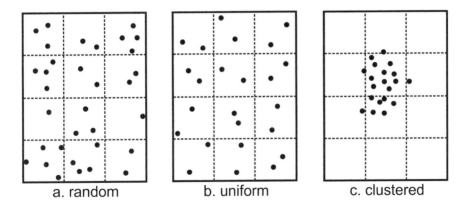

a. random   b. uniform   c. clustered

FIGURE 4.1   Three fundamental types of point pattern: random (a), uniform (b), and clustered (c).

by raindrops. A *uniform pattern* implies a degree of evenness in the point distribution. The entire space is taken up by points separated by a roughly constant distance (Figure 4.1b). It must be noted that uniform distribution is not the same as systematic distribution, as the direction of one point relative to another has a degree of randomness. A uniform pattern usually results from a certain kind of external intervention. For instance, the planting of saplings in an orchard at a constant distance will result in a uniform pattern of trees later on.

The *clustered pattern* is characterised by some points within a locality being in much closer proximity to one another than that allowed in the space, creating voids in the 2D space (Figure 4.1c). A clustered pattern is the most difficult to define quantitatively among the three patterns shown in Figure 4.1. This pattern can be formed out of a number of spatial processes, such as parent-offspring. In a natural forest, the offspring of a mature tree is located very close to the mother plant, where seeds are dropped and have germinated. When multiple clusters are present, there are no clear-cut criteria as to which cluster a point should belong to or how many members one cluster should comprise. A commonly used criterion of judgment is the Euclidean distance, with a shorter distance implying a higher probability of membership. Some points may belong to one cluster under one criterion, but they can be split into two clusters if the criterion becomes more stringent. Point observations can be grouped into a certain number of clusters specified by the analyst during clustering analysis.

The above types of spatial pattern apply to points of the same identity (e.g., univariate). When two types of point entities are involved in formatting a spatial pattern, it can be described as independent, attractive, or repulsive (Figure 4.2). The *independent pattern* can be described as the superimposition of one random pattern over another. As both variables are random in their spatial distribution, none is related to another in any predictable manner. In the *attractive pattern*, points of one identity lie in closer proximity to those of another identity, as well as members of their own identity, than members in another group. Jointly, they form a clustered pattern. In the *repulsive pattern*, points of the same identity always stick to their own clusters,

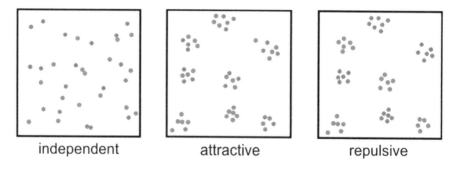

FIGURE 4.2   Three point patterns formed by two types of points.

which are distant from members of a different identity in other clusters. The two groups never mingle spatially, and the distance between clusters is variable.

4.1.3   POISSON PROCESS BEHIND RANDOM PATTERNS

A random point pattern can reveal quite a lot about the process behind its formation. It can be formed only under a uniform setting. In other words, all locations inside a study area have an equal chance of receiving an event. Furthermore, it also means total independence, in that the selection of a location by one point event in no way influences or interferes with the selection of the same location by other point events. Take raindrops as an example: the falling of one raindrop at a spot does not preclude or affect the falling of another raindrop at the same spot. Furthermore, the presence of an event at a location does not preclude or affect the presence of other events at the same spot. The probability of a spot on the ground receiving a raindrop is not affected by the preceding raindrops falling on the same spot. This is known as *complete spatial randomness* (CSR), in which the occurrence of an event at any point is independent of other events. Every event has exactly the same probability of occurrence as every other event, and the likelihood of occurrence is equal over the whole area of study. This implies that the study area is completely homogeneous.

In some cases, a process can still be random even if the area to receive it is heterogeneous because the two are completely independent of each other. For instance, the fall of raindrops on a mountain slope can be regarded as spatially random, but the area that receives them may be covered by diverse types of vegetation, such as trees, shrubs, grasses, and even lichens. Thus, the random distribution is not subject to the external environment. It is the formation of raindrops in the atmosphere and the weather conditions during the fall that jointly dictate the nature of the spatial pattern of raindrops. The surface cover on which they fall plays no role in the formation of the spatial pattern, even though raindrops may exert an influence on the surface cover some time later.

4.1.4   FORMATION OF CLUSTERED PATTERNS

Clustered patterns are commonly associated with dwellings in rural areas and trees in a forest. These point features stick together spatially to form a community or as a consequence of environmental heterogeneity. Roads are the main factors affecting dwelling clustering, while soil fertility, moisture content, and nutrients, as well as the local environmental settings (e.g., slope orientation), all play a role in the formation of a clustered plant pattern. Moisture-loving species are likely to be clustered around a wetland. Clustered patterns are a manifestation of the lack of uniformity in those variables that exert an influence on the formed pattern. If a site has been occupied by a tree already, it is unlikely to be occupied by another tree, as a result of unfavourable competition for light, water, and nutrients. However, once a plant becomes established somewhere, its offspring will likely become established nearby, forming a clustered pattern (Figure 4.3a). In this formation, topographic variability may also play a role, as slope gradient and orientation are decisive parameters in the redistribution of moisture and energy over the landscape. Other processes involved in

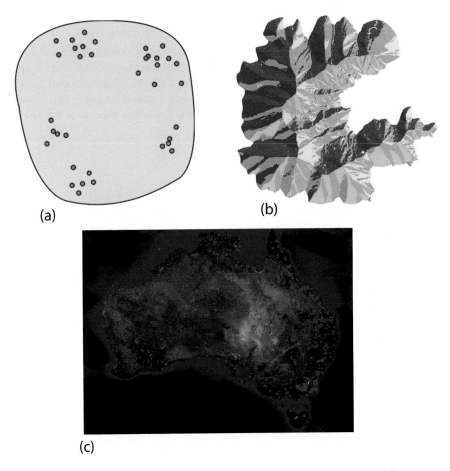

(a)

(b)

(c)

FIGURE 4.3 Clustered point patterns at three scales, formed by different spatial processes: (a) slope-scale parent–offspring relationship of redwood seedlings, (b) catchment-scale slope aspect formed by topographic relief (source: Dexter, 2007), and (c) continental-scale bushfires in Australia in January 2020 (source: screen capture from FIRMS).

the formation of a clustered pattern may include diffusion and competition. A short distance of seed dispersal is more likely to result in a clustered pattern. On the other hand, a longer distance of seed dispersal can result in a spatial distribution leaning towards the random pattern. However, the same cannot be said of competition. Weak competition can result in more randomness in the distribution than stronger competition.

Beyond the slope scale, clustering can also take place at the watershed level, and even on the regional scale. For instance, pixels facing the same slope orientation are likely to be clustered if represented in raster format (Figure 4.3b). Regional clustering of events, as in the spatial distribution of wildfires in Australia, may be attributed to the similarity in the environmental settings, such as abundant dry fuels (Figure 4.3c). The high concentration of fires near the eastern coast of Australia is likely caused by the availability of highly combustible fuels in a hot and dry climate.

When a clustered pattern is formed, a number of questions can be asked about it and the potential factors responsible for its formation. For instance, do clusters merely manifest a priori heterogeneity in the study area (e.g., sunlit slopes and shady slopes)? Do they have anything to do with other specific features of interest geographically, such as proximity to transport? Statistically, an observed point pattern can be regarded as the outcome (e.g., realisation) of a spatial process. This pattern can be characterised mathematically to capture the important aspects of the formation process in *descriptive point pattern analysis*.

4.1.5 Measures of Point Dispersion

Point dispersion can be easily studied via *point density* or the number of points per unit area. Point density is an important descriptive indicator of point pattern, in which the number of observations per unit area is analysed by specifying an aggregation unit. It defines the neighbourhood, within which all the available observations are counted as the neighbours. They are used in deriving the point density by dividing the observed population within the neighbourhood by its area. A commonly used neighbourhood is a circle, easily defined by a radius (r), with an area of πr^2. A higher density indicates that points are closer to one another, while a lower density suggests a higher degree of dispersion. However, this indicator is very crude, as it says nothing about how the points are distributed, only their proximity.

This drawback can be overcome with point patterns (or dispersion of points), which can be quantified using a number of methods. Two of the most commonly used ones are the quadrat method and the distance method. A quadrat can be circular or rectangular in shape. The former tends to be spatially scattered, the latter contiguous (Figure 4.4). In the scattered quadrat method, the same quadrat is repeatedly and randomly overlaid with the observed point features distributed over the study area each time the number of point features enclosed in the quadrat is enumerated. Space is thus partitioned into a varying number of quadrats that are selected randomly, uniformly, and independently, as the placement of the circular quadrats is not mutually exclusive. The enclosure of one point in one scattered quadrat does not preclude its inclusion in other multiple quadrats, which conforms to uniformity. The placement of one quadrat is independent of that of the next. Thus, it is possible for some

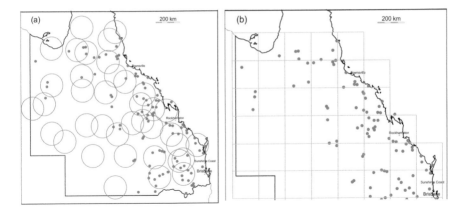

FIGURE 4.4 Comparison of scattered quadrats (a) and continuous quadrats (b) used in point pattern analysis of wildfire hotspots observed 24 hours before 11 March 2021 in Queensland, Australia (data source: MyFireWatch).

quadrats to overlap, and the same points can fall into multiple quadrats (Figure 4.4a). This method is more scientific than continuous quadrats.

In comparison, the contiguous quadrat method makes use of grid quadrats overlaid on the study area (Figure 4.4b). All quadrats are of the same size, shape, and orientation, but they never overlap with each other. This method is less desirable than scattered quadrats because a point observation can fall into one of the quadrats only, no matter how the grid quadrat is placed. If it falls into one quadrat, it cannot fall into other quadrats at the same time, which violates the independence requirement of statistics. An array of square quadrats can be superimposed on the study area repeatedly, and the number of points falling within each quadrat is then counted. The results of the tally can still vary if the grid quadrats are placed differently. This randomness can be overcome by repeating the placement in different manners a number of times. Nevertheless, this method should be avoided, as it does not conform to the statistical requirements of some tests.

4.1.6 Nearest Neighbour Analysis

In *nearest neighbour analysis*, point dispersion is quantified based on the distance between a point and its closest neighbours. After the distances to all the considered neighbouring points have been calculated, their mean is determined (Equation 4.1) and used to indicate the compactness of the distribution of the neighbouring points considered. Nearest neighbour analysis requires the random selection of a number of points lying within a certain distance from their neighbours under the assumption that each of them has an equal chance of being selected, and that the selection of any one point in no way influences the selection of other points.

$$d = \frac{1}{n} \sum_{i=1}^{n} d_i \qquad (4.1)$$

where d stands for the mean nearest neighbour distance and n refers to the number of sampled neighbouring points used in the calculation.

Nearest neighbour analysis can be performed not only on point data, but also on polygon data. In the latter case, it is the location of the polygon centroid that is analysed (i.e., polygons are treated just as point data). Nevertheless, the results must be interpreted differently. A large distance between polygons may not suggest dispersion, as polygons are always contiguous and have an area. A larger mean distance can suggest a larger polygon size, which naturally makes the centroids of neighbouring polygons further apart from each other. When nearest neighbour analysis is performed on polygons of the same identity, such as forest patches, a large nearest neighbour distance may suggest that the landscape is not particularly fragmented. More details on nearest neighbour analysis will be presented in Section 4.4.3.

4.1.7 HOTSPOT ANALYSIS

Although point density can indicate the existence of clusters in the data, it cannot reveal whether or not the clusters are statistically significant. This limitation can be overcome by *hotspot analysis*, a special point pattern analysis that aims to identify statistically significant clusters. Also known as Getis-Ord Gi* (G-I-star), hotspot analysis examines each event in the dataset in the spatial context of its neighbours (Getis and Ord, 1992). A statistically significant hotspot is found if a high attribute value is surrounded by points of also a high value. Similarly low values are grouped and identified as clusters of low values, or cold spots. Whether a cluster is a statistically significant hotspot or a cold spot is evaluated using the Getis-Ord Gi* statistic calculated using Equation (4.2):

$$
G_i^* = \frac{\sum_{j=1}^{n} w_{i,j} x_j - \bar{X} \sum_{j=1}^{n} w_{i,j}}{S \sqrt{\dfrac{n \sum_{j=1}^{n} w_{i,j}^2 - \left(\sum_{j=1}^{n} w_{i,j} \right)^2}{n-1}}}
\tag{4.2}
$$

where x_j denotes the attribute value of observation j, w_{ij} stands for the weight between observations i and j, n represents the total number of observations, and $\bar{X}$ and S refer to the mean (Equation 2.11) and standard deviation (Equation 2.13) of all the observations. Similar to Gj(d), G_i^* is a Z-score statistic whose calculation has been presented in Section 3.4.4.

Hotspot analysis outputs two statistical parameters:, a Z-score and a p-value (Figure 4.5). The former indicates the intensity of clustering. The higher the Z-score, the more intense the clustering, and vice versa. A Z-score close to 0 equates to no spatial clustering. The p-value reveals the probability of spatial clustering. The combination of a high Z-score and a low p-value suggests a significant hotspot. Conversely, a low negative Z-score combined with a low p-value indicates a cold spot.

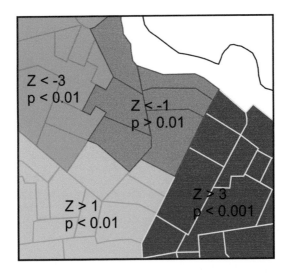

FIGURE 4.5   Two numerical outputs of hotspot analysis in each observation unit (polygon). Their combination can reveal either a hotspot or a cold spot, depending on the exact value. A polygon with a high Z-value but a low p-value is labelled as a hotspot. Conversely, a polygon with a low negative Z-value but a small p-value is identified as a significant cold spot.

Hotspot analysis can be performed on both point and polygon data. Point data such as spots of crashes can be analysed only after they have been aggregated spatially to the polygon level. Spatial aggregation renders the attribute of point observations enumerated over an area. Suitable candidates of polygon data for spatial aggregation are census tracks, administrative boundaries, and customised grids (Figure 4.6). All those point observations falling into the same grid or the same census unit will be aggregated into one value. In order to generate reliable analytical results, the number of observations in each areal unit should not fall below 30, below which the results are not statistically reliable. Also, any observation units that do not have any data (either 0 or missing) should be excluded from the analysis as they can be mistakenly construed as cold spots.

Hotspot analysis has found applications in various fields, one of which is transport, to identify accident-prone hotspots that can reveal defects in road design. Another field is epidemiology, to identify statistically significant clusters of infection cases of communicable diseases. The identified hotspots of infection should receive more attention and efforts to bring their community transmission under control. In combating crimes, hotspot analysis can be used to identify clusters of burglaries and even homicides, so that the identified crime hotspots can be patrolled more frequently by police to deter potential perpetrators.

4.1.8   INFERENTIAL ANALYSIS OF POINT PATTERNS

The *inferential analysis* differs from *descriptive analysis* in that only a small sample is available for analysis, and it aims to yield some information about the population

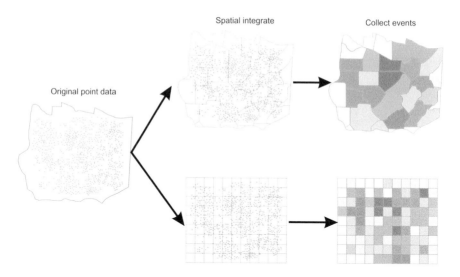

FIGURE 4.6   Spatial aggregation of point data into areas using irregularly shaped census tracks (top) and regular grids (bottom) before hotspot analysis can be performed on them.

from the available sample. Inferences can be made about sample variation or uncertainty, and variation of the results if the same study is repeated, and whether the sample obeys a certain kind of spatial distribution collectively, all via statistical tests such as Chi-square (χ^2). The inferential spatial analysis enables the identification of the plausible range of the true value about the population, such as its mean. Key to the inferential spatial analysis is the statistical test of the *null hypothesis* (H_0). It postulates a typical statement to the effect of negativity, opposite to the alternative hypothesis (H_1) which motivates undertaking the test in the first place. Chi-square tests whether the samples are distributed in the Chi-square manner statistically. All hypothesis tests must involve two critical statistical parameters of confidence and p-value. The confidence interval indicates the level at which the statistical parameters of the samples generated from repeated and independent sampling represent that of the population. It is usually set rather high, such as 95% or 99%. The p-value represents the likelihood of yielding a value more extreme than the observed in case the alternative hypothesis is rejected. If p = 0.02, then the chance of getting a value more extreme than that observed is 2%.

In order to illustrate how to test if the spatial pattern of a point distribution is random or not, let us examine the example of the wildfire hotspots in Queensland, Australia (Figure 4.4a). In this example, the scattered quadrat method is used to test the distribution of the hotspots rather than describing them. In total, the number of placements sums to 31 (n = 31), excluding the seven incomplete quadrats that are not placed fully within the boundary. The number of quadrats is counted according to the number of wildfire hotspots enclosed in each of them (Table 4.1, column 2). In the χ^2 test, the null hypothesis (H_0) states that the observed pattern is not significantly

TABLE 4.1

Chi-square (χ^2) Testing of the Wildfire Hotspots Pattern in Queensland, Australia Shown in Figure 4.4a

[1] No. of obs/ quadrat	[2] Obs frequency Oi	[3] Merged	[4] Probability	[5] Expected frequency Ei	[6] Merged	[7] (Oi–Ei)²/Ei
0	6	6	0.040	1.237	1.237	18.342
1	7	7	0.128	3.977	3.977	2.297
2	4	5	0.207	6.414	13.308	5.187
3	1		0.222	6.894		
4	4	7	0.179	5.561	9.151	0.506
5	3		0.116	3.590		
6	3	6	0.062	1.934	3.326	2.149
7	2		0.029	0.896		
8	0		0.012	0.366		
9	1		0.004	0.130		
Total	31		1.000	30.999		28.481

different from CSR, and the alternative hypothesis (H_1) states that the observed pattern is significantly different from CSR. It is important to select a proper significance level α, usually 5%, which means that 95% of the times the hypothesis is correct. The entire process of calculation is presented in Table 4.1, with the main steps explained below.

Step 1: List the number of hotspots enclosed in each of the qualified quadrats (column 1 in Table 4.1).

Step 2: Count the number of quadrats containing each of the possible number of hotspots shown in column 1, and place the counts in column 2; aggregate all the quadrats if their observed frequency falls below 5. For instance, numbers 2 and 3 are merged, as each contains less than 5 observations. This merging does not alter the total number of complete quadrats (31). It only reduces the number of categories from 10 to 5.

Step 3: Calculate the mean number of hotspots per quadrat (λ):
λ = number of total hotspots (n)/number of complete quadrats (k) = 100/31 = 3.226 (note: an observation falling on an offshore island is excluded).

Step 4: Calculate the probability p(x) of a quadrat with x hotspots according to Equation (4.3) (column 4):

$$p(x) = \frac{\lambda^x}{e^\lambda \cdot x!} \tag{4.3}$$

where $x = 0, 1, 2, ..., 9$; e = 2.7183, so
$p(0) = 3.226^0/(2.7183^{3.226} \times 0!) = 0.040$;
$p(1) = 3.226/(2.7183^{3.226} \times 1) = 0.128$;
$p(2) = 3.226^2/(2.7183^{3.226} \times 2 \times 1) = 0.207$;
......
$p(9) = 3.226^9/(2.7183^{3.226} \times 9!) = 0.004$.

Step 5: Calculate the expected frequency by multiplying the probability calculated in step 4 (column 4) by the total number of the observed complete quadrats (31) (column 5). For instance: $0.0399 \times 31 = 1.2369$ (note: the column total still remains at 31, the same as that of column 2. The expected frequency is also aggregated in a manner identical to that of the observed frequency).

Step 6: Calculate the χ^2 value using the equation given in column 5 (column 7), then derive the column sum (28.481), for instance:

$$\left(O_i - E_i\right)^2 / E_i = \left(6 - 1.237\right)^2 / 1.237 = 18.342.$$

Step 7: Search for the theoretical χ^2 value at the significance level of α = 0.05 in Table 4.2 using a degree of freedom of 5–1–1 = 3 (5 = number of categories after merging) ($\chi^2_{3, 0.05} = 7.81$). Compare the observed and theoretical χ^2 values.

Since $\chi^2_{3, 0.05}$ (7.81) is <28.481, H_0 is rejected. The observed pattern is significantly different from CSR.

TABLE 4.2
χ^2 Values at Different Degrees of Freedom and p-values.

Degree of freedom (df)	χ^2 value										
1	0.004	0.02	0.06	0.15	0.46	1.07	1.64	2.71	3.84	6.63	10.83
2	0.10	0.21	0.45	0.71	1.39	2.41	3.22	4.61	5.99	9.21	13.82
3	0.35	0.58	1.01	1.42	2.37	3.66	4.64	6.25	7.81	11.34	16.27
4	0.71	1.06	1.65	2.20	3.36	4.88	5.99	7.78	9.49	13.28	18.47
5	1.14	1.61	2.34	3.00	4.35	6.06	7.29	9.24	11.07	15.09	20.52
6	1.63	2.20	3.07	3.83	5.35	7.23	8.56	10.64	12.59	16.81	22.46
7	2.17	2.83	3.82	4.67	6.35	8.38	9.80	12.02	14.07	18.48	24.32
8	2.73	3.49	4.59	5.53	7.34	9.52	11.03	13.36	15.51	20.09	26.12
9	3.32	4.17	5.38	6.39	8.34	10.66	12.24	14.68	16.92	21.67	27.88
10	3.94	4.87	6.18	7.27	9.34	11.78	13.44	15.99	18.31	23.21	29.59
p-value (probability)	0.95	0.90	0.80	0.70	0.50	0.30	0.20	0.10	0.05	0.01	0.001

4.1.9 KERNEL DENSITY ANALYSIS

The conventional point density analysis discussed in Section 4.1.5 becomes problematic when too many points overlap heavily with each other. If they are too close to one another, then it is impossible to appreciate their spatial density clearly (Figure 4.7a). This problem can be circumvented by making use of *kernel density of points*, also known as kernel estimation. It can shed light on the intensity of clustering. In kernel density estimation, the point pattern is analysed by counting the number of events per unit area within a moving quadrat or "window". Point events within the moving window are weighted in terms of their distance from the point in question. This method is able to yield smooth estimates of univariate probability densities from a sample of points $\hat{f}(x, y)$ (Figure 4.7b). It is calculated as:

$$\hat{f}(x, y) = \frac{1}{nh_x h_y} \sum_i^n k\left[\frac{x - x_i}{h_x}, \frac{y - y_i}{h_y}\right] \tag{4.4}$$

where $k\left[\dfrac{x - x_i}{h_x}, \dfrac{y - y_i}{h_y}\right]$ represents the kernel weighting function, in which

$$h_x = \sigma_x \left(\frac{2}{3n}\right)^{\frac{1}{6}} \tag{4.5}$$

where n stands for the number of points enclosed in bands h_x and h_y, the kernel bandwidth in the x and y directions, respectively; σ_x denotes the standard deviation of all the enclosed observations within the bandwidth.

Kernel density estimation is effective at showing the degree of spatial clustering when numerous points overlap each other (Figure 4.8a). In this case, it is difficult to identify individual points. In comparison, if represented in contour form, kernel

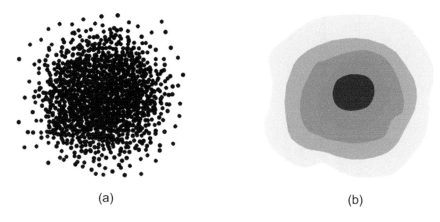

(a)   (b)

FIGURE 4.7   The point pattern is lost when so many points fall on top of each other (a). The kernel density estimate plot is much better at showing the intensity of clustering (b).

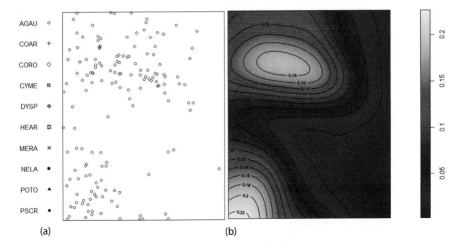

(a)   (b)

FIGURE 4.8   Spatial distribution of kauri trees (a), and their density visualised using kernel density contours (b).

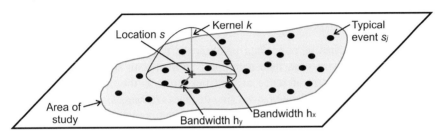

FIGURE 4.9   Kernel estimation of point density at location s (+) in which the bandwidth (green line) must be specified.

density estimation is able to show the level of clustering quantitatively and the spatial pattern clearly (Figure 4.8b). The key to the success of kernel density analysis is the selection of appropriate bandwidths h_x and h_y (Figure 4.9). A proper bandwidth is able to preserve the pattern maximally while still showing the pattern. h_x and h_y are virtually the radii of the bumps in the respective direction. Either too conservative a bandwidth or an over-generous bandwidth will not be able to reveal the intensity of spatial clustering genuinely. A larger bandwidth causes more generalisation to the spatial pattern, and the kernel density estimate will be noisy and less genuine (Figure 4.10a). Conversely, too conservative a bandwidth may not show the pattern adequately (Figure 4.10c).

There are no rules governing how an appropriate bandwidth can be determined. Prior to implementing the analysis, a series of small "bumps" (2D probability distribution) centred on each observation need to be averaged to yield an estimate of the probability density. A practical approach towards its determination is to run the analysis at a number of bandwidths, from which the appropriate one can be determined (Figure 4.10b).

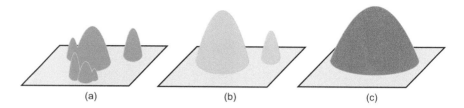

FIGURE 4.10 Impact of bandwidths on kernel density patterns: (a) too conservative a band-width, resulting in numerous small spikes, (b) appropriate, and (c) too broad a bandwidth, resulting in only one spike.

Kernel density is different from ordinary point density, in that the calculated ker-nel density around each observation is based on a quadratic formula with the maxi-mum value at the centre (where the point observation is located, e.g., s in Figure 4.9), and it tapers to 0 at close to the search radius (Figure 4.10), whereas in ordinary density the same value applies to the entire search neighbourhood and the density is spatially homogeneous.

4.1.10 SECOND-ORDER ANALYSIS OF POINT PATTERNS

Second-order analysis of point data involves the derivation of a function for the observed point pattern first, then a comparison of it with some kind of standard pattern that serves as a reference or yardstick, such as the most commonly used CSR. Point density is calculated at varying distances or ranges, so it is a function of range (distance) that is used to identify neighbouring points. One of the most com-monly used methods of calculating range-based point density is Ripley's K-function, a widely used second-order statistic of point data. It requires the calculation of point density at multiple ranges or distances, within which the number of observations is compared with the expected pattern under the CSR distribution. It can detect depar-ture from CSR over different ranges (r). Similar to the Moran's I function, it is able to yield information on the observed point distribution at the user-specified scale, and the distribution of a collection of point data. Since the K-function is virtually a multi-distance spatial clustering analysis, the user can determine whether the dis-tribution is dispersed, clustered, or random at different ranges of analysis. With this knowledge, it is thus possible to select an optimal scale to study the phenomenon of interest. Ripley's K-function is expressed as:

$$\hat{K}(r) = \frac{1}{\hat{\lambda}} \sum_{i=1}^{N} \sum_{j=1(i \neq j)}^{N} \frac{w(S_i, S_j)^{-1} i(\|S_i - S_j\| \leq r)}{N, t > 0} \quad (4.6)$$

where r represents the step size, N denotes the number of points enclosed in the area formed by r, $\hat{\lambda}$ stands for the estimated point density calculated as N/A (A = size of the study area), and w(S_i, S_j) refers to the portion of the circumference of a circle centred at S_i, passing through S_j, within the study area A.

The Ripley's K-function $\hat{K}(r)$ describes the distribution of 2D points that marginally have a fixed intensity (λ), but jointly are independent of each other (e.g., raindrops). As such, it can be used to yield summary information about a point pattern (e.g., its distribution parameters), test hypotheses about the pattern, and even fit the observed patterns of distribution with models.

$\hat{K}(r)$ is estimated from K(r) using Equation (4.7):

$$K(r)=E(N_r)/\lambda \qquad (4.7)$$

where N_r is the number of events (points) at range r and λ stands for the number of events per unit area (e.g., density). The reference distribution is $K_{CSR}(r) = \pi r^2$. If K(d) > K_{CSR}(d), there is an excess of points nearby, and the pattern leans towards clustering. If K(r) < K_{CSR}(r), there is less likely a chance for other observations to occur nearby, indicating a dispersed pattern of distribution.

Ripley's K-function curves are able to indicate the general pattern of distribution for the points under study (Figure 4.11). The curve of the uniform distribution lies at the bottom, while a clustered distribution has a curve at the top. Comparison of an observed curve with these reference curves is able to shed light on the nature of distribution for the observed points. This comparison is able to reveal the general type of distribution (Figure 4.12). If the observed curve lies below the expected random distribution curve, it suggests a dispersed pattern. If it lies above the random distribution line, then the distribution is clustered. If it runs parallel to the line of the expected distribution within the envelope, then the distribution is random.

The derived curve of $K(r)$ may involve a degree of uncertainty or variability. The range of uncertainty is commonly indicated via *envelope*. This refers to the level of confidence, represented as a band between the highest and the lowest values (e.g.,

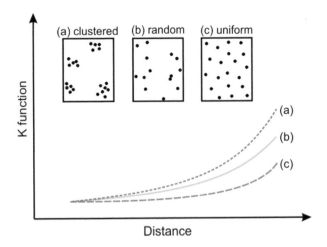

FIGURE 4.11   Three Ripley's K-function curves indicative of three types of point patterns.

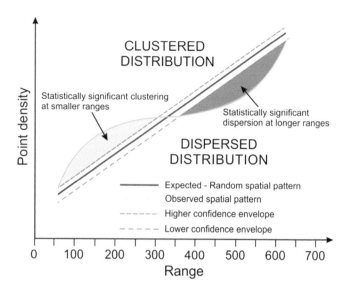

FIGURE 4.12   Interpretation of Ripley's K-function.

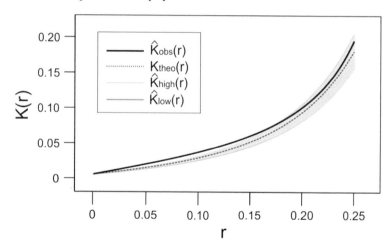

FIGURE 4.13   The concept of envelope in analysing point patterns in which the simulation approach is used to identify the highest and the lowest values (e.g., envelope) of $K(r)$ at a given range r.

envelope) of K(r) at a given distance. As shown in Figure 4.13, the envelope is rather narrow at a small range, but becomes increasingly widened at a large range.

As shown in Figures 4.11 and 4.13, the $K(r)$ function curves are non-linear, which makes it rather difficult to judge whether an observed distribution curve is parallel to the theoretical distribution curves. Nevertheless, the subtle or slight departure of the observed curve from the theoretical ones over different ranges is much easier to perceive and judge if the theoretical curves are straight, as shown in Figure 4.12.

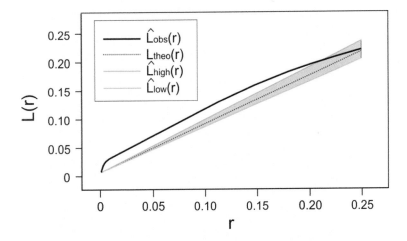

FIGURE 4.14   The *L(r)* transformation of *K(r)*, which shows a linear relationship between the degree of clustering and range with the envelope (shaded) shown. *L(r)* facilitates judgement of the deviation of an observed distribution from the theoretical pattern of distribution over different ranges.

Thus, *K(r)* has been transformed into various linear forms, one of which is *L(d)* calculated as:

$$L(r) = \sqrt{\frac{k(r)}{\pi}} = \sqrt{\frac{A \sum_{i=1}^{N} \sum_{j=1, i \neq j}^{N} k(i, j)}{\pi N (N-1)}} \tag{4.8}$$

where *A* = area, *N* = the number of points, *r* = range or distance, and k(i, j) is the weight. It has a value of 1 if the distance between *i* and *j* is ≤*d*, or 0 otherwise. This transformation is good at producing a straight line with a slope of 1 (45°) that passes the origin, so it is much easier to judge the nature of point patterns at different ranges via the comparison of the curves of the observed distribution with it than a quadratic curve (Figure 4.14).

4.2   FRACTALS AND SPATIAL ANALYSIS

4.2.1   LINE/SURFACE COMPLEXITY

In reality, there is frequently a need to quantify linear and areal features. For instance, it is well known that a river channel is more meandering or sinuous in its lower stretches at a low elevation, but is much straighter in its upper stretches (Figure 4.15). The sinuosity of a river channel can indicate not only its age, but also the surrounding landscape and even the underlying geology. Similarly, a surface of a coastal plain is nearly flat, whereas a mountainous region can experience tremendous topographic relief. How can we compare landscapes with different levels of surface relief? Is it possible to come up with a quantitative measure so that linear and

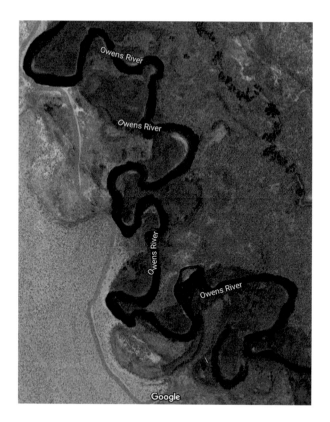

FIGURE 4.15   This section of the Owens River channel in California is so curved that a quantitative measure is needed to indicate its complexity (image credit: screen capture from Google Maps).

areal geographic entities can be compared with each other objectively? These questions can be answered by a number of indices covered in this section.

The complexity of a linear feature can be defined by the ratio of the shortest length to its actual length. This ratio can be quantified from the direct (straight) line length (distance) of the starting and ending points divided by the actual length (D_{actual}) between the same two points (Equation 4.10). This calculation can be easily accomplished in a GIS as all linear features are represented by numerous tiny segments of straight lines, from which the actual length can be summed up. The smaller the ratio, the more contorted the feature. Conversely, the closer the ratio is to 1, the straighter and geometrically simpler the river channel is. In the comparison, the reference length is the straight-line length or the shortest Euclidean distance (D_{direct}) calculated from the plane coordinates of the two points (Equation 2.4).

$$\text{Ratio} = D_{\text{direct}} / D_{\text{actual}} \tag{4.10}$$

Although simple to derive, this ratio is rather crude as it does not indicate how and where the line is curved.

In comparison to lines, surface features have properties that are much more complex and difficult to quantify and compare with each other. In the comparison, the commonly used reference surface is flat without any relief. Surface relief can be measured using the standard deviation of elevations, or their range. Although easy to calculate, these parameters lack a definite meaning. For instance, how much more contorted is a surface with a standard deviation in height that is twice that of another? In fact, the standard deviation of heights or the range of relief does not help us understand and appreciate surface variability intuitively. The solution to this difficulty lies in the fractal dimension.

4.2.2 FUNDAMENTALS OF FRACTAL GEOMETRY

Fractal geometry can be defined as a shape composed of parts that are similar to the whole in some way. The geometrical or physical structure of a fractal feature may be irregularly shaped or fragmented at all scales of measurement, namely scale-independence or geometric invariance under certain transformations, which is central to the understanding of the fractal concept. It can manifest in two fundamental forms, *self-similarity* and *self-affinity*. Self-similarity or exact fractal means that each segment of a shape is geometrically identical or similar to the whole, either statistically or verbatim. In other words, a spatial entity is composed of *n* copies of itself with possible rotation and translation (Figure 4.16). Each copy is scaled down by the same ratio in all Cartesian coordinates from the whole. In reality, such exact fractals are exemplified by snowflakes, but are not commonly associated with geographic or spatial entities that behave more like self-affinity fractals. In comparison with self-similarity entities, self-affinity or *statistically self-similar* features are more commonly associated with spatial entities. In other words, a portion of an entity looks similar to the whole if enlarged and/or reduced to the appropriate scale statistically. For instance, they appear to be the same if the length along different directions is scaled down or up by varying factors. Statistical self-similarity is exemplified by

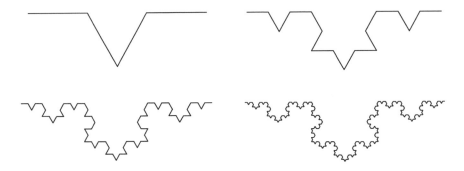

FIGURE 4.16 Self-similarity fractals, in which the whole is made up of the same unit that has been resized and rotated differently.

coastal lines or shorelines in geomorphology. The fractal dimension of linear and areal features is able to shed light on their geometric complexity quantitatively.

4.2.3 FRACTAL DIMENSION AND ITS DETERMINATION

Also called the Hausdorff-Besicovitch dimension, *fractal dimension* is a real number indicative of the degree of irregularity or complexity of a spatial entity that can be linear or areal in its topologic dimension. A linear feature can have a fractal dimension between 1 and 2. A fractal dimension of 1 means that it is straight. A dimension close to 2 indicates that the line is so curved and plane-filling that it occupies virtually the entire 2D space. Similarly, a surface can have a fractal dimension between 2 and 3, depending on its complexity. A dimension of 3 implies that the surface is so crumpled that it fills up virtually the entire 3D space. Therefore, the value of fractal dimension can realistically and objectively reflect the complexity of linear and areal spatial entities.

The determination of fractal dimension relies on regression analysis, in which the log-log plot of step size and length (area) is fitted with a straight line whose slope represents the fractal dimension. The goodness of fit of regression analysis indicates whether or not the geographical entity under study behaves like a fractal. Some common methods of calculating fractal dimension are presented in Table 4.3. These methods have different applicability and may yield drastically differing dimensions for the same spatial entities. Of the eight methods listed in the table, only the first is applicable to linear features, while the remaining ones are suitable for areal features. Of the areal algorithms, the walking dividers or line dividers, variogram, and box-counting methods are used very commonly. The variogram method yielded a dimension of 2.16 for a mountainous area (Roy et al. 1987). This dimension is almost identical to the 2.17 calculated using the walking dividers method. Both figures are slightly larger than the 2.09 obtained using the box-counting method. In fact, the box-counting and the area-perimeter methods produced the most divergent dimensions among a coastline, contours, and the outline of an island (Goodchild, 1982).

The precise determination of fractal dimension of spatial entities has been facilitated by the wide and easy availability of data in digital format, be they linear or areal. In particular, the representation of a surface by a digital elevation model (DEM) at a regular grid enables the fractal dimension of the Earth's surface to be determined easily. Fractal dimension can be derived using a few computer systems, one of which is FRAGSTATS (see Section 4.4.3).

The main use of fractal dimension in spatial analysis is related to the quantification of topography, polygons, and coastlines. Fractal analysis has been successfully used to characterise highly irregular linear features, various types of landforms, and urban land use (Purevtseren et al., 2018), and to describe habitat fragmentation and delineate landform regions statistically (Gao and Xia, 1996). One novel application of fractal analysis is to simulate a terrain with a known dimension against which hypotheses can be tested. However, it is very difficult to link a fractal dimension to the underlying geomorphic processes because there is no one-to-one correspondence between the two. More success has been achieved in descriptive spatial analysis of

TABLE 4.3
Methods Commonly Used to Estimate Fractal Dimension (source: Gao and Xia, 1997)

Method	Algorithm	Estimate of D	Source
Stream Number-Stream Length	$D = \log R_b / \log R_l$ R_b = bifurcation ratio R_l = stream length ratio		La Barbera & Rosso (1989)
Isarithm (Line-divider)	$L(\lambda) = k\lambda^{1-D}$ λ-step size $L(\lambda)$-total length	Plot $\log L$ against $\log \lambda$ $D = 1-B$	Mandelbrot (1967) Goodchild (1980) Shelberg et al. (1982)
Variogram	$2\gamma(d) = kd^{4-2D}$ d-sampling interval $2\gamma(d)$ - incremental variogram	Plot $\log\gamma(d)$ against $\log d$ $D = 2-B/2$ for profiles $D = 3-B/2$ for surfaces	Mark & Aronson (1984) Roy et al. (1987)
Box-Counting	$N = kl^{-D}$ l-cell size, N-average number of adjacencies	Plot $\log N$ against $\log l$ $D = -B$	Goodchild (1982) Shelberg et al. (1983)
Area-Perimeter	$A = kP^{2/D}$ A-estimated area P-estimated perimeter	Plot $\log A$ against $\log P$ $D = 2/B$	Kent & Wong (1982)
Power Spectrum	$P(\omega) = \omega^{2D-5}$ ω-frequency $P(\omega)$-power	Plot $\log P(\omega)$ against $\log\omega$ $D = (5-B)/2$ for profile $D = (8-B)/2$ for surface	Turcotte (1987) Goff (1990)
Triangular Prism	$A(\lambda) = k\lambda^{2-D}$ λ-step size $A(\lambda)$-total area	Plot $\log A$ against Log (step-size^2) $D = 2-B$	Clarke (1986) Jaggi et al. (1993)
Size-Frequency	$N(A>a) = ka^{-D/2}$ $N(A>a)$-no. of islands above size a	Plot $\log N(A>a)$ against $\log a$ $D = -2B$	Kent & Wong (1982)

D = fractal dimension; k = constant; B = slope of the regression line.

urban agglomerations and patterns, and in evaluating the urbanisation process, such as the degree of spatial absorption, change, and continuity in urban morphogenesis (Frankhauser, 1998).

4.3 SHAPE ANALYSIS OF POLYGONS

Shape analysis attempts to understand the geometric properties of and the spatial relationship between a group of 2D spatial entities. So far, it has been exploited by anthropologists to quantify the shape of gorilla skulls (Harmon, 2007). However, in this section shape analysis is restricted to planar forms, usually performed on polygon data in 2D, in which either a singular entity or a group of similar entities is analysed collectively. It is significant to perform shape analysis, as it is a prerequisite for better comprehending the physical process underpinning the formation of the observed shape. For instance, the shape of a kiwi habitat can be very revealing of the spatial range of the bird's foraging activities. The habitat shape can also reveal the traversal distance of animals and their interaction with the environment. A circularly shaped habitat may indicate traverse in all directions with minimal topographic hindrance. On the other hand, a highly elongated habitat may result from the presence of barriers (e.g., a creek). In order to facilitate the comparison of polygons (habitats), there is a need to come up with a quantitative measure of their shape, even though this measure may not be perfect.

4.3.1 Desirable Qualities of Shape Measures

In order to produce a reliable indication, the derived shape index must meet certain qualities, such as uniqueness, rotation invariance, and scale-invariance (Clark, 1981).

(1) *Uniqueness* refers to the quality a unique shape should correspond to in a unique index value, and vice versa. This quality is difficult to achieve as by default, all shapes are inherently 2D and are hardly quantifiable by a number. Thus, some shape indices (e.g., the simple ratio of the shortest radius to the longest) fail to meet this quality.

(2) *Rotation invariance* means that the same shape should yield the same index value regardless of its orientation. This is important in spatial analysis, as the same object can be captured in different directions on a photograph or a satellite image. The shape of a polygon does not vary with its orientation. For certain polygons, such as kiwi habitat, the direction is fixed. In other cases, the polygon can be rotated, such as the dispersal area of plant seeds. Whenever the predominant wind changes direction, the direction of dispersal may change, causing the shape to orient differently. As long as the shape of a polygon remains the same, its quantitative descriptor should have the same value regardless of its orientation. This quality is easier to achieve than uniqueness, as the commonly used indices are all direction-independent.

(3) *Scale-invariance* means that the same shape should have the same index value regardless of the scale at which the spatial entity is captured or measured. This is particularly relevant if the shape is mapped from remote sensing images of different spatial resolutions. They cause the same polygon to appear to be very large or small, depending on the scale at which it is mapped. Scale-invariance means that the shape should be dimensionless. However, this quality cannot always be met, subject to the shape index used.

(4) *Parsimony* means that the number of descriptors of a shape should be kept to a minimum. Naturally, more descriptors will add more information about the shape of a polygon, but not all descriptors are equally important or effective. Only those most revealing descriptors should be retained.

(5) *Independence* means that no matter how many descriptors are retained, they must all be independent of each other. In other words, the deletion of one of them should not impact other retained descriptors: they do not need to be re-calculated.

Of the five qualities, the first three are more important than the last two. These qualities are all related to the actual value of the shape index. The last two are concerned with the shape indices themselves, not with how shape should be calculated or interpreted.

Up to now, polygon shape has been quantified using a variety of indicators, including the elongation ratio, the circularity ratio, and the aspect ratio, calculated using a number of methods. They fall into three general categories: outline-based, compactness-based, and comparison to standard shapes (Table 4.4). All shape analysis algorithms can be grouped into two categories in terms of whether they focus on the perimeter or the whole area, or whether they describe the original polygon in terms of scalar measurements or through structural descriptors (Pavlidis, 1978).

4.3.2 Outline-based Shape Analysis

Outline-based shape analysis focuses on the outline or boundary of a shape, while its interior is completely ignored (e.g., whether it has a hole or not does not affect the quantified shape). There are two approaches under this category: transfer of the 2D shape to 1D boundary presentation, and Fourier radius expansion. Both require unrolling the outline and expressing the discrete points on the perimeter as polar coordinates (Figure 4.17). Once unrolled, it is relatively easy to describe those points using the Fourier series. The split of the outline takes place at some arbitrarily selected position. The rolled-out outline curve is then decomposed into a spectrum of harmonically related trigonometric (e.g., sine and cosine) curves. Even a complex shape can still be precisely described by an appropriate number of Fourier coefficients (harmonics). Fourier analysis has various versions with different complexities, one of which is called Fast Fourier Transform. It can reduce a large number of coefficients to two computationally independent coefficients per harmonic. However, the result of shape analysis using this method is highly dependent on the position at which the outline is split and rolled out (e.g. the starting point of the digitised shape outline or phase). It is also computationally intensive and complex. This approach is applicable only to individual polygons without any holes.

Furthermore, Fourier methods of analysis are sensitive to whether the captured shape outline has been smoothed or not (Haines and Crampton, 2000). If it contains a large number of spurious, high-frequency noises, then they can distort or even corrupt the analysis. In addition, Fourier analysis results are sensitive to the placement of the starting position of the digitised outline, which can unduly affect the proper interpretation of the results. This problem is particularly acute when there is no unambiguously defined, biologically homologous point on the outline that is suitable for serving as the starting point.

TABLE 4.4
Comparison of Commonly Used Descriptive Shape Indices and Their Best Uses

Method	Formula	Pros	Cons	Best uses
P/A ratio	$\dfrac{P_{ij}}{2\sqrt{\pi a_{ij}}}$	Simple, revealing compactness	No information on shape, just a degree of deviation from circular shape	For a group of polygons (e.g., land cover patches)
Outline-based	Fourier harmonics	Robust, accurate	Subject to noise and point of splitting, complex	Individual shapes with no holes
Axis ratio	R_{short}/R_{long}	Simple, easy to calculate	General shape in two directions, imprecise	For individual polygons (e.g., shape of a city)
Radii average	$100\displaystyle\sum_{i=1}^{n}\left\|\dfrac{r_i}{\sum_i^n r_i}-\dfrac{1}{n}\right\|$	Deviation in radius (reference shape: circle)	Subject to the number of sectors dividing the shape	For individual polygons only
Major–minor ratio	L_{min}/L_{max}	Simple to calculate, measures elongation	Too crude, shape in two primary directions	The shape of rain shields
Fractal dimension	$Ln\,(p_{ij})/Ln\,(a_{ij})$	Precise indicator of shape complexity	Complex to calculate	Both individual and collective polygons possible
Compactness based on moment of inertia	$C_{MI}=\dfrac{A^2}{2\pi I_g}$ (Ig centroid of polygon)	Able to handle shapes with holes and multiple parts, results additive	Highly complex	Regionalisation of a large number of polygons
Gravelius compactness coefficient	$GC=\dfrac{P_r}{2\sqrt{\pi A_r}}$	Robust, not subject to DEM grid size or the basin scale	Complex in calculating basin surface area	Quantifying the shape of natural drainage catchments

4.3.3 COMPACTNESS-BASED MEASURES

Alternatively, outline-based measures can make use of a reference shape that can be easily defined, such as a circle. It is chosen as the reference shape because it is the most compact, in that the area enclosed by a given circumference is the maximum. A circle is also advantageous in that its area depends solely on one

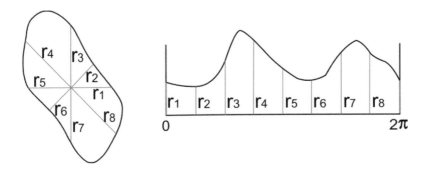

FIGURE 4.17 The outline approach to quantifying shape by unrolling it at some arbitrarily selected starting point that is going to affect the phase of the unrolled curve.

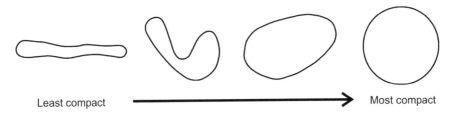

Least compact ————————————————————→ Most compact

FIGURE 4.18 The outline approach to quantifying shape based on the perimeter-to-area ratio. All four shapes have the same perimeter, but their compactness increases from left to the right.

parameter, the radius. A shape index can be expressed as the perimeter-to-area ratio:

$$\text{Perimeter-to-area shape} = \frac{P_{ij}}{2\sqrt{\pi \cdot a_{ij}}} \tag{4.11}$$

where P_{ij} stands for the perimeter of the jth polygon of type i and a_{ij} represents its area. All circles have a perimeter-to-area ratio of 1, regardless of their size and orientation. Thus, they possess the scale-invariant and rotation invariant qualities. The smaller the ratio, the less area is enclosed by the perimeter, indicating an increasingly elongated and skinny shape (Figure 4.18). Although very simple to calculate, this ratio is unable to reveal the exact shape. Instead, it indicates the degree of deviation from the reference shape. This ratio faces a major limit, in that two totally different shapes can have the same ratio. Thus, it fails to meet the desired quality of uniqueness.

The compactness ratio is applicable to both individual polygons and groups of polygons of the same type, such as forested polygons in a landscape. When multiple polygons are the targets of analysis, the mean shape index of all polygons of the same type can be calculated by summing the shape index and dividing the sum by the number of polygons:

$$\text{Mean shape index} = \frac{1}{n_i} \sum_{j=1}^{n_i} \left[\frac{P_{ij}}{2\sqrt{\pi \cdot a_{ij}}} \right] \tag{4.12}$$

where n_i is the total number of polygons of type i. With this mean shape index, it is possible to compare different types of polygons or habitats quantitatively.

A special version of the perimeter-to-area ratio is known as the *Gravelius compactness coefficient* (GC), defined as the ratio of the basin perimeter to the circumference of a circle having the same surface area as the basin. It is calculated as:

$$GC = \frac{P_r}{2\sqrt{\pi \cdot A_r}} \tag{4.13}$$

where P_r and A_r refer to the perimeter and the area of the basin calculated from a DEM at the relative resolution R_r respectively. GC is not subject to the DEM grid size or the basin scale, meeting the desired scale-invariant quality. This robust shape index can be used to compare basins in hydrological studies (Sassolas-Serrayet et al., 2018).

Instead of relying on the perimeter, the compactness index proposed by Li et al. (2013) makes use of the moment of inertia (MI). It is calculated as:

$$C_{MI} = \frac{A^2}{2\pi I_g} \tag{4.14}$$

where I_g stands for the polygon centroid and A refers to the area of the shape. Dissimilar to other compactness-based shape measures, C_{MI} is applicable to polygons of any shape, even those having holes and multiple parts, although the determination of the centroid in the case of holes is a rather complex and computationally intensive operation. MI can be calculated using the trapezium-based approach developed by Li et al. (2013). This improved the computation efficiency for shapes represented in either vector or raster format. This measure is robust, and the calculated shape index value is not sensitive to the uncertainty of the digitised shape outline. The calculated value is stable, and does not vary with polygon size and spatial resolution, so is scale-invariant. This compactness-based measure is particularly suited to shape analysis in regionalisation problems involving a large number of polygons, such as census data, in which a large number of shapes need to be aggregated into a few larger regions because the index value is additive. Whenever new polygons are added to a group of existing ones, there is no need to recalculate the shape index of all polygons. Instead, the compactness of the newly added shapes can be calculated and then added to the existing shape index value.

The fractal dimension of shape is also based on the ratio of area to perimeter (Table 4.4). The only difference between this measure and others based on compactness is the natural logarithmic transformation of both area and perimeter. The dimension is the slope of the regression equation of the two parameters.

4.3.4 COMPARISON TO A STANDARD SHAPE

A slight variation of the perimeter-to-area ratio shape index is the average radii index, or comparison to a standard shape. This index is able to indicate shape by comparing the derived results to some known standard shape, such as a circle. This radial line approach of quantifying a shape, developed by Boyce and Clark (1964), concentrates on the deviation of the outline from a common reference point in certain representative directions. This reference point is commonly served by the polygon centroid, which can be conveniently calculated from all the coordinates of the perimeter. The general shape is determined by drawing a number of radial lines and statistically analysing their lengths to derive the absolute deviation (Figure 4.19). The shape will be circular if the deviation equals 0. The calculation of average radii follows five steps:

(1) Determine the polygon centroid. This can be achieved by averaging all the horizontal and vertical coordinates of vertices defining the outline, respectively.

(2) Draw n equally spaced radials from the centroid to the shape's perimeter, and measure the radial distance (r_i) from the centroid to the outline for each of the radial lines.

(3) Convert the measured absolute radii to a percentage by dividing them by the sum of all radii $\left(\dfrac{r_i}{\sum_i^n r_i} \times 100\% \right)$ (n = the number of radials, n = 16 in Figure 4.19).

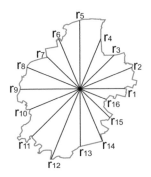

Formula	r_1	r_2	r_3	r_4	r_5	r_6	r_7	r_8	r_9	r_{10}	r_{11}	r_{12}	r_{13}	r_{14}	r_{15}	r_{16}	Σ
r_i	6.7	8.4	7.2	8.7	10.4	8.2	6.9	8.5	8.9	8.1	10.2	11.4	8.7	8.6	6.2	4.2	131.3
r_i (%)	5.10	6.40	5.48	6.63	7.92	6.25	5.26	6.47	6.78	6.17	7.77	8.68	6.63	6.55	4.72	3.20	100.00
$\left\| \dfrac{r_i}{\sum_i^n r_i} - \dfrac{1}{n} \right\|$	1.15	0.15	0.77	0.38	1.67	0.00	0.99	0.22	0.53	0.08	1.52	2.43	0.38	0.30	1.53	3.05	15.15

FIGURE 4.19 An example of analysing the shape of Mount Albert, an inner suburb of Auckland, New Zealand based on 16 radii.

(4) Calculate the theoretical mean radius by dividing 1 by n, and express the ratio as a percentage $\left(\dfrac{1}{n} \times 100\% = 1/16 = 6.25\%\right)$.

(5) Subtract the observed percentage obtained in step 3 from the theoretic mean radii (%) calculated in step 4.

(6) Add up all the absolute differences obtained in step 5, or:

$$\text{Boyce and Clark index} = 100 \sum_{i=1}^{N} \left| \frac{r_i}{\sum_i^n r_i} - \frac{1}{n} \right| \tag{4.15}$$

The final result is a measure of variability of the radial lengths from the reference point (centroid) to the theoretical radius. A circle has a value of 0. Indicating primarily the compactness of a shape, this measure sheds no light on the shape, such as where the largest variation in radial length occurs. Neither does it reveal anything about the direction in which the polygon is indented or convex, hence it is unable to indicate the genuine shape. This method is easy to implement and the calculation is simple and straightforward, but the derived index value varies with the number of radial lines and their orientation. This method is applicable for a singular polygon that must be convex-shaped. Otherwise, the radial line in a given direction can intersect with the perimeter more than once, resulting in more than one radius in the same direction.

A simpler version of this method is the *major–minor ratio*. Instead of n radials, the radius is measured twice in only two crucial directions: the maximum length L_{max} along the major axis, and the minimum length L_{min} along the minor axis. The minimum–maximum ratio of L_{min}/L_{max} measures the contrast between the shortest and the longest dimension of a polygon, or elongation (Wentz, 2010). It ignores the variation in other directions. Thus, it is much cruder than the Boyce and Clark shape index. Both share the same trait of being applicable to singular polygons without holes.

4.3.5 PRACTICAL APPLICATIONS

So far, shape analysis has found wide applications in geography, such as evaluation of the effects of urbanisation, characterisation of drainage basins in hydrology, quantification of the shape of drumlins and coral atolls in geomorphology, and examination of patterns of geologic formations (Wentz, 2020). In particular, it has been used to quantify the shape of tropical cyclone rain shields (Matyas, 2007). Quantitative shape metrics of tropical cyclones can help us better understand the influence of various factors on shaping tropical cyclones that produce complex rainfall patterns. These metrics may include the area–perimeter ratio and the minimum–maximum ratio. They enable the differentiation of cyclones via stepwise discriminant analysis. The area–perimeter ratio provides a simple measure of storm compactness, and the major–minor axis ratio is able to shed light on the shape of the rain shield (circular versus elongated). After the shape of the rain fields has been quantified using these indices, they can be used to model rain intensity and facilitate a more reliable prediction of the spatial distribution of cyclone-induced rainfall and the locations where flood-producing rainfall events are going to occur (Matyas, 2007).

In hydrology, the shape of a drainage catchment as measured by the GC coefficient is related to the Hack's coefficient, and gives a physical meaning to data dispersion published in previous studies about catchment length and area (Sassolas-Serrayet et al., 2018). Analysis of a wide range of catchments reveals a GC value that varies as a continuum, typically between 1.2 and 2.1. In landscape ecology, the shape of hypothetical habitats (e.g., linear, branching, rectangular, and square) is found to affect the dispersal and abundance of invasive populations (Cumming, 2002). The same habitat will be invaded faster, and the invasion may ultimately result in more abundant invasive species in the habitat if it has a more geometrically complex shape. Such findings have profound implications for the proper design of restorative habitat in that attention should be paid not only to habitat fragmentation and connectivity but also to their shape.

4.4 ANALYSIS OF AREAL DATA AND FRAGSTATS

Areal data are commonly associated with land use and land cover patches commonly derived from satellite imagery. These polygon data have different shapes and sizes, and may be oriented differently as well. Some of them may share a common boundary, while others can be scattered widely in a study area. Areal data can also be socioeconomic, enumerated over a census unit area. Depending on their nature, some of the data have to be processed before any analysis can proceed. For instance, per capita income has to be generalised from the ratio scale to the ordinal scale of high, medium, and low before it is possible to examine whether this variable is spatially clustered or random in its distribution. All the polygons are non-overlapping and spatially exhaustive with no holes in the landscape, even though it is possible for some areal units not to have data associated with them (e.g., missing data). *Area pattern analysis* is mostly descriptive, in that a general index is produced to describe all the polygons, or only certain types of them that meet the specified criteria. A pattern can be generated not only for patch (polygon) shape, but also its size, perimeter, and its relationship with other adjacent polygons, both individually and collectively.

4.4.1 Objects of Study

Earlier, we discussed spatial analysis concentrated on individual patches or polygons. In this section, the focus shifts to polygons as a group. Statistical information is generated from all polygons, or polygons of a given type in the case of multiple polygon identities. Some typical questions that can be addressed with area pattern analysis include the following:

(1) How to come up with quantitative indices to describe a landscape comprising multiple polygons of various types?

(2) How does one landscape compare with another quantitatively?

Comparisons can be made in terms of polygon size and its variation, polygon shape, and polygon perimeter (known as edge), distance from other polygons, as well as polygon identity in relation to that of other polygons. Metrics of analysis can be easily derived using existing powerful computer packages, owing to the wide availability of polygon data in digital format. However, attention must be paid to the proper interpretation of the obtained quantitative results, otherwise the entire purpose of area pattern analysis is lost.

4.4.2 REPRESENTATIVE METRICS

In landscape ecology, not all polygons are of the same type. These polygons can be studied at three levels – patch level, class level, and landscape level – with the scale gradually broadening. *Patch*, also called *ecotope* and *biotope*, is the basic element of a landscape. It is represented as a polygon in vector format. At the polygon level, each patch is treated as an individual observation or unit of analysis. As such, all those statistics covered in Sections 4.2 and 4.3 are also applicable to them, such as shape, fractal dimension, nearest neighbour distance, and proximity. However, some of these indices may be available only in the statistical sense (e.g., they apply to all polygons, instead of individual ones).

Class refers to all patches having the same identity or attribute value, such as forest polygons and grassland polygons. At the class level, the same set of metrics can be derived as at the polygon level. Moreover, the derived results can be weighted by polygon area for a group of polygons. The influence of each individual polygon on the final results is proportional to its dominance in the landscape, hence large polygons exert a more significant influence on the final results than small ones. Class-level indices may include the largest patch index, patch density, mean patch size, and area-weighted mean edge contrast and shape index. Since all the patches may have different identities, additional indices can also be computed based on their identities at the class level, such as landscape similarity and edge contrast (Table 4.5).

Finally, at the *landscape* scale, all polygons are treated as members of the same class and are analysed jointly regardless of their identity. Here, landscape refers to a heterogeneous land area composed of a cluster of interacting ecosystems that are repeated in a similar form throughout the area of study. In addition to all those aforementioned metrics, more can be calculated at this level, including fractal dimension, landscape similarity index, Shannon's diversity index, Shannon's evenness index (see Table 4.5), area-weighted mean shape index (Equation 4.16), and the contagion index (Equation 4.17). Contagion describes the probability that two randomly chosen adjacent polygons have the same identity.

$$\text{Area Weighted Mean Shape Index} = \sum_{j=1}^{n} \left[\frac{p_{ij}}{2\sqrt{\pi \cdot a_{ij}}} \cdot \frac{a_{ij}}{\sum_{j=1}^{n} a_{ij}} \right] \quad (4.16)$$

TABLE 4.5

Landscape-level Metrics Commonly Used for Characterising Area Patterns

Index	Formula	Explanation
No. of patches	n	n = number of patches
Total patch area	$\sum a_i/10{,}000$	a_i = area (ha) of patches
Mean patch size	$\sum a_i/n$	a_i = area (ha) of patches, n = number of patches
Edge (total, density, contrast)	$1{,}000 \sum e/A$	e = total length (m) of patch edges, A = total landscape area
Mean shape*	$\dfrac{1}{n} \sum_{j=1}^{n} \dfrac{p_j}{\sqrt{\pi a_j}}$	p_j = perimeter of patch j, a_j = area of patch j, n = number of patches
Fractal dimension	$2 \, \mathrm{Ln}\,(p_{ij})/\mathrm{Ln}\,(a_{ij})$	p_{ij} = perimeter of patch j of identity i, a_{ij} = area of patch j of cover identity i
Landscape similarity	$100 \sum_{j=1}^{n} a_{ij}/A$	A = total landscape area
Area weighted mean patch fractal dimension	$\sum_{i=1}^{m} \sum_{j=1}^{n} \left(\dfrac{2\mathrm{Ln}\left(0.25 p_{ij}\right)}{\mathrm{Ln}\left(a_{ij}\right)} \dfrac{a_{ij}}{A} \right)$	m = number of patch types, n = number of patches of a class, p_{ij} = perimeter of patch ij, a_{ij} = area of patch ij, A = total landscape area
Shannon's diversity	$-\sum P_i \cdot \mathrm{Ln}(P_i)$	P_i = proportion (n/N) of type i polygons (n) out of the total number of polygons (N)
Shannon's evenness	$-\sum(P_i \cdot \mathrm{Ln}(P_i))/\mathrm{Ln}(m)$	m = number of polygons of different land cover types, 0–1

*Can be area-weighted (see Equation 4.16).

where p_{ij} and a_{ij} represent the perimeter and area of patch j with a cover identity i, and n denotes the total number of polygons with a cover identity i.

$$\text{Contagion} = 100 \left[1 + \frac{\sum_{i=1}^{m}\sum_{k=1}^{m} P_i \dfrac{g_{ik}}{\sum_{k=1}^{m} g_{ik}} \cdot \ln\left(P_i\right)\dfrac{g_{ik}}{\sum_{k=1}^{m} g_{ik}}}{2\ln\left(m\right)} \right] \quad (4.17)$$

where P_i represents the proportion of patch type i in the landscape, g_{ik} refers to the number of adjacencies between pixels of patch types i and k, and m is the number of possible patch types. This equation is suitable for raster data. The vector version of contagion is calculated as:

$$C_u(d) = 1 + \frac{\sum_{i=1}^{m}\sum_{j=1}^{m} p_{ij}(d)\ln\left(p_{ij}(d)\right)}{2\ln(m)} \quad (4.18)$$

where d denotes the distance between patch type i and type j.

4.4.3 FRAGSTATS

FRAGSTATS by McGarigal and Marks (1994) is a computer system designed for characterising area pattern in general and quantifying landscape in particular. This spatial pattern analysis system is able to quantify landscape structure and yields a comprehensive range of landscape metrics. This versatile system can be run almost completely automatically and requires little technical training to use. In order to run spatial analysis, the user just needs to tick the box in front of each index to ensure the inclusion of the calculated metrics in the output (Figure 4.20). Metrics can be derived at the patch level (vector version) or the cell level (raster version). The advent of this system considerably eased spatial analysis of polygon data as it recognises spatial data saved in the ArcGIS format. The flawless integration of FRAGSTATS with ArcGIS rids the necessity of compatibility in data format.

FRAGSTATS comes in two versions: vector and raster. The vector version is known as Patch Analyst. It accepts ArcGIS polygon coverage or shape files, and facilitates spatial pattern analysis of polygon data and modelling of patch attributes (Figure 4.20). The analytical results produced are saved in vector format. The raster version of FRAGSTATS is called Patch Grid, also an extension of ArcGIS. Patch Grid is used when the input land cover map is in raster format (e.g., produced from raster imagery directly via image classification). It accepts image files in various

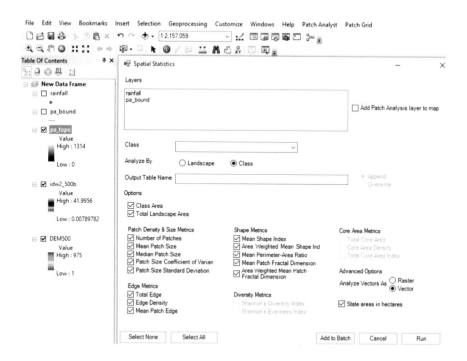

FIGURE 4.20   The interface of FRAGSTATS in deriving spatial indices at the landscape, class, and patch levels.

formats, including ERDAS and IDRISI, plus other common graphic formats such as TIFF and JPEG. There is no need to convert the raster map into vector format. The latest version, 4.2, can work with 8-, 16-, and 32-bit integer grids.

A large number of structural and functional metrics can be derived at the landscape, class, and patch (polygon) levels. Common functional metrics include measures of core area, edge contrast, and isolation. They are able to account for depth-of-edge effects. Edge contrast can show the difference in magnitude between adjoining patch types. Isolation can reveal ecological neighbourhood size and similarity between different types of patches. Prior to running FRAGSTATS, the analyst is given the option of sampling the landscape for analysis at the sub-landscape level. Exhaustive sampling may be carried out using user-supplied tiles (sub-landscape), uniform tiles, a moving window, or randomly generated focal points around which a window (sub-landscape) is placed. The analyst can specify the size and shape of the window. Provisions are also available for handling missing data (e.g., unclassified cells in a raster input), background (classified, but anonymous), and border (buffer of classified cells outside the designated landscape boundary).

4.4.4 Application in Quantifying Urban Sprawl

It is possible to simulate urban growth under different scenarios (Torrens, 2006). The simulated outcomes in different scenarios must be compared with each other quantitatively in order to reveal their differences. This comparison is based on the derived metrics of polygon data. They are able to describe the simulated pattern of land cover polygons precisely. The level of urbanisation can be described from the perspective of land use composition by examining the presence and the quantity of different patch types (e.g., urban versus non-urban). Next, the spatial configuration can be described. For instance, how are these patches distributed spatially inside the study area? Are there any distinct spatial agglomerations among each type of patches? These questions can be answered by a number of descriptive indices jointly, such as contagion, perimeter-area fractal dimension (PAFRAC), and the interspersion and juxtaposition index (IJI). Contagion can be calculated using Equation (4.17). A small contagion value is indicative of a landscape comprising a large number of small and dispersed polygons (in vector format) or clusters of cells (in raster format). High contagion values indicate more compact landscapes. PAFRAC measures the extent to which patches fill a landscape (e.g., patch shape complexity), and is calculated as:

$$
PAFRAC = \frac{2}{N \sum_{i=1}^{m} \sum_{j=1}^{n} \ln p_{ij} \cdot \ln a_{ij} - \sum_{i=1}^{m} \sum_{j=1}^{n} \ln p_{ij} \cdot \sum_{i=1}^{m} \sum_{j=1}^{n} \ln a_{ij}} {N \sum_{i=1}^{m} \sum_{j=1}^{n} \ln p_{ij}^{2} - \left(\sum_{i=1}^{m} \sum_{j=1}^{n} \ln p_{ij} \right)^{2}} \tag{4.19}
$$

Where a_{ij} refers to the area (m²) of patch ij with a perimeter of p_{ij} (m); N denotes the total number of patches in the landscape. The actual determination of PAFRAC is based on the slope of the regression relationship of the log-transformed patch area (a) against the log-transformed patch perimeter (p).

The IJI index measures adjacency on a patch-by-patch basis. It is calculated as:

$$IJI = \frac{-100 \sum_{i=1}^{m} \sum_{k=i+1}^{m} \frac{e_{ik}}{E} \cdot Ln\left(\frac{e_{ik}}{E}\right)}{Ln\left[0.5m(m-1)\right]} \quad (4.20)$$

where m represents the number of patch types, e_{ik} stands for the total length of edges between patch types i and k in the landscape, and E denotes the total length of edges in the landscape, excluding the background. High IJI values (e.g., close to 100%) suggest that patch types in the landscape are well interspersed with an equal adjacency between each other, and the landscape under study has a relatively high degree of homogeneity. When the patches are poorly interspersed, they will have a low IJI value close to 0, and patch type adjacencies will have a disproportionate distribution.

The joint consideration of the above three indices should paint a comprehensive and informative picture about the level of urban sprawl and its spatial properties. As shown in Table 4.6, the number of patches in the general growth scenario (14,375) is twice the combined number of patches in the other two scenarios. This mode of urban sprawl will cause the resultant landscape to be much more fragmented than the other two modes of growth. This disparity can also be appreciated from the smaller fractal dimension. The contagion value of this mode of growth is almost the same as that of the Midwestern example, both being much lower than the polycentricity mode. Judging from this index value, this mode of growth seems to result in a less sprawling city. The much higher IJI index of general growth suggests that the resultant city is much more homogeneous than that formed by the other two modes of growth.

4.5 ANALYSIS OF DIRECTIONS

Analysis of directions plays a vital role in a number of applications, such as where to site a wind turbine to harvest wind energy. The mounting of solar panels on a building roof to capture solar radiation must be guided by the results of directional data

TABLE 4.6

Fractal and Landscape Metrics of Polygons Used to Compare Urban Sprawl in Three Scenarios of Growth

Metric	General growth	Polycentricity	Midwestern example
No. of patches	14,375	3,066	3,782
PAFRAC	1.5305	1.5321	1.5479
Contagion (%)	48	65	45
IJI (%)	54	37	20.15

Source: Torrens (2006).

analysis. Even the landing of aeroplanes at an airport can benefit from information derived from directional data analysis. Common directional data include wind direction and ocean current circulation. They are known as *directed*, as they are indicative of the direction of movement (Schuenemeyer, 1984). Directions can also be *undirected*, merely indicating the bearing or orientation from a reference direction, such as the orientation of a fault line and slope aspect. For undirected directions, the orientation of θ is the same as θ + 180°. These two types of directional data are represented differently. Undirected directional data only show the bearing without any magnitude attached to them. They can be represented either categorically (e.g., north) or continuously with a range of 0–360°. In contrast, directed directional data have to be represented as a vector, with the vector length proportional to the velocity or pace of movements. Directions can be 2D or 3D in space. Some undirected line segments, such as landslide chutes, have a 3D direction, while wind direction and ocean current direction are mostly planar. This section covers only planar directions, and 3D directions are treated as if they were 2D.

4.5.1 Reference Systems of Directions

Just as locational data need a proper reference system, directed directional data (vector) also require a proper reference system so that results obtained from different studies by different authors are directly comparable to one another. Directions are conveniently represented in the polar system, in which two parameters are needed to define a directional vector uniquely: (ρ, θ), in which ρ refers to the magnitude of the vector or radius, and θ represents the angle measured counter-clockwise from due east (the L-axis in Figure 4.21). Thus, its potential value falls within a range of 0–360°, with 0° being the same as 360°. In Figure 4.21, the green line represents a radial length of 4 (*r*) and an angular coordinate of 60° (θ) at B. The blue line represents a vector with a magnitude of 2 (*r*) and an angular coordinate of 290° at C.

It must be emphasised that this polar system is only suitable for representing directed directional data. For undirected data, the information to be conveyed is simply the orientation, while the magnitude (e.g., landslide path length) is ignored or

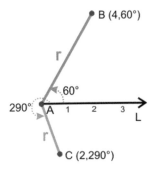

FIGURE 4.21 The polar coordinate system with polar axis *L* for representing vector directions (A is the pole).

not considered. In the raster environment, it is defined by the relative height of cells within a 3 by 3 window (see Section 2.3.5 for more details).

4.5.2 DESCRIPTIVE MEASURES

Directed directional data can be analysed statistically in the same manner as any other types of spatial data, such as the derivation of mean and standard deviation. However, since the possible range of directional observations is confined to 0–360°, there is some uniqueness to the calculation of the statistical parameters. Figure 4.22 illustrates how two directions are summed to derive the resultant vector. Each direction must be projected to the two primary directions of easting and northing. Only the directional vectors having the same direction can be arithmetically manipulated. The final direction is calculated from the ratio of the summed vector lengths in the two primary directions. The arctan of the two angles is then calculated as the final direction. The detailed calculation can be expressed as:

$$X_R = X_A + X_B = \rho_A \cos\theta_A + \rho_B \cos\theta_B \tag{4.21}$$

$$Y_R = Y_A + Y_B = \rho_A \sin\theta_A + \rho_B \sin\theta_B \tag{4.22}$$

$$\rho = \sqrt{X_R^2 + Y_R^2} = \sqrt{\left(\rho_A \cos\theta_A + \rho_B \cos\theta_B\right)^2 + \left(\rho_A \sin\theta_A + \rho_B \sin\theta_B\right)^2} \tag{4.23}$$

$$\theta = \arctan\left(\frac{Y_R}{X_R}\right) = \frac{\rho_A \sin\theta_A + \rho_B \sin\theta_B}{\rho_A \cos\theta_A + \rho_B \cos\theta_B} \tag{4.24}$$

It is important to know the quadrant of the sum as it determines the sign in the calculation, and θ must be properly adjusted by the quadrat it falls into. The adjustment

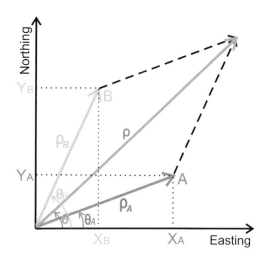

FIGURE 4.22   The resultant vector (ρ, θ) formed by adding two directions, θ_A and θ_B.

to the calculated direction is 0 in quadrant I (numerator >0, denominator >0), 90° in quadrant II (numerator >0, denominator <0), 180° in quadrant III (numerator <0, denominator <0), and 270° in quadrant IV (numerator <0, denominator >0).

The directional mean or the direction of the resultant vector R_0 is calculated as:

$$R_0 = \arctan \frac{\sum_{i=1}^{n} \sin \theta_i}{\sum_{i=1}^{n} \cos \theta_i} \tag{4.25}$$

where θ_i refers to the angle of the ith direction (vector). The variance of all directions is also circular, calculated as:

$$S_0 = 1 - R / n \tag{4.26}$$

where R stands for the resultant vector length $\left(\sqrt{X_R^2 + Y_R^2}\right)$ and n refers to the number of directions used in the calculation. It is possible to test whether a collection of directions is randomly distributed or not using the inferential analysis methods presented in Section 4.1.8.

4.5.3 Schematic Representation

The results of a directional analysis can be represented graphically in a few ways. The first type of diagram is known as *linear histogram* (Figure 4.23). Essentially, it

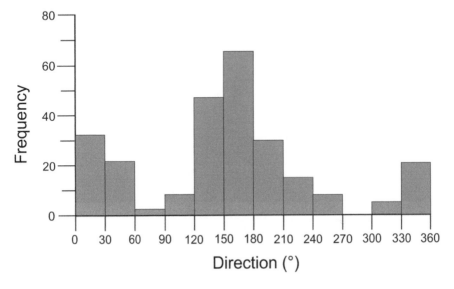

FIGURE 4.23 An unwrapped circular histogram showing the frequency distribution of landslide chute orientation in 12 directional sectors.

is just a plain, unwrapped circular histogram that illustrates the frequency distribution with respect to the direction categorised at various levels. This histogram is no different from ordinary histograms except that the horizontal axis represents the direction. Thus, there is no direct correspondence between the illustrated direction and the actual direction. For instance, the sector of 240–270° does not indicate the actual direction graphically. Thus, it is not so easy to perceive the directional distribution intuitively.

The limitations of the unwrapped circular histogram can be effectively overcome with the *circular histogram*, in which data are grouped according to the pre-determined directional sectors and represented as bars of a uniform width, with the bar length proportional to the frequency value being represented (Figure 4.24). Due to the absence of the vertical axis, concentric circles of a distinct value are drawn to illustrate the magnitude being visualised. Thus, bars tend to overlap each other at a low value. In order to avoid heavy overlapping, an empty circle is drawn at the centre, signifying the reference direction. However, this can cause misinterpretation. There is no room for marking directions in the circular histogram. It is implicitly assumed that a bar oriented in a given direction is synonymous with the direction being depicted. Thus, the circular histogram is rather intuitive.

The *rose diagram* is by far the most commonly accepted form of visualising directional results (Figure 4.25). Instead of using bars of a uniform width, a rose diagram comprises sectors that converge at a single point. Sector radius is proportional

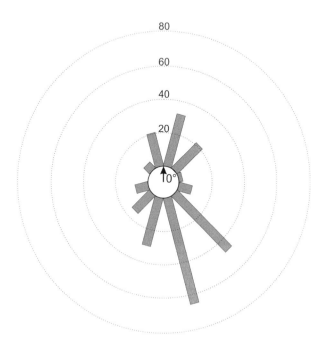

FIGURE 4.24   A circular histogram showing the distribution of observations with direction. The frequency data shown are the same as in Figure 4.23.

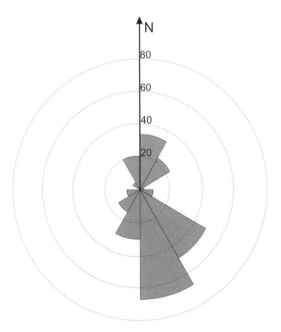

FIGURE 4.25   A rose diagram illustrating the directional distribution of the data shown in Figures 4.23 and 4.24.

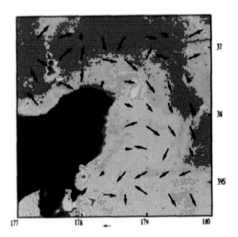

FIGURE 4.26   Direction of ocean currents off the east coast of Hawk's Bay in New Zealand as detected from multi-temporal satellite images. The colour background shows sea surface temperature from which the flow direction was detected (source: Gao and Lythe, 1998).

to class frequency. The area of a sector is proportional to the square of its frequency. The magnitude of a sector is shown in a vertical axis with markers for distinctive values (Figure 4.25), so the possibility of misinterpreting the diagram is minimised. The rose diagram is similar to the circular histogram, in that the sector direction represents the actual direction being depicted.

Directional data analysis is highly restricted in its applications, as it always deals with a pair of points (from to) in the vector format. It is commonly used in meteorology to identify the predominant wind direction and the direction of traffic and channel flow. In oceanography, it is critical to analysing the direction of ocean current circulation for navigation and climate change modelling (Figure 4.26). In hazard studies, it can be used to determine the orientation of landslide chutes and snow avalanche paths, so that the most vulnerable places can be avoided, and remedial measures implemented to mitigate risks.

REVIEW QUESTIONS

1. It can be argued that the three fundamental spatial patterns of random, clustered, and uniform are interchangeable. Discuss the extent to which you agree or disagree with this argument.
2. The random spatial pattern can be indicative of the environment or totally independent of the environment. Use an example each to illustrate whether this pattern can be used to infer the environment responsible for its formation.
3. Clustered patterns can be formed at different scales. How does the formation of a clustered pattern at a local scale differ from that at the regional or global scale?
4. Compare and contrast the methods that are commonly used to analyse spatial patterns of point data.
5. Hotspot analysis is a special kind of point pattern analysis. In which sense is it special?
6. Compare and contrast hotspot analysis of point data with inferential analysis of point pattern.
7. Compare and contrast hotspot analysis with kernel density analysis of point data.
8. What are the main similarities and differences between second-order analysis of point pattern and analysis of spatial pattern using joint count statistic?
9. Which type of features is easier to quantify both individually and collectively, areal features or linear features? What underpins the revelation of their complexity via their quantification?
10. Although fractal dimension is a value that can indicate the complexity of linear and areal features precisely, what problems are commonly encountered in its derivation?
11. Compare and contrast the major philosophies by which the shape of a polygon is quantified. Why no one is perfect?
12. Compare and contrast how perimeter is used to quantify shape in outline-based, compactness-based, and the mean radii-based methods.
13. If you are going to quantify a landscape in ecology, what indices would you use to indicate its degree of fragmentation?

14. What are the common applications of directional analysis in geography? What parameters are usually derived in direction analysis?
15. Compare and contrast direction analysis in 2D with that in 3D.

REFERENCES

Boyce, R. R., and Clark, W. A. V. (1964) The concept of shape in geography. *The Geographical Review*, 54: 561–57.

Clark, M. W. (1981) Quantitative shape analysis: A review. *Mathematical Geology*, 13(4): 303–320.

Clarke, K. C. (1986) Computation of the fractal dimension of topographic surfaces using the triangular prism surface area method. *Computers and Geosciences*, 12: 713–22.

Cumming, G. S. (2002) Habitat shape, species invasions, and reserve design: Insights from simple models. *Conservation Ecology*, 6(1): 3, www.consecol.org/vol6/iss1/art3/.

Dexter, L. R. (2007) *Mapping Impacts Related to the Senecio Franciscanus Greene Phase II* (Final Report), U.S. Forest Service, doi: 10.13140/2.1.4939.0084.

FIRMS. Fire Information for Resource Management System, https://firms.modaps.eosdis .nasa.gov/map/#d:2021-05-03..2021-05-04;@134.3,-26.4,5z.

Frankhauser, P. (1998) The fractal approach. A new tool for the spatial analysis of urban agglomerations. In *Population: An English Selection, Special Issue New Methodological Approaches in the Social Sciences*, 205–40.

Gao, J., and Xia, Z.-G. (1996) Fractals in physical geography. *Progress in Physical Geography*, 20(2): 178–91.

Gao, J., and Lythe, M. (1998) Evaluation of the MCC method in detecting oceanic circulation patterns at a local scale. *Photogrammetric Engineering and Remote Sensing*, 64(4): 301–8.

Getis, A., and Ord, J. K. (1992) The analysis of spatial association by use of distance statistics. *Geographical Analysis*, 24(3): 189–206, doi: 10.1111/j.1538-4632.1992.tb00261.x.

Goff, J. A. (1990) Comments on "Fractal mapping of digital images: Application to the topography of Arizona and Comparison with synthetic images" by J Huang and D.L. Turcotte. *Journal of Geophysical Research*, 95(B4): 5159.

Goodchild, M. F. (1980) Fractals and the accuracy of geographical measures. *Mathematical Geology*, 20: 615–20.

Goodchild, M. F. (1982) The fractal Brownian processes as a terrain simulation model. *Modeling and Simulation*, 13: 1133–7.

Haines, A. J., and Crampton, J. S. (2000) Improvements to the method of Fourier shape analysis as applied in morphometric studies. *Palaeontology*, 43(4): 765–83.

Haining, R., Wise, S., and Ma, J. (1998) Exploratory spatial data analysis in a geographic information system environment. *The Statistician*, 47: 457–69.

Harmon E. H. (2007). The shape of the hominoid proximal femur: A geometric morphometric analysis. *Journal of Anatomy*, 210(2): 170–85, doi: 10.1111/j.1469-7580.2006.00688.x.

Jaggi, S., Quattrochi, D. A., and Lam, N. S. (1993) Implementation and operation of three fractal measurement algorithms for analysis of remote-sensing data. *Computers and Geosciences*, 19: 745–67.

Kent, C., and Wong, J. (1982) An index of littoral zone complexity and its measurement. *Canadian Journal of Fisheries and Aquatic Sciences*, 39: 847–53.

La Barbera, P., and Rosso, R. (1989) On the fractal dimension of stream networks. *Water Resources Research*, 25: 735–41.

Li, W., Goodchild, M. F., and Church, R. (2013) An efficient measure of compactness for two-dimensional shapes and its application in regionalization problems. *International Journal of Geographical Information Science*, 27: 1227–50, doi: 10.1080/13658816.2012.752093.

Mandelbrot, B. B. (1967) How long is the coast of Britain? Statistical self-similarity and fractal dimension. *Science*, 56(3775): 636–38.

Mark, D. M., and Aronson, P. B. (1980) Scale-dependent fractal dimensions of topographic surfaces: An empirical investigation with applications in geomorphology and computer mapping. *Mathematical Geology*, 16: 671–83.

Matyas, C. (2007) Quantifying the shapes of U.S. landfalling tropical cyclone rain shields. *The Professional Geographer*, 59(2): 158–72.

McGarigal, K., and Marks, B. J. (1994). FRAGSTATS spatial pattern analysis program for quantifying landscape structure, retrieved from https://www.umass.edu/landeco/pubs/mcgarigal.marks.1995.pdf.

MyFireWatch. Bushfire map information Australia, https://myfirewatch.landgate.wa.gov.au/map.html.

Pavlidis, P. (1978) A review of algorithms for shape analysis. *Computer Graphics and Image Processing*, 7(2): 243–58.

Purevtseren, M., Tsegmid, B., Indra, M., and Sugar, M. (2018) The fractal geometry of urban land use: The case of Ulaanbaatar City, Mongolia. *Land*, 7(67), doi: 10.3390/land7020067.

Roy, A. G., Gravel, G., and Gauthier, C. (1987) Measuring the dimension of surfaces: A review and appraisal of different methods. *Auto-Carto 8 Proceedings*, Baltimore, MD, March 29-April 3, 1987, pp. 68–77.

Sassolas-Serrayet, T., Cattin, R., and Ferry, M. (2018) The shape of watersheds. *Nature Communication*, 9: 3791, doi: 10.1038/s41467-018-06210-4.

Schuenemeyer, J. H. (1984) Directional data analysis. In *Spatial Statistics and Models*, Gaile, G. L., and Willmott, C. J., (eds.), pp. 253–70, doi: 10.1007/978-94-017-3048-8_14.

Shelberg, M. C., Lam, N., and Moellering, H. (1982) Measuring the fractal dimension of empirical cartographic curves. *Proceedings, Fifth International Symposium on Computer-Assisted Cartography (Auto-Carto 5)*, Washington, DC, 481–90.

Shelberg, M. C., Lam, N., and Moellering, H. (1983) Measuring the fractal dimensions of surfaces. *Proceedings of the Sixth International Symposium on Computer-Assisted Cartography (Auto-Carto 6)*, Ottawa, Canada, pp. 319–28.

Torrens, P. M. (2006) Simulating sprawl. *Annals of the Association of American Geographers*, 96(2): 248–75.

Turcotte, D. L. (1987) A fractal interpretation of topography and geoid spectra on the Earth, Moon, Venus and Mars. *Journal of Geophysical Research*, 92: E597–601.

Wentz, E. (2000) A shape definition for geographic applications based on edge, elongation, and perforation. *Geographic Analysis*, 32(2): 95–112.

5 Geostatistics and Spatial Interpolation

5.1 INTRODUCTION

5.1.1 Spatial Interpolation and Geostatistics

Spatial interpolation is a process of estimating attribute values at locations where no ground observations are available. This field of spatial data analysis is theoretically grounded in the First Law of Geography by Tobler (1970). This theoretical grounding has opened up the study of a number of spatial relationships, including spatial dependency and spatial auto-correlation. In particular, it has been widely used in spatial interpolation, in which attribute values at unsampled locations are estimated from those of their neighbouring observations. Spatial interpolation is necessitated by data scarcity, as it is either prohibitively costly to make numerous observations, or impossible to make observations at the desired location due to the nature of random sampling, remoteness, or inaccessibility. Even if the data are spatially irregularly distributed dense points, as with light detection and ranging (LiDAR) point cloud data, it is still necessary to convert them into regular grid lattices via spatial interpolation. Once in the regular grid form, these data allow further information to be generated via spatial analyses. Spatial interpolation can be realised using several methods, including moving averaging, minimum curvature, and kriging. They will be covered in separate sections in this chapter.

Some of the aforementioned spatial interpolators are founded in geostatistics, which has been associated with essential prediction and modelling, in which the rules of prediction are derived statistically based on the past behaviour or the behaviour of the neighbouring observations. Geostatistics is especially powerful in predicting values at unsampled locations and in constructing a surface from a limited number of irregularly spaced observations. To some degree, it represents the application of statistics to the analysis of spatial data, but not spatial modelling. The central focus of geostatistics is on how an observed phenomenon varies spatially within a certain neighbourhood. It has more restrictive applicability than spatial statistics as it can handle only regionalised variables.

5.1.2 Regionalised Variables

Regionalised variables, by default, are spatial in nature. They have a property lying between truly random and completely deterministic. In other words, they are too irregular to be depicted by a mathematical equation, but still obey some kind of spatial distribution that allows their properties to be predicted mathematically. The

DOI: 10.1201/9781003220527-5

degree of similarity in the variable's attribute values at different locations is a function of their spatial proximity. Two observations of the same variable closer to each other are more likely to have a similar value than two observations further distant from each other. Some common examples of regionalised variables in geography and environmental science include elevation, precipitation, air pollutants, suspended sediments, soil pH and moisture content over a slope, and even population density. These variables all are spatially continuous in their distribution, but there can be holes (e.g., islands) amid the space of distribution.

A regionalised variable has the following three traits that deserve special attention:

(1) The spatial correlatedness of the same variable's attribute value at different locations may indicate a cause–effect relationship between spatially adjacent observations, or merely spatial association. For example, the amount of rainfall received at point A may be extremely similar to the amount of rainfall at point B simply because they are very close to each other. However, there is no cause–effect relationship between the two observed values. We cannot say that the high rainfall at point A is caused by the abundant rainfall at point B, or vice versa. However, the relationship is caustic if the variable of study is air pollutant concentration. In this case, the high level of concentration at point A affects the low concentration level at point B. In fact, both the high and low concentrations inevitably affect each other in the process of diffusion.

(2) Another trait is the enumeration unit. Most regionalised variables are point-based features having a topological dimension of 0. It is also possible for the observation unit to be line-based (e.g., traffic volume along a road, and channel discharge) or area-based (e.g., population density in a census tract). Some of the line-based observations may have a cause–effect relationship between them. For instance, the traffic volume of a street is affected by the traffic volume in another adjacent road or street. If a nearby street has a traffic accident, another street in the neighbourhood will become more congested if the traffic of the affected street is diverted to it. Similarly, population density in one census tract is related to that in another. If a suburb in a neighbourhood has a high density, its neighbouring suburbs are likely to have a similar density because of the same zoning regulations. Area-enumerated data can still be analysed using the methods presented in this chapter after they have been converted to point-based data using the centroid of the enumeration units as the observation location. However, linear data cannot be analysed in the same way as point and/or areal data. They have to be analysed through the road network.

(3) The measurement scale of a regionalised variable attribute can be nominal, ordinal, interval, or ratio (see Section 3.1.3). Of these scales, only ratio data such as mortality/fertility rate and pollution concentration level are numerical values depicting the quality of a regionalised variable quantitatively. They are the most detailed describer of the quality, and can be manipulated via spatial interpolation.

In order to be analysed using the geostatistical methods presented in this chapter, a regionalised variable must fulfil the following two criteria (Bárdossy, 2017):

(1) *Data homogeneity*: The data should involve only one variable collected using the same method and to the same reliability level.
(2) *Additivity*: The variable's mean calculated using Equation (2.11) should have the same meaning as Z(x, y). In reality, some variables such as hydraulic conductivity are clearly not additive. They have to be transformed into additive variables before they can be analysed geostatistically.

As illustrated in Figure 5.1, a regionalised variable Z(x, y) can be decomposed into three parts: (a) a structural component m(x, y) associated with a constant mean value, (b) a spatially correlated random component ε'(x, y), and (c) a random noise or residual error ε''. Mathematically, their relationship can be described using Equation (5.1):

$$Z(x, y) = m(x, y) + \varepsilon'(x, y) + \varepsilon'' \tag{5.1}$$

where both $\varepsilon'(x, y)$ and ε'' are assumed to have a uniform statistical property over the whole sample area and $m(x, y)$ may not be present in all cases. If m(x, y) = 0, the data are termed *stationary*. If not, m(x, y) is known as *drift*, indicating the presence of a global trend in the spatial distribution of the data. Drift falls into different types, the commonest being linear and quadratic. They will be discussed in detail in Section 5.2, under "trend surface analysis".

5.1.3 VARIOGRAM AND SEMI-VARIOGRAM

In order to explore how the attribute value of a regionalised variable behaves spatially so as to predict its value at unsampled locations, it is necessary to study its variance over a spatial range. Spatial variance can be calculated using Equation (2.12) after some modifications:

$$\gamma(h) = \frac{1}{2n(h)} \sum_{i=1}^{n} \left[z(x_i) - z(x_i + h) \right]^2 \tag{5.2}$$

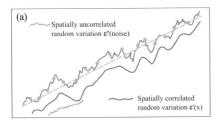 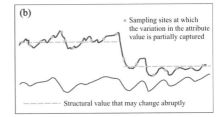

FIGURE 5.1 Composition of a regionalised variable: a "structural" value that may change abruptly at a boundary (a) or vary with a constant trend (b) and the spatially correlated random variation ε'.

where *h* is called separation or lag distance. It defines all possible pairs of observations separated by *h*; *n(h)* stands for the number of paired observations with a separation of *h*. Naturally, as *h* increases, the possibility of pairing two observations among all the observations diminishes (see Figure 3.9). Since two pairs can be established between any two observations (A and B, and B and A), this double-counting must be discounted by dividing the variance by 2; *γ(h)* is known as the *semi-variogram*, a diagram illustrating the variance or mean squared difference in value between two adjoining observations separated by a lag *h* graphically. According to the definition in Equation (5.2), the semi-variogram has the following four properties (Bárdossy, 2017):

- γ(h) always ≥ 0.
- When h = 0, γ(h) = 0.
- γ(h) = γ(−h).
- Given that the regionalised variable is spatially continuous (apart from holes), the semi-variance increases with the length of *h* to a certain extent.

5.1.4 STRUCTURE OF SEMI-VARIOGRAM

For a given set of observations, their *semi-variogram* (Figure 5.2) illustrates how the variability of the attribute value behaves spatially. Initially, as the lag distance (*h*) rises, the semi-variance increases steadily. However, after a certain distance, it reaches a plateau and fluctuates around the maximum value. This diagram can be characterised with the following four traits:

- *γ(h)* increases rapidly with *h* if it is very small.

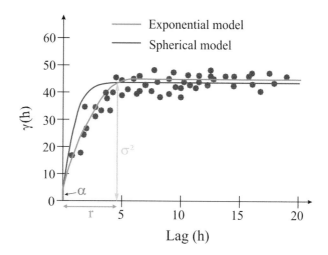

FIGURE 5.2 Structure of a typical semi-variogram with three parameters of range (*r*), sill (σ^2), and nugget effect (*α*) clearly marked.

- At a large *h*, the variance does not necessarily change predictably because of the absence of spatial dependence between neighbouring data points.
- The curve does not pass through the origin.
- For areal-based observations, a similar diagram can be constructed for them by assuming the centroid of each polygon to be the point of observations.

The curve of a semi-variogram has three critical parameters: range, sill, and nugget effect. *Range* (*r*) refers to the lag distance at which the variance reaches the plateau. Range is significant in that it indicates the spatial scale beyond which the neighbouring observations no longer exert any influence on the observation under consideration. All the neighbouring observations within the range obey Tobler's First Law of Geography. Namely, the closer they are, the more similar their values are to that of the observation in question. *Sill* (σ^2) represents the plateau variance that is first reached at range *r*. The *nugget effect* (α) refers to the vertical jump of the variance from 0 at the origin to the variance at an extremely small separation. Theoretically, $\gamma(0)$ should equal 0 at h = 0 (e.g., comparison of the observed value with itself), but in reality, it does not (see Figure 5.2). There are two sources for this discrepancy: measurement inaccuracy and the micro-scale variation of the measured variable, namely:

$$\text{nugget effect} = \text{error variance} + \text{micro variance} \tag{5.3}$$

Error variance is caused by sampling inaccuracy. For instance, the measured attribute value can deviate from the true value slightly for various reasons (e.g., equipment inaccuracy). Because of this inaccuracy, two measurements at the same location are likely to differ from each other within a small range. Micro-variance refers to the short-scale variability or variance of the variable at a scale smaller than the minimum sampling distance. The nugget effect may be expressed relatively. Relative nugget effect is defined as the ratio of the nugget effect to the sill, or α/σ^2, usually expressed as a percentage.

Given that all the observations are confined to the 2D space, pairs of observations can be formed either horizontally or vertically. This brings out the directional dimension of semi-variograms. Directionally, there are two types of semi-variogram, *isotropic* and *anisotropic*. The former refers to a semi-variogram that does not vary with the direction of vector *h*, but only its length, because direction exerts no influence on the observed attribute value. The latter indicates that the observed values are affected by direction to a certain degree, such as the latitudinal distribution of temperatures or the concentration of air pollutants along the predominant wind direction. In reality, isotropic regionalised variables are more common than anisotropic ones. Anisotropy can be further divided into two types in terms of the relative variations in sill and range: *geometric anisotropy* and *zonal anisotropy* (Figure 5.3). The former means that the sill value varies only with the lag distance, but not with direction. In other words, the sill of adjoining observations remains unchanged as long as the distance between them does not change. In contrast, *zonal anisotropy* means just the opposite, in that only sill changes with direction, while range remains

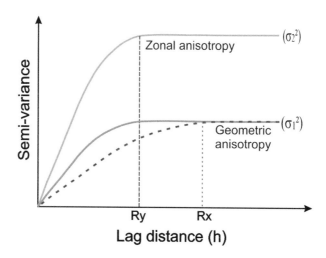

FIGURE 5.3 Three semi-variogram curves illustrating zonal and geometric anisotropy. The green and blue curves show the same sill (σ_I^2) over different ranges, indicating geometric anisotropy. The green and cyan curves show a different sill (σ_2^2) at the same range, indicating zonal anisotropy.

constant (e.g., at R_y). Geometric anisotropy can be transformed to an isotropic one via rotating the ellipsoid of the anisotropic random functions, followed by shrinkage if necessary. This transformation is defined by two parameters: the angle between the x coordinate and the main axis of the anisotropy (ellipse), and the ratio of the two orthogonal ranges corresponding to the maximum and the minimum variability.

5.1.5 Semi-variogram Models

The observed distribution of $\gamma(h)$ with h must be fitted with a theoretical model that can be precisely described mathematically before the semi-variogram can be of any use for subsequent spatial interpolation (to be covered in Section 5.5). A number of theoretical models can serve this purpose. Some of the most commonly used ones are spherical (Equation 5.4), exponential (Equation 5.5), and Gaussian (Equation 5.6).

$$\gamma(h) = \begin{cases} \sigma^2 \left(\dfrac{3h}{2r} - \dfrac{h^3}{2r^3} \right) \\ \sigma^2 \end{cases} \tag{5.4}$$

$$\gamma(h) = \sigma^2 \left(1 - e^{-3h/r} \right) \tag{5.5}$$

$$\gamma(h) = \sigma^2 \left(1 - e^{-3h^2/r^2} \right) \tag{5.6}$$

where r stands for the semi-variogram range and σ^2 refers to the variance of the semi-variogram. All of the above equations do not factor in the nugget effect (e.g., h = 0, γ(h) ≠ 0). After taking the nugget effect into consideration, they will become much more complex. For instance, the Gaussian model can be rewritten as:

$$\gamma(h) = \begin{cases} \alpha + \left(\sigma^2 - \alpha\right)\left(1 - e^{-\frac{3h}{r}}\right)(h > 0) \\ 0(h = 0) \end{cases} \tag{5.7}$$

It is worthwhile to note that there are many other models in use (e.g., circular and rational quadratic), but these three are the most common. More models can be constructed from a linear combination of these three. These models are suitable for isotropic processes, but not anisotropic semi-variograms. Figure 5.4 illustrates a comparison of the three models. The main difference between them lies in the range and the semi-variance within the range. The exponential model reaches the highest variance up to a range of 3, beyond which it is surpassed by the Gaussian model. The spherical model generally achieves a lower variance except at a lag <1.7. At a longer lag, there is almost no difference between them.

Which model is the best? There is no simple answer to this question. The best model can be judged visually by comparing the theoretical curve with the observed distribution of $\gamma(h)$. As shown in Figure 5.2, some models may be more accurate at a short h, while others follow the distribution trend more closely at a large h. In either case, there is no quantitative indication of the goodness of fit, so the choice of a particular semi-variogram model rests with the analyst.

5.2 TREND SURFACE INTERPOLATION

Trend surface analysis is an analytical method aiming to identify the general tendency of change in the attribute value of a variable over space. This attribute value

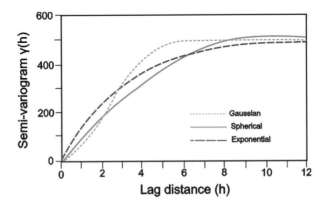

FIGURE 5.4 Comparison of three theoretical models to approximate a semi-variogram with h.

can be decomposed into two parts resulting from processes operating at two scales (Unwin, 1978). The first part, also the broad-scale component, is a trend that varies smoothly over space. The second component is the local-scale variations or residuals that stem from random fluctuations and inaccurate measurements. Thus, they are spatially unpredictable. It is the first component that can be addressed using trend surface analysis. Mathematically, a trend surface can be expressed as a polynomial equation:

$$Z(X,Y) = \sum_{i=1}^{n} b_i X^r Y^s + u_i \tag{5.8}$$

where $Z(X, Y)$ refers to the attribute value of a spatially varying variable, X and Y are its coordinates in the horizontal (easting) and vertical (northing) directions, respectively, b_i stands for the coefficients, r and s refer to the order of the polynomial equation, and u_i denotes the random error or residual, the discrepancy between the observed value and that shown in the fitted trend. The exact form of Equation (5.8) varies with r and s, the polynomial order. In its simplest form, $r = 0$, $s = 0$, $Z(X, Y) = b_0$, which is a constant that does not vary spatially. In the first order ($r = 1$, $s = 1$), the surface is represented linearly as:

$$Z(X,Y) = b_0 + b_1 X + b_2 Y + u_i \tag{5.9}$$

In the second order ($r = 2$, $s = 2$):

$$Z(X,Y) = b_0 + b_1 X + b_2 Y + b_3 X^2 + b_4 XY + b_5 Y^2 + u_i \tag{5.10}$$

As the order rises, the polynomial equation becomes more complex, and the trend surface represented by it becomes increasingly curved (Figure 5.5). Regardless of the order, however, all curves can encapsulate only the general variation of the trend. There is no room for capturing local variations. Thus, the established trend surface is not aligned precisely with all the original values. This is because all the trend surface equations are resolved through the principle of the least squared criterion of

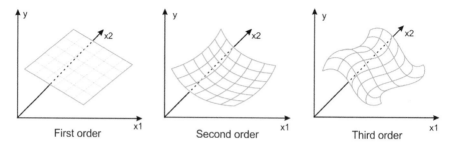

First order   Second order   Third order

FIGURE 5.5 The nature of a trend surface in relation to the order of the polynomial equation. As the order rises, the surface becomes increasingly curved.

the goodness of fit, through which an optimal set of coefficients (b_i) is determined among all the values.

A trend surface is commonly established via regression analysis based on the least squared principle, and involves three inherent assumptions about the data used:

(1) All the residuals must have a normal distribution with a mean centred around 0 (e.g., expectation = 0).
(2) They are spatially uncorrelated with each other. This assumption guarantees that the significance of the fitted surface can be tested, although it is frequently violated in reality for a wide range of spatial data, such as rainfall and topography, as spatial auto-correlation of residuals always exists to a varying extent.
(3) They must be measured with a negligible error at precisely known locations, and their quantity must exceed the number of orders. This assumption ensures that the coefficients are solvable.

The exact number of data points needed to determine b_i varies with the polynomial order. At least three data points are needed to resolve Equation (5.9), and six to resolve Equation (5.10). It must be emphasised that this stated number is the bare minimum that is unable to indicate the goodness-of-fit genuinely, because the fit will always be perfect under all circumstances, even if the observed attribute values are incorrect (e.g., violation of assumption 1 in the preceding list). In practice, it is necessary to make use of more data points than these bare minima.

The reliability of the fitted trend is judged against the coefficient of regression (R^2). It represents a ratio of the corrected sum of squares of the computed trend to the corrected sum of squares of the observed values. A larger R^2 value is synonymous with a better goodness-of-fit. The significance of the established trend surface is determined by analysis of variance (ANOVA). This can be achieved by dividing the total number of observations into two sections, one of which is used for constructing the trend surface, and the other for checking residuals to validate the equation. At least four questions can be asked of the established trend surface regarding its statistical significance (Unwin, 1978):

(1) Is the overall fitted trend significantly different from zero?
(2) Is the trend surface with an order of $k+1$ significantly better than that of k order?
(3) Is there any coefficient in the trend surface that is significantly different from 0?
(4) Spatially, where are the observations well fitted to the trend surface, and where should the fitted trend be treated with caution?

Some of these questions can be answered based on the outcome of the F-test, in which the null hypothesis (H_0) states that all the coefficients in the trend surface equation equal a constant (e.g., the trend is flat or does not exist). The alternative hypothesis (H_1) claims that some of the coefficients in the equation are not 0. The degree of freedom in the test is $n–1$ (n = the number of observations used in establishing the trend surface equation).

Once established and properly validated, a trend surface can be used to estimate values at any location within the convex hull of all the observations used in its establishment. Trend surface analysis is the default choice of spatial interpolation when only scarce data are available and their trend overwhelms the local random variation (u_i). These data must be spatially continuous, such as rainfall. Whenever this is not the case, trend surface analysis cannot be used, or can be used only after special restrictions have been imposed (such as the exclusion of lakes devoid of people in population density estimation). By incorporating an additional geographic information system (GIS) layer in the analysis, it is possible to exclude such holes from consideration. If the data are enumerated over an area (such as population in a census tract), they are ill suited to trend surface analysis unless the data are construed as acquired at the centroid of polygons so as to eliminate the impact of varying size and shape of the observation units. In order to generate a genuine and realistic trend surface, the area under study must be well represented by observations, especially near the border and in corner areas.

Trend surface analysis is advantageous in that it is easy to understand, and it can capture broad features with a low-order trend surface. However, it is disadvantageous in three respects:

(1) The established trend surface is highly susceptible to outliers in the input data. The solutions to Equations (5.9) and (5.10) are based on the assumption that all the observed values are correct. Whenever this assumption is violated, software will make every effort to accommodate the mistake in the established trend, even if the value is vastly different from the remaining values of the group (e.g., an outlier).

(2) There is a high degree of uncertainty in the results beyond the spatial extent covered by the data. In other words, trend surface analysis is competent only for observations within the convex hull of all the data. As such, trend surface analysis is ill suited to predicting values beyond this spatial extent (e.g., extrapolation).

(3) Residuals are not spatially independent of each other. A large residual likely lies next to other large residuals.

The above drawbacks can be overcome with other methods of spatial interpolation that will be presented in Section 5.5.

As shown in Table 5.1, a total of 14 observations was used in constructing a trend surface of both the first and second orders. However, there is no difference in the mean residual (0) between them. This seems to suggest that the order of the trend surface does not make much difference to the interpolated results. Nevertheless, a closer inspection of the root-mean-square error (RMSE) reveals that the first order is about 22% less accurate than the second order, suggesting that the NO_2 values are not distributed as a linear plane. Of the 14 residuals, seven are negative and seven are positive. The extreme residuals are as large as 22.54, slightly lower than the average

TABLE 5.1

Annual Mean Concentration of Vehicle-yielded NO_2 in Auckland in 2011 Interpolated Using First- and Second-order Trend Surface Analysis in Comparison with the Observed Values

Coordinates	Observed concentration	First-order Estimated	First-order Residual	Second-order Estimated	Second-order Residual
1,753,733, 5,919,769	21.48	22.08	−0.60	26.24	−4.76
1,753,071, 5,935,353	29.53	28.58	0.95	27.04	2.50
1,755,982, 5,924,003	28.81	25.21	3.60	28.29	0.52
1,748,363, 5,926,715	30.96	27.66	3.30	28.17	2.79
1,753,733, 5,927,310	13.74	25.47	−11.72	27.54	−13.80
1,747,383, 5,926,384	15.88	21.58	−5.70	20.43	−4.55
1,742,356, 5,926,648	16.23	19.03	−2.81	10.41	5.82
1,747,118, 5,912,427	10.63	15.34	−4.71	11.52	−0.89
1,762,041, 5,906,461	29.91	20.58	9.33	27.16	2.74
1,774,661, 5,896,142	18.55	22.85	−4.30	18.59	−0.05
1,771,460, 5,912,361	16.52	28.28	−11.76	19.65	−3.13
1,757,371, 5,919,968	46.43	24.11	22.32	28.37	18.06
1,756,246, 5,913,022	20.80	20.43	0.37	26.25	−5.45
1,764,382, 5,915,073	27.37	25.64	1.73	27.18	0.19
Mean	23.34	23.34	0.00	23.34	0.00
RMSE			8.32		6.85

Unit: $\mu g \cdot m^{-3}$.

concentration, but as low as −11.76. In the second order, this relativity is reversed, and the maximal positive residual is lowered to 18.06, but the maximal negative residual worsens to −13.80. Furthermore, the negative residuals in the first order are clustered, but they are more widely scattered in the second order. This phenomenon is accounted for by the plane surface of the trend. In areas where the NO_2 concentration is much lower than that suggested by the trend (e.g., near urban green fields or areas devoid of truck traffic), they are all underestimated. In contrast, the fitting of the NO_2 concentration with a curved surface enables a closer match between the observed and estimated concentrations, resulting in a lower RMSE and less clustering of underestimates. Thus, second-order trend surface analysis is superior to its first-order counterpart in interpolating the spatial distribution of vehicle-generated annual NO_2 concentration of Auckland.

5.3 MOVING AVERAGING

Variously known as weighted averaging and inverse distance weighting (IDW), *moving averaging* is a spatial analysis method of estimating the value at an unsampled location or any location of desire from its nearby observations based on their weight.

It is commonly defined as an inverse function of the distance to the neighbouring observation in question. Mathematically, the interpolated value can be expressed as a weighted, linear combination of all the considered neighbouring observations' attribute value, or:

$$Y_p = \sum_{i=1}^{n} W_i Y_i = \frac{\sum_{i=1}^{n} Y_i d_i^{-m}}{\sum_{i=1}^{n} d_i^{-m}} \tag{5.11}$$

where Y_p denotes the attribute value at the point of interest (p) to be estimated, m is known as the power of inverse distance weighting, n stands for the number of nearest neighbouring observations considered in the estimation, and d refers to the distance of a neighbouring observation to p.

Moving averaging interpolation can be accomplished by first calculating the Euclidean distance between the point in question and those neighbouring points selected for the averaging using Equation (2.4), as shown in Figure 5.6b, then using a value of either 1 or 2 for m. Thus:

$$m=1, Y_p = \left(138/2.236 + 152/2.693 + 143/1.581\right)/\left(1/2.236 + 1/2.693 + 1/1.581\right) = 143.76 \,(\text{mm});$$

$$m=2, Y_p = \left(138/2.236^2 + 152/2.693^2 + 143/1.581^2\right)/\left(1/2.236^2 + 1/2.693^2 + 1/1.581^2\right) = 143.33 \,(\text{mm}).$$

The above calculations are easy to understand, in that adjacent observations will more likely share a similar value to the one in question than distant ones. The relationships between all the observed values are related inversely to their distance, as stated in Tobler's First Law of Geography. Thus, closer neighbours exert a more profound influence on the attribute value to be estimated than distant ones. The numerator in Equation (5.11) is a way of standardising Y_p so that it is not artificially

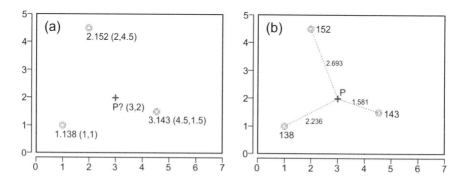

FIGURE 5.6 Location of three adjoining rain gauges and their recorded rainfall readings (a), and the Euclidean distance between the point in question and three neighbouring observations (b).

inflated after the interpolation. While conceptually intuitive and easy to understand, this interpolator faces three uncertainties:

(1) The exact value of m remains unknown. In practice, it is commonly taken as 1 or 2, without any theoretical justification for either option. There is no guidance about which choice is better. If two analysts perform the same analysis using the same dataset and one uses $m = 1$ and another uses $m = 2$, then the two interpolated results will differ from each other slightly, violating the scientific tenet that the analytical results should be objective and exactly replicable no matter who is performing the analysis. As shown in the examples in Figure 5.6a, a difference of 0.43 mm exists between $m = 1$ and $m = 2$.

(2) The number of neighbouring observations (n) that should be used in an interpolation remains unknown. Theoretically, if more neighbouring observations are used, the estimated value will be more reliable. However, this is not true if a neighbouring observation is too distant from p, as it may have virtually no bearing on the value to be estimated, and thus should be excluded from consideration. There is no theoretical guidance as what is the appropriate number of the nearest neighbouring observations that should be taken into consideration in the estimation. In practice, it is commonly taken as 10, purely due to convenience, without any justification.

(3) This quantity-based or distance-based selection of the nearest neighbouring observations ignores their spatial distribution. In reality, they may be clustered in certain directions (Figure 5.7). In this case, the distance-based selection of neighbouring observations will lead to an uneven representation of samples in different sectors. The influence of neighbouring observations in certain directions may be absent, while the influence in other directions may be overrepresented by the selected neighbouring observations, which may critically degrade the reliability of the interpolated value. The influence of clustering in the observation data can be minimised by limiting the number of neighbouring points in each direction after the 2D space is partitioned into four quadrants at the point of interest (Figure 5.7). A maximum number of points can be selected from each quadrant (such as $n = 5$). This search strategy is able to minimise the clustering effect, but cannot deal with the trend issue in the data.

The limitation of the quantity- or distance-based strategy for selecting neighbouring observations can be overcome by using the alternative area-based search strategy by specifying the neighbourhood within which all observations will be searched and automatically selected. This search area can be elliptical in shape and have different sizes and orientations (Figure 5.7). The exact shape of the elliptical search neighbourhood depends on the specified radius in the horizontal and vertical directions that can be rotated to different orientations. Nevertheless, the theoretical grounding of a suitable radius of the search for neighbouring observations still remains unresolved.

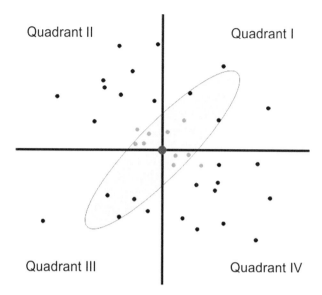

FIGURE 5.7 Partitioning of the 2D space centred around the point of interest into four quadrants, from which neighbouring observations are selected. Blue dots represent the ten nearest neighbouring observations to be selected in a circular search window; the rotated elliptical search window is able to remedy the clustering effect in selecting the same number of neighbouring points, but from more quadrants.

All of the calculations in Equation (5.11) assume that Y_p is spatially independent of each other. However, this may not be always true. If a planar trend exists in the observations, moving averaging will not be able to handle it by itself. However, it can be resolved simply by running a trend surface analysis on the data first, and the established trend is then subtracted from all the observations. Afterwards, the usual process of moving averaging interpolation can be performed on the calculated differences, as described previously.

Moving averaging is simple and easy to understand. However, the interpolated results are affected by the order of the weight function (m) and the window size from which the neighbouring observations are drawn, and hence are susceptible to clustering in the data points. The accuracy of spatial interpolation cannot be revealed by Equation (5.11). If the interpolated results represent a topographic surface, then the interpolated elevations can be contoured and visualised to indicate their reasonableness. The presence of pseudo-contours indicates unreliable interpolation. Furthermore, residuals can be plotted out to show the exact trend and locations of inaccurate interpolations. However, the generation of a quantitative measure of interpolation accuracy still requires additional efforts. It is commonly determined by dividing the available dataset into two parts, one of which, known as the checkpoints, is excluded from being used in the interpolation. Instead, the checkpoints are reserved for validating the interpolated results. The accuracy of interpolation is commonly assessed by comparing the observed and interpolated values at these points independently and expressed as RMSE (Equation 2.15). While it is relatively easy to calculate RMSE, what RMSE is considered acceptable remains unknown. There are no standards about what accuracy of interpolation is acceptable.

Table 5.2 presents a comparison of the major properties of a global interpolator exemplified by trend surface analysis with those of a local interpolator exemplified by moving averaging. The former makes use of one global polynomial equation from all of the observations available. Once constructed, this equation is used to interpolate values at any location within the convex hull of all the observations. Thus, the accuracy of interpolation is low, as the constructed surface rarely passes through the raw data. If a location falls outside the covered spatial extent, its interpolated value will even be less accurate. This deficiency is overcome with the local interpolator, which makes use of only a limited number of observations closest to the location in question. As such, the accuracy tends to be higher, even though it is subject to the number of neighbouring observations used, and how they are selected and weighted. Ideally, the location at which each value is to be interpolated should fall within the convex hull of all the observations. Even if it falls outside the hull slightly, reasonable accuracy can still be achieved. The interpolated result will be more reliable if the data do not have any drift in them. Otherwise, the trend can be detected first using the global interpolator, then subtracted from all the observations (Table 5.2). The difference values are then analysed using the local interpolator, and the interpolated results are subsequently added back to the trend to derive the final outcome of interpolation.

5.4   MINIMUM CURVATURE

Minimum curvature is a special interpolator, in that it is able to interpolate a group of spatial observations into a regular grid lattice that must have an odd number of dimensions. As such, it is commonly used to construct digital elevation models

TABLE 5.2
Comparison of Major Properties of a Global Interpolator with a Local One.

Items	Global interpolator	Local interpolator
Example	Trend surface analysis	IDW
Format	Polynomial	Weighted averaging
No. of observations used	All	A subset of neighbouring observations
Computation	Low to moderate	Intensive
Accuracy	Low	High
Ability to extrapolate	Limited	Reasonable
Assumption	None	No drift
Best use	Scarce observations available	Huge, dense data available
Relationship	Unable to address local variations	Global first, local later in the presence of drift

(DEMs) from elevational data acquired using a LiDAR scanner. This interpolator is based on the curvature at location (i, j) expressed as *C(i, j)*, and the total squared curvature in the discrete 2D space is expressed as:

$$C = \sum_{i=1}^{I} \sum_{j=1}^{J} \left(C_{i,j} \right)^2 \tag{5.12}$$

where *I* and *J* refer to the total number of rows and the total number of columns of the gridded DEM to be constructed, respectively.

The sum of *C* is minimised if the following functions are made to equal zero (Brigg, 1974):

$$\frac{\partial C}{\partial u_{i,j}} \left(i = 1, 2, \ldots, I; j = 1, 2, \ldots, J \right) \tag{5.13}$$

A number of equations are needed to spell out a set of relationships between neighbouring grid cell values, a unique relationship for each cell. In two dimensions, the simplest approximation of the curvature *C(i,j)* at (x_i, y_i) is expressed as:

$$C_{i,j} = \left(u_{i+1,j} + u_{i-1,j} + u_{i,j+1} + u_{i,j-1} - 4u_{i,j} \right) \big/ h^2 \tag{5.14}$$

where *h* stands for the DEM grid size and *u* is the polynomial function. The above generic equation is applicable to non-border grid cells only. It must be modified for cells near the edges and rows from the edges or near the corners. For instance, at an edge (j = 1):

$$C_{i,j} = \left(u_{i+1,j} + u_{i-1,j} - 2u_{i,j} \right) \big/ h^2 \tag{5.15}$$

The border condition dictates that:

$$u_{i+2,j} + u_{i,j+2} + u_{i-2,j} + u_{i,j-2}$$
$$+ 2\left(u_{i+1,j+1} + u_{i-1,j+1} + u_{i+1,j-1} + u_{i-1,j-1} \right) \tag{5.16}$$
$$- 8\left(u_{i+1,j} + u_{i-1,j} + u_{i,j-1} + u_{i,j+1} \right) + 20u_{i,j} = 0$$

If j=1 (e.g., an edge), the above equation becomes:

$$u_{i-2,j} + u_{i+2,j} + u_{i,j+2} + u_{i-1,j+1} + u_{i+1,j+1}) - 4\left(u_{i-1,j} + u_{i,j+1} + u_{i+1,j} \right) + 7u_{i,j} = 0 \tag{5.17}$$

Both equations are resolved iteratively. The $u_{i,j}$ at the *p*+1 iteration is expressed as:

$$u_{i,j}^{p+1} = \left[4(u_{i-1,j}^{p} + u_{i,j+1}^{p} + u_{i+1,j}^{p}) - \left(u_{i-2,j}^{p} + u_{i+2,j}^{p} + u_{i,j+2}^{p} + u_{i-1,j+1}^{p} + u_{i+1,j+1}^{p} \right) \right] \big/ 7 \tag{5.18}$$

The resolution of these equations requires initial values that are assumed to be the same as those of the nearest observation for those grid cells that do not coincide with a grid point (e.g., the grid cell in which the observation falls), or a weighted sum of several neighbouring observations. This process is terminated once the set number of iterations has been reached or the specified minimum curvature has been met.

Measuring smoothness, $\sum \left(c_{i,j}\right)^2$ varies with h and the precision of approximating $C_{i,j}$. The interpolated grid-point surface should be as smooth as any other grid-point surface, if not smoother, at a given h, because the linear Equations 5.14–5.18 are deduced in accordance with the principle of minimum curvature. As illustrated in Figure 5.8, the contours of the interpolated values at the grids are smooth and there are no artificial contours, even though the contour interval is set at a rather fine level of 1 mgal. Thus, the best use of minimum curvature interpolation lies in the creation of regular grid DEMs. It has a strong capability of handling uncertainties caused by missing observations, and is able to achieve an accuracy comparable to that of kriging (to be covered in Section 5.5), even though the accuracy is highly sensitive to DEM grid size. However, the interpolated contours along a line are not as good as a one-dimensional spline (Brigg, 1974).

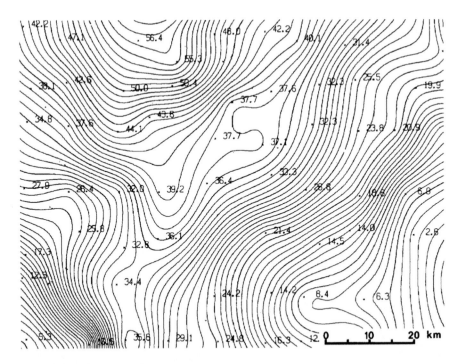

FIGURE 5.8 A contoured gravity field interpolated using minimum curvature (grid spacing = 1.85 km; a total of 900 grids are interpolated) (source: Brigg, 1974, with permission).

5.5 KRIGING

Grounded in geostatistics, *kriging* was named after a South African statistician and mining engineer called Daniel Krige (1951), who developed the geostatistical method as early as the 1950s. Since then, it has found wide applications in spatial analysis, particularly spatial interpolation. This suite of interpolators is based on semi-variograms, so they are the best at estimating values for regionalised variables. Since its inception, kriging has evolved into several types, including ordinary kriging (OK), universal kriging (UK), block kriging, and co-kriging. Of these types, ordinary kriging is the most common, and is critical and fundamental to understanding other types of kriging. As a geostatistical interpolator, ordinary kriging, also called *punctual kriging*, can be described as the best linear unbiased estimator. It is the "best" in the sense that the variance of the interpolation residuals is minimised. "Linear" means that the estimated values are obtained via a linear and weighted combination of the adjacent observations. "Unbiased" implies that the residuals of interpolation have a mean value equal to 0. The selection of a particular type of kriging interpolator depends on the nature of the data being analysed.

5.5.1 ORDINARY (PUNCTUAL) KRIGING

Ordinary kriging is suitable for stationary data (e.g., absence of drift), or m(x, y) = 0 in Equation (5.1). It addresses the ε'(x, y) component of a regionalised variable only, thereby assuming that the data in use have no structural components. If this is not the case, then ordinary kriging must be preceded by trend surface analysis, through which drift can be identified and then removed. After this, the stationary residuals are kriged, and the final interpolation result can be obtained by adding the estimated residuals to the drift. This whole process is known as *universal kriging*.

The mathematical expression of ordinary kriging resembles exactly weighted averaging, or:

$$\hat{z}(B) = \sum_{i=1}^{n} w_i z(x_i) \tag{5.19}$$

where w_i stands for the weight of the ith neighbouring observation whose observed value is Z(x_i). This equation resembles the first half of Equation (5.11), the only difference being how w_i is treated. In kriging, the weight for a particular neighbouring observation is determined from the semi-variogram of the neighbouring observations' value instead of being a universal distance-decay function. In other words, weights are tailor-made, calculated specifically for every neighbouring observation individually based on the constructed semi-variogram. Thus, the amount of computation is rather intensive. The semi-variogram can be constructed using two approaches, local and global. *Local kriging* means that the semi-variogram is constructed from a subset of the observations near the point in question. Numerous such semi-variograms have to be constructed during the interpolation. Global implementation of kriging means that all of the observations are used to construct one semi-variogram that is used universally for determining the weights at all the points of interest. This manner of implementation is simpler and requires far less computation.

Mathematically, the equations to determine the weights can be expressed as a matrix:

$$
\begin{bmatrix}
\gamma(Z_1-Z_1) & \gamma(Z_1-Z_2) & \dots & \gamma(Z_1-Z_n) & 1 \\
\gamma(Z_2-Z_1) & \gamma(Z_2-Z_2) & \dots & \gamma(Z_2-Z_n) & 1 \\
\dots\dots & \dots\dots & & \dots\dots & \dots \\
\gamma(Z_n-Z_1) & \gamma(Z_n-Z_2) & \dots & \gamma(Z_n-Z_n) & 1 \\
1 & 1 & \dots & 1 & 0
\end{bmatrix}
\begin{bmatrix}
W_1 \\ W_2 \\ \dots \\ W_n \\ \mu
\end{bmatrix}
=
\begin{bmatrix}
\gamma(X_1-X) \\ \gamma(X_2-X) \\ \dots\dots \\ \gamma(X_n-X) \\ 1
\end{bmatrix}
\tag{5.20}
$$

where γ stands for the semi-variance value obtainable from the constructed semi-variogram, Z_i refers to the observed value at location i (i = 1, 2,…, n), X_i represents the location of the ith neighbouring observed value, X denotes the location at which the attribute value is to be estimated, and W_i refers to the weight for the ith neighbouring observation. Of these variables, only W_i is unknown. There are n+1 equations in Equation (5.20), compared to n weights to be determined. The last calculation in Equation (5.20) is inserted to constrain the sum of all weights to 1 so that the estimated value is not artificially inflated during interpolation. Its functionality is equivalent to the denominator of Equation (5.11). Thus, a slack variable (μ) is inserted into the first n equations to ensure that all the n calculated weights are exact and unique.

How kriging works is illustrated in Figure 5.6a, in which the horizontal coordinates of all the neighbouring observations are known already, as in all spatial analyses, in addition to the observed values. The entire process of estimation is accomplished in six major steps:

(1) Calculate the Euclidean distance between every potential pair of neighbouring observations, including the location at which the value is to be estimated from their geographic coordinates (Figure 5.9) using Equation (2.4).

(2) Construct a semi-variogram based on all the observations near P. Since the lag (h) has a small range, the semi-variogram is approximated as a linear model (Figure 5.10).

(3) Convert the Euclidean distance between neighbouring observations calculated in step 1 to a semi-variance value from the semi-variogram constructed in step 2 (see Figure 5.10). This can be achieved by drawing a vertical line perpendicular to the horizontal axis, and then a horizontal line in parallel to the horizontal axis from the intersection of the vertical line with the semi-variogram curve. The intersection of this horizontal line with the vertical axis indicates the estimated γ(h) value corresponding to the given range.

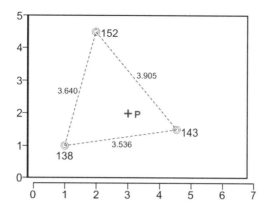

FIGURE 5.9 The first step of kriging is to calculate the distance between any two neighbouring observations in order to determine their weights.

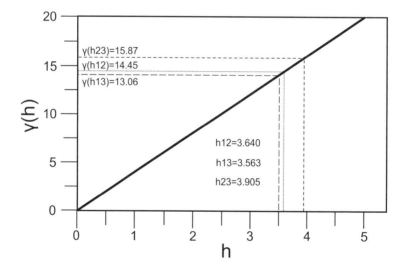

FIGURE 5.10 Determination of the semi-variance value from the constructed semi-variogram based on h (distance between neighbouring observations). In this case, a linear model is adopted for the semi-variogram because the observations are not distant enough from each other to reach the sill.

(4) Plug all the semi-variance values into Equation (5.20) to determine w_i:

$$
\begin{cases}
w_1 \times 0 + w_2 \times 14.45 + w_3 \times 13.06 + \lambda = 9.20 \\
w_1 \times 14.45 + w_2 \times 0 + w_3 \times 15.87 + \lambda = 9.92 \\
w_1 \times 13.06 + w_2 \times 15.87 + w_3 \times 0 + \lambda = 6.46 \\
\qquad\qquad w_1 + w_2 + w_3 = 1.00
\end{cases}
\tag{5.21}
$$

Equation (5.21) can be rewritten in matrix format:

$$\begin{bmatrix} 0 & 14.45 & 13.06 & 1 \\ 14.45 & 0 & 15.87 & 1 \\ 13.06 & 15.87 & 0 & 1 \\ 1 & 1 & 1 & 0 \end{bmatrix} \begin{bmatrix} w_1 \\ w_2 \\ w_3 \\ \lambda \end{bmatrix} = \begin{bmatrix} 9.20 \\ 9.92 \\ 6.46 \\ 1.00 \end{bmatrix} \quad (5.22)$$

The multi-linear equations can be resolved iteratively using computer software. The answer is:

$$\begin{pmatrix} w_1 \\ w_2 \\ w_3 \\ \lambda \end{pmatrix} = \begin{pmatrix} 0.239 \\ 0.282 \\ 0.479 \\ -1.134 \end{pmatrix}$$

(5) Plug all the calculated w_i into Equation (5.19) to generate the unbiased estimate at the point of interest (P):

$$\hat{Y}_p = \sum_{i=1}^{3} w_i Y_i = 0.239 \times 138 + 0.282 \times 152 + 0.479 \times 143 = 144.34.$$

(6) Estimate the uncertainty associated with the interpolated result from the variance values.

Dissimilar to moving averaging, kriging allows the reliability of the estimate to be assessed based, again, on the variance values multiplied by the weights, namely:

$$s_\varepsilon^2 = \sum_{i=1}^{3} w_i \gamma(h_{ip}) = 0.239 \times 9.20 + 0.282 \times 9.92 + 0.479 \times 6.46 = 8.09.$$

Thus, the final outcome is expressed as $144.34 \pm \sqrt[2]{8.09} = 144.34 \pm 2.84 \, (\text{mm})$, so the actual value at the location of interest can range from 141.50 to 147.18 mm.

Table 5.3 compares ordinary kriging with moving averaging in terms of a number of major properties. In general, ordinary kriging surmounts all the deficiencies of moving averaging identified in Section 5.3. For instance, there is no longer any need to specify the power of distance decay. The exact number of neighbouring observations that should be considered in interpolation is automatically decided by the software, as those observation points located beyond the range are excluded from consideration. Their weights are determined individually from the constructed semi-variogram objectively without analyst intervention, so the interpolated results are completely replicable as long as the same input parameters are used (e.g., whether the nugget effect should be considered in the semi-variogram curve and the semi-variogram

TABLE 5.3
Comparison of the Main Properties of Moving Averaging and Ordinary Kriging

Interpolator	Moving averaging	Ordinary kriging
Number of neighbouring points used (n)	Arbitrarily set	The software decides based on the constructed semi-variogram
Search neighbourhood	Circular or elliptical	Variable and automatic (machine decides)
Weights	Universal, inverse distance (m = 1 or 2); sum ≠ 0	Tailored, sum = 1
Computation	Moderate	Intensive
Accuracy	Moderate	High
Replicability	No (subject to n and m values)	Yes for the same set of semi-variogram parameters
Ability to handle drift	No	No
Reliability of interpolated value	Unknown	Known (σ^2), accuracy subject to the semi-variogram model used

model adopted). Because of this, ordinary kriging-interpolated results are more reliable than those from moving averaging. More importantly, ordinary kriging is able to yield an indication of the reliability of the interpolated value via the weighted semi-variance. The only downside of ordinary kriging is its high intensity of computation, which can be easily handled by today's powerful desktop computers. It must be noted that both are local interpolators that are unable to deal with drifts in the dataset.

5.5.2 A Comparative Evaluation

This section will comparatively evaluate the three local interpolators: moving averaging, minimum curvature, and ordinary kriging, including their ability to handle uncertainties stemming from missing elevations in the input data. *Topographic uncertainty* refers to the degree at which the elevation at the point of interest can be predicted from those of its neighbours. It falls into distinctive levels in different parts of mountainous terrain. Uncertainty reaches the maximum level at extreme heights such as pits and peaks (Figure 5.11, A). If these extreme values are not captured in the input data, the interpolated elevations are subject to a rather high level of uncertainty as they are virtually extrapolated from the nearby lower elevations (e.g., extrapolation of attribute values). The uncertainty drops to the moderate level for elevations along a ridge, whose curvature remains unknown, but elevations along it still behave predictably (Figure 5.11, B). In other words, they are predictable in one direction, but less so in another. Those elevations lying in the middle of a slope have the least uncertainty as they can be estimated from their downslope and upslope neighbours at a high degree of confidence, especially if the slope has a linear profile

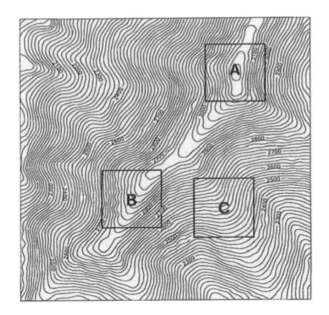

FIGURE 5.11   Level of topographic uncertainty at three geomorphic positions. A = peak, B = ridge, and C = mid-slope. Sampled elevations are deliberately removed from these blocks to create topographic uncertainties, which are the highest at A, but the lowest at C (source: Gao, 2001).

(Figure 5.11, C). Topographic uncertainty is deliberately introduced to an input dataset sampled along contours by removing 136, 127, and 123 sample points around a summit (A), along a ridge (B), and in the middle of a slope (C) (Figure 5.11), respectively. They account for 5.11%, 3.21%, and 7.13% of the total sampled points.

The capability of ordinary kriging in handling topographic uncertainties in relation to the other two local interpolators is assessed by interpolating the newly created datasets containing deliberately removed points at strategic locations to a DEM of 50 by 50 grid cells using the three interpolators. The interpolation settings are the decay power (m = 2) and the number of nearest neighbouring elevations (n = 10) for moving averaging. They are searched within a radius of 34.79 mm using the circular search neighbourhood. This search strategy disregards the distribution of the neighbouring points. The two interpolation parameters for minimum curvature are curvature (0.005) and the number of iterations (500). The interpolated DEM dimension is reduced to 49 by 49 to fulfil the requirement. In ordinary kriging, the linear semivariogram model without drift is adopted without considering anisotropy because the topographic units under test are micro-scaled, and do not exhibit any apparent directional patterns.

Judged against RMSE, all three interpolators become less competent as the uncertainty level rises (Table 5.4). The same finding still holds true if judged against the standard deviation of residuals, and their mean (except medium uncertainty, which is an absolute value only). Of the three interpolators, minimum curvature is the most

TABLE 5.4
Comparison of Three Interpolators in Handling Topographic Uncertainties Caused by Missing Observations

Level of uncertainty	Interpolator	RMSE (m)	Standard deviation (m)	Mean (m)	No. of residuals <3 m	No. of residuals <6.1 m
Low (mid-slope)	MA	16.76	14.57	8.39	103	91
	OK	10.72	8.87	6.07	76	52
	MC	5.19	3.63	3.72	62	34
Medium (ridge)	MA	18.08	17.89	−46.37	114	97
	OK	11.82	11.58	−36.31	97	68
	MC	3.82	2.84	−7.97	63	14
High (peak or pit)	MA	25.16	12.97	1.08	134	122
	OK	19.58	11.99	−0.34	119	101
	MC	10.32	7.44	−5.75	96	67

Source: Gao (2001).
Results are based on the statistics of interpolation residuals.
Contour interval = 6.1 m.
MA = moving averaging; OK = ordinary kriging; MC = minimum curvature.

capable of handling topographic uncertainties. It always achieves the lowest RMSE, which is less than half the next smallest value. In comparison, ordinary kriging is moderately capable of handling uncertainties, achieving a RMSE that is only slightly smaller than that of moving averaging. It is the least capable interpolator, as the interpolated elevation depends directly on the neighbouring observations' heights. When they are missing, more distant neighbours whose elevations bear less resemblance to that of the point in question are used, resulting in a less reliable estimate.

5.5.3 SIMPLE KRIGING

Simple kriging is a special type of kriging in which the local mean is taken as the population mean that is known. It is hence also called *kriging with known mean*. It is termed simple kriging because the model has the simplest mathematical formula. As an unbiased estimator, simple kriging is analogous to linear regression. The main differences between simple kriging and ordinary kriging are twofold:

(1) The variable or random function $Z(x, y)$ is known or can be assumed. It is integrated into the model to improve the estimate of $Z(x, y)$ at a location. The expectation remains constant throughout the region, namely, $u(x, y) = $ constant.

(2) $Z(x, y)$ is second-order stationary with the covariance function $Cov(Z(x, y), Z(x+h, y+h))$ known. In reality, there is little difference between simple kriging and ordinary kriging in terms of accuracy.

In practice, the covariance function $C(h)$ does not exist, so the first step of simple kriging is to estimate either the covariance or the corresponding variogram:

$$Z(x, y) = u(x, y) + \varepsilon'(x, y) \tag{5.23}$$

where $u(x, y)$ is a known constant (in reality, it is unknown). Its implementation also consists of five steps:

(1) Estimation of an appropriate covariogram using all the data.
(2) Estimation of the covariance between any pair of observations at lag h from the covariogram.
(3) Estimation of the mean $u(x, y)$ in Equation (5.23) using all the observations in the dataset.
(4) Calculation of the difference between the observed value and the mean to derive $\varepsilon'(x, y)$ at every sampling position, then performing ordinary kriging on it to yield the unbiased exact value.
(5) Addition of the interpolated value to the mean obtained in step 3 to generate the simple kriging prediction at all sampled locations.

In spite of its simplicity, simple kriging suffers from limited applicability and can produce only suboptimal results if used beyond the embedded assumptions (Olea, 1999).

5.5.4   Universal Kriging

The inability of ordinary kriging to deal with non-stationary data with a drift can be overcome by using *universal kriging*. It combines trend surface analysis with ordinary kriging, each addressing a major component in Equation (5.1). The unknown value is estimated via a linear combination of a smoothly varying trend $m(x, y)$ and a deterministic function $\varepsilon'(x, y)$. The local trend (or drift) is continuous and represents a gradually varying surface that can be described mathematically as a second-order polynomial. It can be determined using trend surface analysis. The deterministic function $\varepsilon'(x, y)$ is handled using ordinary kriging. Universal kriging is almost identical to simple kriging except for the treatment of $u(x, y)$. In universal kriging, the drift parameters are estimated iteratively in five steps (Bárdossy, 2017):

(1) Specify the type of drift or the order of the polynomial.
(2) Select a theoretical variogram $\gamma(h)$, and calculate the drift coefficients.
(3) Calculate the residuals as in step 4 of simple kriging.
(4) Construct a semi-variogram based on the residuals.
(5) Compare the theoretical and the constructed semi-variograms to see whether they are similar to each other. If yes, terminate the iteration. If not, then repeat steps 2–4 with a new theoretical semi-variogram fitted to the experimental data until they are close to each other.

The implementation of universal kriging is almost identical to moving averaging for spatial datasets with a trend. The operation comprises three steps:

(1) The drift is estimated and removed (the equivalent of trend surface).
(2) Stationary residuals are kriged.
(3) The estimated residuals are combined with the drift to obtain the estimates of a surface.

As an example, the capability of moving averaging, ordinary kriging, and universal kriging in estimating the spatial distribution of vehicle-originated NO_2 concentration was comparatively assessed using 107 observations acquired via a monitoring network along major roads with passive diffusion tubes. They were widely distributed over an area of 533.1 km² (observation density: 0.2 observations km⁻²) in central Auckland, New Zealand. A power of 2 was specified for inverse distance in moving averaging. An exponential correlation structure was assumed and selected in ordinary kriging, and the mean and variogram parameters were determined by maximum likelihood. The interpolated results were output at a grid size of 50 m, and their quality was assessed against the observed values at eight fixed monitoring stations that were not used in the interpolation (Table 5.5). The accuracy of interpolation was judged by regression coefficient (R^2) and RMSE. In addition, Leave-One-Out Cross-Validation (LOOCV) R^2 and RMSE were also calculated for comparison.

As shown in Table 5.5, moving averaging was the worst interpolator, followed by ordinary kriging. Universal kriging was the most accurate interpolator regardless of the accuracy criteria used. However, both moving averaging and ordinary kriging had a very similar accuracy, while universal kriging was much more accurate than both of them. A further inspection of the residuals at the eight stations showed that moving averaging overestimated all the eight values consistently. In comparison, ordinary kriging resulted in only one negative residual, the same as universal kriging. At some stations, both ordinary kriging and universal kriging predicted a similar concentration that was much higher than the observed concentration, higher than that from moving averaging. Universal kriging was more robust in extrapolation than ordinary kriging. It was more accurate than ordinary kriging because it could handle any trend in the 107 data points that was ignored by both moving averaging and ordinary kriging. All three interpolators tended to overestimate the concentration, an outcome likely attributable to the lower NO_2 concentration in the validation dataset than that in the interpolation dataset. The fixed monitoring stations were located in areas distant from main roads, where the concentration was much higher.

5.5.5   BLOCK KRIGING AND CO-KRIGING

All of the aforementioned kriging methods are generically known as *point kriging*, as the observations used in the interpolation are made at points. Point kriging estimates the value of the attribute at unsampled points or on a grid. If the samples are collected with high spatial uncertainty due to Global Positioning System (GPS) unreliability, it may be desirable to average the local observations over an area by

TABLE 5.5

Comparison of Accuracy in Interpolating Mean NO$_2$ Concentrations (μg·m^{-3}) to 50 m Grids at Eight Hold-out Fixed Monitoring Stations

Monitoring station	Observed value	Moving averaging		Ordinary kriging		Universal kriging	
		Estimated	Residual	Estimated	Residual	Estimated	Residual
1	7.80	10.34	2.54	17.12	9.32	17.76	9.96
2	12.30	19.33	7.03	21.47	9.17	21.35	9.05
3	27.10	45.33	18.23	40.45	13.35	35.01	7.91
4	6.00	32.09	26.09	28.49	22.49	20.67	14.67
5	23.70	30.95	7.25	30.80	7.10	24.89	1.19
6	26.00	42.06	16.06	40.77	14.77	38.30	12.30
7	43.20	48.49	5.29	42.05	-1.15	41.46	-1.74
8	19.90	30.70	10.60	28.99	9.09	29.47	9.57
R^2		0.65		0.69		0.83	
RMSE		13.80		12.29		9.40	
LOOCV R^2		0.30		0.32		0.67	
LOOCV RMSE		8.44		8.26		5.72	

Source: Modified from Ma et al. (2019), with permission.

overlaying them with a lattice grid (Figure 5.12). All those observations falling into the same grid cells are averaged, and the averaged values are then kriged. In this way, point kriging becomes *block kriging*, a generalised version of kriging that makes use of values averaged over an area instead of at particular points. Similarly, the block kriging results are applicable to the block level only. In block kriging, the estimated average value of the rectangular blocks is centred on the grid. It has been reported that the estimates generated by block kriging closely resemble those from punctual kriging (Oliver and Webster, 1990). Although block kriging generates smoother contours, it is not a perfect interpolator (e.g., an area may not be divided into blocks neatly due to its irregular shape).

Co-kriging is an extension of ordinary kriging by taking into account a second or co-variable in addition to the one considered in ordinary kriging. The two are usually combined linearly in the following fashion:

$$\hat{z}_1\left(x_0\right) = \sum_{i=1}^{n} \lambda_{1i} z_i\left(x_i\right) + \sum_{j=1}^{p} \lambda_{2j} z_2\left(x_j\right) \qquad (5.24)$$

where λ_1 and λ_2 represent the semi-variograms of the first and second variables, respectively, and n and p denote the total number of observations in the two datasets. Co-kriging requires the computation of a cross-variogram that is fitted with a theoretical model, just like in ordinary kriging. It is used to evaluate how the covariance of auxiliary variable 1 and the co-variable changes with the increasing lags

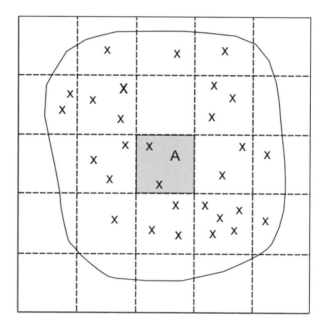

FIGURE 5.12   The concept of block kriging, in which observations within a block are averaged to minimise uncertainty in their location.

between observations. This diagram can also reveal how the spatial structure of the co-variable varies with the auxiliary variable. The cross-variogram can characterise whether these two variables are spatially correlated with each other or have a similar spatial structure.

In order for co-kriging to function well, the co-variable must be closely related to the target variable. Otherwise, there will be little improvement in the interpolated results. Such a conclusion is based on the use of scare observations (e.g., only 117 random observations) (Knotters et al., 1995). It remains unknown whether the same finding still holds true with a huge number of observations.

REVIEW QUESTIONS

1. What are the main characteristics of a regionalised variable? How is it commonly taken advantage of in spatial analysis?
2. Define (semi-)variogram. Why is it so important in certain spatial analysis?
3. What is anisotropy? Why is it important to differentiate it into zonal and geometric in spatial analysis?
4. What is the role of semi-variogram in trend surface analysis? Based on this role, discuss the accuracy of trend surface analysis in interpolating the spatial distribution of rainfall.
5. Discuss why it is important to have a spatially balanced distribution of samples in spatial interpolation regardless of the interpolator used.
6. What are the main problems with spatial interpolation based on moving averaging or inverse distance weighting? Suggest some means by which they can be overcome.
7. What are the strengths and limitations of minimum curvature in comparison with inverse distance weighting?
8. Compare kriging with inverse distance weighting, paying particular attention to the advantages of the former over the latter.
9. What is the relationship between universal kriging, ordinary kriging and trend surface analysis?
10. In which way is blocking kriging preferable to ordinary kriging?

REFERENCES

Bárdossy, A. (2017) *Introduction to Geostatistics*. University of Stuttgart.

Briggs, I. C. (1974) Machine contouring using minimum curvature. *Geophysics*, 39(1): 39–48.

Gao, J. (2001) Construction of regular grid DEMs from digitized contour lines: A comparison of three interpolators. *Geographic Information Sciences*, 7: 8–15.

Knotters, M., Brus, D. J., and Oude Voshaar, J. H. (1995) A comparison of kriging, co-kriging and kriging combined with regression for spatial interpolation of horizon depth with censored observations. *GeoDerma*, 67(3–4): 227–46.

Krige, D. G. (1951) A statistical approach to some mine valuations and allied problems at the Witwatersrand, Master's thesis, University of Witwatersrand, p. 136.

Ma, X., Longley, I., Gao, J., Kachhara, A., and Salmond, J. (2019) A site-optimised multi-scale GIS based land use regression model for simulating local scale patterns in air pollution. *Science of the Total Environment*, 685: 134–49.

Olea, R. A. (1999). *Geostatistics for Engineers and Earth Scientists (Chapter 2:* Simple kriging). New York: Kluwer Academic Publishers, pp. 7–30.

Oliver, M. A., and Webster, R. (1990) Kriging: A method of interpolation for geographical information systems. *International Journal of Geographical Information Systems*, 4(3): 313–32.

Tobler, W. (1970) A computer movie simulating urban growth in the Detroit region. *Economic Geography*, 46(Supplement): 234–40.

Unwin, D. J. (1978) *Concepts and Techniques in Modern Geography Number 5: An Introduction to Trend Surface Analysis*, p. 40.

6 Spatial Modelling

6.1 FUNDAMENTALS OF MODELLING

6.1.1 MODELS AND TYPES

A model is an abstraction and simplified representation of certain selected aspects of reality that are deemed crucial to understanding problems in a given domain of application. This is because the world we live in is so complex that it is impossible to represent all of its aspects. The selectively represented aspects may be factual, or just unknown potential that cannot be verified (e.g., likelihood of landslides). Apart from depicting the real world, models can also be used to understand how the world changes, namely the processes underpinning constantly evolving phenomena, such as the diffusion of air pollutants along a transport corridor. *Spatial models* are a subset of models concerned about spatial relationships and spatial dependencies among the modelled variables. They are different from statistical and mathematical models in that they involve intrinsically spatial variables, or at least some of them are spatial. Spatial models can be categorised into a variety of types, depending on the criteria used. They range from the status to the purpose of models. Spatial models can be deterministic or stochastic, statistical or mathematical in terms of the way the decision rules are enacted, descriptive or predictive in terms of the functionality of modelling, or heterogeneous and homogeneous in terms of the nature of the modelling space (Figure 6.1).

All spatial models can be depicted mathematically or logically. The mathematical representation of a model captures reality within the framework of a mathematical apparatus so as to better explore the properties of that reality. The variable of interest is a mathematical function of its influencing factors, either linear or non-linear. Linear models are expressed as the arithmetical summation of all the variables concerned. Non-linear models may take the form of exponential or even logarithmic functions between the dependent and explanatory (independent) variables. Linear models are commonly exemplified by the geographic regression models that are constructed for the dependent variable from all the explanatory variables in a database. Non-linear models are quite common in real life, such as rainfall-runoff models and distance-decay models. Statistical models involve a certain number of statistical components. They can be further divided into two types, *probability models* and *deterministic models*. The former have at least one stochastic process represented by one or more random variables. Their outputs are also random. Deterministic models do not contain any random variables. The output is unique for a given set of inputs.

Descriptive models are mainly concerned with describing and possibly explaining a spatial pattern produced from spatial analysis. *Predictive models* aim to forecast the outcome of a spatial variable or phenomenon in the future. They are able to determine the spatial distribution of the dependent variable and its spatial variation

DOI: 10.1201/9781003220527-6

189

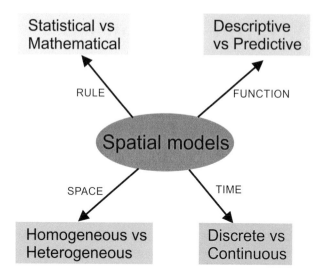

FIGURE 6.1   Major types of spatial models classified using four criteria.

involving both causes and effects. In order for a predictive model to function well, it must include all appropriate and essential parameters, their values, and spatial and temporal variations in relation to the dependent variable being modelled. Predictive models are more powerful than descriptive models in that they can produce predictions about the future distributions or patterns instead of merely describing what has already been observed.

In describing a model, the terms heterogeneous or homogeneous, discrete or continuous all refer to the modelling space (Figure 6.1). In homogeneous models, spatial variation is ignored, leading to a homogeneous space, or at the modelling scale, space is treated as internally homogeneous. Dissimilar to homogeneous models, heterogeneous models' space can be continuous gradients or patches of the environment. An extra layer representing this spatial variation must be present in the modelling process. It can be a layer of impedance or friction, such as traffic volume or slope gradient, which affects travel speed. In continuous models, both space and time are regarded as a continuum. In discrete models, space is non-continuous, comprising non-overlapping, discrete cells or grids. Discrete models are implemented in the raster environment, in which space can be incremented by one cell at a time.

Models can be classified as theoretical or empirical, based on the way they are constructed. *Theoretical models* are physical-based, mathematically describable, and universally applicable (Achinstein, 1965). Physical models describe the relationship between the dependent and independent variables mathematically. However, the use of distributed, physically based models is hampered by the lack of a framework for efficient parameterisation and evaluation of a priori parameters in spatial modelling. For this reason, they are not commonly used. In contrast, *empirical models*, such as those established via regression analysis, rely on experience and intuition gained from the past. They generally have narrow applicability as they are likely to be local, case-specific, and area-specific (e.g., specific to the area for which they are developed). Caution needs to be exercised in applying published empirical models. They can be regional if the external variation can be ignored or if the models are constructed with data collected at the regional level. Thus, a new model needs to be established from scratch for every untested area or case. Empirical models are exemplified by the Hydrologic Model developed by the Hydrological Engineering Center of the US Army Corps of Engineers for predicting runoff associated with a rainfall event. It is expressed as:

$$Q = \frac{\left(P - \dfrac{200}{CN} - 10\right)^2}{P + \dfrac{800}{CN} - 10} \tag{6.1}$$

where Q = runoff (unit: inch), P = rainfall (unit: inch), and CN = the gridded curve number for calculating spatially distributed infiltration. Curve numbers have been tabulated by the US Soil Conservation Service based on land use, treatment, and soil hydrology. This model takes into account land use, soil, land cover types, and hydrologic soil group in determining storm runoff.

In comparison, semi-empirical models are more applicable if they have been extensively parameterised and calibrated using data collected in multiple places, such as the allometric models for calculating the above-ground carbon stock of trees. These models are applicable to areas that share the same or similar traits as the area for which they are initially developed.

In terms of the domain of applications, spatial models can be land use (e.g., urban expansion models), hydrological (e.g., groundwater contamination models), ecological (e.g., wildfire spread models), transportation, and environmental (e.g., air pollutant diffusion and the universal soil erosion models). These models can be hierarchical, as in the Land Use/Transport Interaction models (Simmonds and Feldman, 2011). They aim to understand how the interaction of land use (e.g., where people live) with transport affects people's choice of transport modes based on transport infrastructure and provision. They are hierarchical, in that they are made up of more models at a lower hierarchy that can be activity-based (Figure 6.2).

All of the aforementioned models differ from each other in the model elements involved. In addition, models can also be differentiated by the dimension (unit) of the data to be analysed. If the data are collected along with linear features, they can be studied using *network models*. Network models are good at modelling movements (e.g., flows and accumulation) of the dependent variable in the vector format. Network models have been commonly used in transport, river hydraulic, and wildfire spread modelling.

All of the aforementioned spatial models share one common trait, in that they are all invariably spatially explicit and are usually implemented in the grid environment, except network models. As such, they offer the following strengths:

- Rule-based, impossible mechanisms are ruled out.
- They are useful for parameter testing and quick identification of qualitative changes in system behaviour over a range of parameter values.

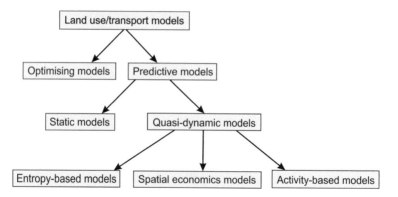

FIGURE 6.2 Sub-models under the hierarchical land use/transport interaction models (source: Simmonds and Feldman, 2011, with permission).

- They are easy to implement, and provide rapid feedback, tremendous flexibility, and relative ease of simulation.
- Emerging phenomena can be visualised as an array of cells, unattainable by other models.

However, none of these models have a temporal component except *spatiotemporal models*. These models involve variables that are dynamic in both space and time and are temporally explicit. They are good at exploring time-varying processes or scenarios. Undertaking spatiotemporal modelling is so different from the non-temporal spatial modelling covered in this chapter that it will be presented separately in Chapter 7.

6.1.2 STATIC VERSUS DYNAMIC MODELS

In terms of the state of the dependent variable, a model can be static or dynamic. *Static models* are algebraic equations between a dependent variable and a set of independent variables. These models provide a "snapshot" of the dependent variable in response to the specified set of input conditions. In a static model, the dependent variable does not change with time, which is absent from the model (Table 6.1). Static models are time-invariant. The same output will be generated whenever the same set of inputs is fed to the model. In addition, all static models have a unique characteristic of being neighbourhood-independent: the variable under modelling is affected by all the independent variables at the location in question, but not their neighbours. There may be or may not be a cause–effect correlation between values at adjoining locations. For instance, the amount of rainfall at one location is not affected by its location nor its neighbours. In contrast, the landslide potential of a cell depends on where it is located. The risk of landslide damage is much higher at a spot lying in the path of a landslide chute, caused by the mobilised debris travelling at a high speed downslope.

A special case of static modelling is called the *spatial diagnostic model*. It focuses on phenomena that are time-invariant, or temporally static (static at a given time), but can still

TABLE 6.1
Comparison of the Main Features of Static and Dynamic Models

Items	Static models	Dynamic models
Nature	Predictive, scenario analysis	Predictive, outcome depending on model run and initial conditions
Variables	Input and output, spatial model, and model parameters (values)	State, rate, and initial conditions; diverse (model, parameters, time increment), rate of change, and neighbourhood effect
Spatial interaction	None	Strong
Treatment of time	Implicit, snapshot	Explicit (time-dependent and increment)
Output	Subject to inputs only	Subject to input, time, previous run, and initial conditions
Nature of variables	Time-invariant (stable)	Temporally changing and dynamic
Format	Categorical, spatial	Quantitative, spatial
Expression	State = F(I, P) I = inputs; P = parameters	State = F(I$_t$, P$_t$, t) I = inputs or forcing functions; P = parameters; t = time
Examples	Empirical: $$Q = \frac{\left(P - \frac{200}{CN} - 10\right)^2}{P + \frac{800}{CN} - 10}$$ Q = runoff (unit: inch); P = rainfall (unit: inch) Semi-empirical: AGC = 2.7 × 10^{-3} (CH × DBH^2)$^{1.19}$ AGC = above-ground carbon; CH = canopy height; DBH = diameter at breast height	Time distribution of excess rainfall: $$\frac{dFa}{dt} = \frac{S^2 d\frac{P}{dt}}{\left(P + 0.8S\right)^2}$$ S = 1,000/CN = −10 Exponential vegetation growth: dB/dt = rB(K–B)/K K = carrying capacity; intrinsic rate of natural increase (c–m); c = growing rate constant; m = mortality rate or decay rate constant; B = scalar describing the stage of the available vegetation
Exemplary applications	Landslide potential; areas to be flooded by the anticipated sea-level rise	Fire burn area; meadow degraded to bare grounds under external disturbances

vary from time to time. Thus, the dependent variable is spatially heterogeneous, but temporally steady (e.g., time-sliced). The modelled value or potential at a particular location or site is contributed by a number of variables whose importance to the dependent variable may be variable. Spatial diagnostic models may be used to study the spatial probability distribution, spatial contiguity, or spatial similarities of the dependent variable.

In comparison with static models, which are unable to show spatial processes due to the absence of the temporal component, *dynamic models* have an explicit component of time. At least one of the independent variables has an attribute that varies with time. The output value depends not only on the inputs, but also on the dependent variable itself, such as the amount of surface runoff accumulated over a slope (Table 6.1). Moreover, the output may also be affected by the initial conditions, as well as the outcome from a previous run. The dependent variable in a dynamic model changes either spatially or temporally. A *temporally dynamic model* contains at least one factor that varies with time.

6.1.3 Spatial Modelling

Spatial modelling refers to a series of mathematical or logical operations in which a spatial attribute (e.g., state or spatial extent) is estimated or predicted from a number of influential, independent variables based on their relationship with the variable being modelled or the dependent variable. The modelled results are yielded from the simultaneous consideration of multiple variables and their interactions (Figure 6.3). Essential to spatial modelling is knowledge of past experience and behaviour, or the mathematical relationship with other factors whose spatial behaviour is known. The relationship between the independent variables and the variable being modelled can take the form of a mathematical or statistical equation. By default, all spatial modelling must have a spatial component and focus on spatial variations, as at least some of the input layers are in map form covering the same area of study.

Spatial modelling can be carried out at a single level or multiple levels. Multilevel modelling is suitable for data that are somewhat hierarchical or clustered, such as

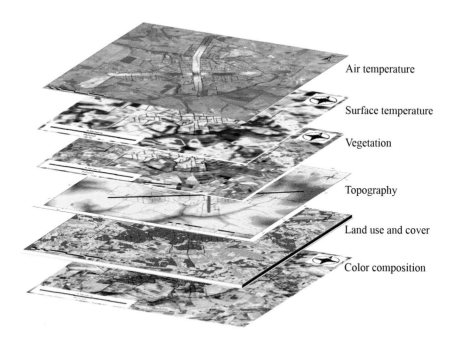

FIGURE 6.3 The concept of spatial modelling, in which the dependent variable is estimated from a number of input variables at the same location (source: Dorigon et al., 2019, with permission).

pupils within schools, households within neighbourhoods, survey respondents within cities, or occasions involving individuals observed multiple times. Spatial modelling is widely practised in a variety of fields, ranging from socioeconomic or political to health and environmental modelling. In urban geography, spatial modelling has been used to predict urban sprawl. In environmental science, spatial modelling has been exploited to model the spatial distribution of air pollutants and the exposure of inhabitants to them in their commutes to and from school at the individual and collective levels (Ma et al., 2020). In epidemiology, spatial modelling can be applied to demonstrate how a communicable disease spreads out spatially.

It is significant to perform spatial modelling, as it can yield new insights into a spatial problem that cannot be gained otherwise. It is able to make predictions about what is going to happen to a spatial phenomenon in the future (e.g., urban sprawl), so that proactive measures can be implemented to contain the event or properly address its negative impacts before it gets out of control. Spatial modelling can support some design processes in which the user is allowed to experiment with a replica. It also enables investigations into what-if scenarios. Through spatial modelling, we are able to understand changes and dynamics and test sensitivity and confidence.

6.1.4 SPATIAL ANALYSIS OR SPATIAL MODELLING?

Spatial analysis, as discussed in the previous chapters, is different from spatial modelling in four main ways:

(a) The target of study is different. The target of spatial analysis is tangible, observable, and in existence. In contrast, the target of modelling is not normally observable, and may not exist at present.

(b) The number of variables being studied is different. Spatial analysis usually focuses on one variable that is studied from a collection of observations. The analytical results are not related directly to the number of variables being analysed. It is also possible to involve more than one variable to study their relationship, but the results of analysis are always non-spatial. In spatial modelling, the output of the dependent variable is produced from the independent variables, and it is always spatial.

(c) The results produced have different degrees of predictability. In spatial analysis, the results are mostly replicable no matter who is performing the analysis (Table 6.2). There are no rules involved, and no state changes in the variables. As long as the same given inputs are analysed, the same results will be produced, all being time-invariant. They are simple and static, showing only one state in time (Figure 6.4). In spatial modelling, on the other hand, the output depends not only on the influence of cells in other layers, but also on cells in the same layer in the neighbourhood. The relationship between the input and output variables can be complex, and may not even be known. Spatial interactions among neighbouring cells are a strong feature that differentiates spatial analysis from spatial modelling. However, it is not always present in all types of spatial modelling (Table 6.2).

TABLE 6.2
Comparison of the Main Properties of Spatial Analysis and Spatial Modelling

Items	Spatial analysis	Spatial modelling
Nature	Descriptive measures, cause–effect	Predictive outcome; logical expressions possible
Functionality	Answers "What?" and "Why?"	Answers "What?" and "How?"
Inputs	Spatial layers of a fixed nature	Spatial layers, spatial model and model parameters; neighbourhood
Spatial interaction	None	Possible in spatial dynamic modelling
Time	Irrelevant (time stable)	Sliced or incremental
Output	Fixed and replicable, non-spatial (numerical)	Fixed and replicable, could be time-dependent; always spatial
Format	Quantitative, non-spatial	Categorical/quantitative, spatial
Examples	Bushfire pattern, channel network density, watershed geometry	Landslide potential, groundwater contamination risk, watershed runoff, fire spread, urban sprawl

people / lane
  0 - 200
201 - 300
301 - 400
401 - 500
500 <

FIGURE 6.4   An example of spatial analysis of the carrying capacity of road networks in Santa Barbara (source: Cova and Church, 1997).

(d) The inputs are different. In spatial analysis, only the layers needed are input. In spatial modelling, additional information on analytical parameters may be required in addition to the layers themselves, such as variable weights, neighbourhood size, time increment, as well as the number of model runs.

In terms of functionality, spatial analysis is able to search for patterns and anomalies from which hypotheses about their formation may be generated and tested. Analysis is able to reveal what would otherwise be invisible, and answers the questions of

"What?" and "Why?", but cannot explain how the observed pattern is formed. The solution to this question is spatial modelling, which is able to link forms with processes. Spatial modelling can help us better understand how a process is formed.

Spatial analysis can be implemented as simple scenario analysis, or as the same analysis repeated under different external conditions, with the results compared to identify the differences. For instance, the coastal area to be inundated by anticipated sea-level rise can be analysed in several scenarios, such as global warming by 0.5°C and by 1°C. This application is considered to be spatial *analysis*, as the results are a mathematical combination of all the variables considered, none of which change with time, and the results are always replicable. Spatial *modelling*, in contrast, is more sophisticated, in that the factors under consideration can change with time. Usually, the same model is re-run multiple times, and in each run time is incremented. The outcome of the current run is subject to the output of a previous run. For instance, whether an area will be burned in a bushfire is subject not only to the wind direction and fuels, but also to its proximity to an area (cell) that is on fire. A spot may be safe in the first few runs, but may ignite a few runs later. Thus, both the modelled process and outcome are time-dependent.

In the above two extreme examples, the disparity between spatial analysis and spatial modelling is distinctive and clear-cut. However, not all cases have such a definite distinction. Take the modelling of best sites as an example. In order to determine the potential areas suitable for a rubbish dump, a number of spatial variables related to the spatial issue being considered are analysed, such as distance to roads, aquifers, and residential areas, topography, and land cover. Any areas meeting the specified requirements are the candidate sites. This modelling involves a series of spatial overlay analyses and is called landfill site modelling. In this case, spatial modelling can be regarded as a special instance of sophisticated spatial analyses involving several data layers. Apart from this indistinctiveness, the two are becoming increasingly integrated and inseparable as spatial modelling becomes highly complex and demands ever more inputs, some of which can be obtained only through spatial analysis (Figure 6.4). For instance, hydrological modelling requires information on watershed parameters that can only be derived from spatial analysis. In this example, spatial analysis feeds inputs to spatial modelling, and both are essential in the same system, as exemplified by the TRansportation Analysis and SIMulation System (TRANSIMS) based on individual vehicles (Smith et al., 1995). The successful running of this model requires spatial analysis preceding spatial modelling because the outcome of spatial analysis is fed to spatial modelling as the input. Without the results from spatial analysis, it is impossible to simulate traffic for entire cities and second-by-second movements of every person and vehicle through the transportation network.

6.2 NATURE OF SPATIAL MODELLING

6.2.1 VARIABLES IN SPATIAL MODELLING

All the variables in a spatial model without exception fall into two broad categories of independent and dependent. Independent variables, also known as explanatory variables, can be of any type, either spatial (e.g., precipitation) or non-spatial (e.g., rate of change) in nature, depending on the dependent variable being modelled. The number and type of independent variables that should be considered in spatial modelling requires specialist knowledge to determine. To be included in a model, they must all affect the dependent variable directly or indirectly, even though the exact manner of influence may remain unknown in some cases. Selecting the best independent variables is such a complex topic that it will be covered separately in Section 6.3.1. In reality, some independent variables themselves may be dependent on each other. Each independent variable, if spatial in nature, may be spatially correlated as well. For instance, slope orientation affects vegetation distribution, so the inclusion of both topography and vegetation in modelling slope susceptibility to landslides may cause double counting of the same independent variable. How to avoid this issue is so complex that it will be addressed separately in Section 6.3.2. The relationship between dependent and independent variables is usually established via regression analysis, either linear or non-linear, subject to the plotted relationship between them. In some cases, the relationship may have already been established by consideration of other factors (e.g., watershed discharge and its parameters), and can be used without modification.

By default, the dependent variable is mostly two-dimensional in the form of f(x, y), restricted to a surface in most spatial modelling. Common dependent variables in spatial modelling include landslide potential, habitat suitability, risk of bushfires, and earthquake hazards. In reality, the modelled variable can be 3D in nature in the form of f(x, y, z), filling the whole 3D space, such as air pollutant distribution in the atmosphere, underground dispersion of surface hazards (e.g., oil spillage in aquifers), and distribution of soil moisture concentration and pH value over a slope. Naturally, 3D modelling is much more sophisticated and challenging than 2D modelling, as most geographic variables can be easily represented in map format, which is inherently 2D. This challenge is commonly addressed by reducing 3D modelling to 2D modelling by replicating the same modelling at multiple representative slices of height/depth.

Both the dependent and the explanatory variables can have a temporal dimension when the phenomenon being modelled changes with time (e.g., diffusion of air pollutants, accumulation of surface runoff, and dispersal of plant seeds). The temporal variation of both variables is regarded as an additional dimension if the dependent variable is temporally dynamic, so 3D becomes 4D: f(x, y, z, t). Time is treated differently in different types of spatial modelling. In static modelling, the modelled phenomenon is assumed to be time-invariant, or temporally static at a given moment, but can still vary from time to time, so time is implicit. Even in spatial dynamic modelling, time is not treated explicitly. Instead, it is incremented from one run to the next.

6.2.2   Types of Spatial Modelling

In terms of the nature of the phenomena being modelled, spatial modelling can be classified into three types – static, predictive, and exploratory (Figure 6.5) – with the last two being dynamic. *Static modelling* merely treats all the input variables as they are. Statically modelled results are available at every point in space

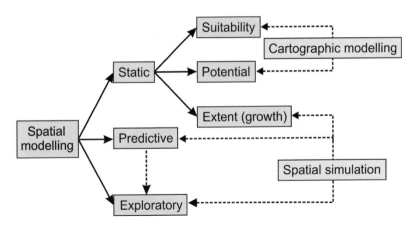

FIGURE 6.5 Types of spatial modelling and their relationships. Green boxes show methods of implementing spatial modelling.

(e.g., hazards and earthquake damage) and can be spatially continuous. The dependent variable is spatially heterogeneous but temporally steady, as time is not explicitly involved in the modelling. According to the nature of the modelled outcome, spatial modelling falls into three categories: *suitability modelling* (siting), *potential modelling*, and *spatial extent modelling* (Figure 6.5). The first aims to come up with a quantitative index indicative of the general suitability for a particular purpose. It is based on simultaneous consideration of a number of restriction criteria, such as where to site a transmission tower, a bank branch, or a landfill. As shown in Figure 6.3, all the layers involved are overlaid with each other, usually two at a time. Each layer is either treated equally or weighted by assigning a coefficient to it. All the layers depict a certain aspect of reality and show its state at a particular given time. Individually, some input layers may need to be recoded by converting the original value into a binary value first, with 1 reserved for true (e.g., meeting the condition) and 0 false, prior to the overlay analysis. All the layers that have been recoded based on the specified criteria can be multiplied in one operation. Only those cells in the final output that have a value of 1 meet all the requirements and thus are the candidate cells (Figure 6.6). The output of suitability modelling can be binary (suitable versus unsuitable) or graded (most suitable to least suitable) (Figure 6.6a). For more effective visualisation, the originally modelled suitability index expressed as a continuously varying attribute is usually grouped into a few classes (Store and Jokimäki, 2003). The difference between the binary and continuous types of suitability stems from the different treatments of the input layers. If they are turned into binary through recoding, then the output will be spatially isolated and binary. If the influencing factors are not converted into binary but expressed on a continuous scale, then the resultant suitability will be continuous. In order for the multiplication to be successful, all the raster layers must have the same spatial resolution and must have been projected to the same coordinate system and referenced to the same geodatum.

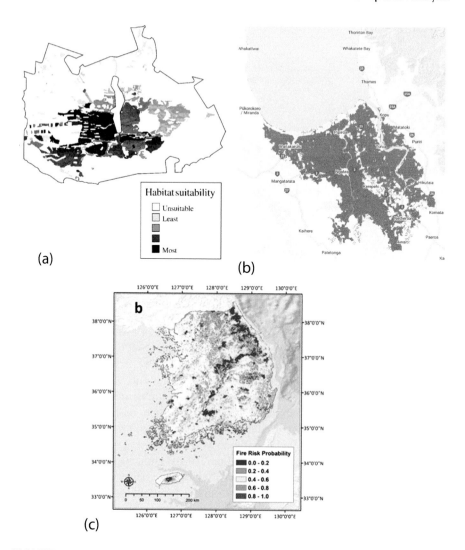

FIGURE 6.6 Comparison of three objectives of static modelling: (a) suitability modelling (source: Store and Jokimäki 2003, with permission); (b) extent modelling – area to be inundated underneath 1 m (red) of the projected sea-level rise in 2050 near Thames, Coromandel in the North Island of New Zealand (source: screen capture from Climate Central, https://coastal.climatecentral.org/map/); (c) potential modelling – the spatial probability of forest fire in the Republic of Korea (source: Lim et al., 2019, open access).

A very important instance of suitability modelling is site suitability, the identification of candidates that meet all the specified requirements through a logical combination of all the variables. Such suitability modelling is carried out for effective spatial allocation, such as where to locate a wind farm or even a McDonald's eatery. Other examples include where to locate a shopping mall to attract the maximum number of shoppers, or a rubbish dump or landfill site to minimise its environmental impacts. This kind of site suitability modelling is basically a process of exclusion, in which all the sites (locations) failing to meet the criteria or the least favoured are not included in the output. Instead of identifying the candidates, suitability modelling can also identify gaps in the existing coverage. Those areas that are not currently served adequately are considered as candidates for further action. This kind of site suitability modelling simply narrows down the areas or sites that can meet the specified conditions or requirements. The candidate sites or locations produced can be regarded as the preliminary results, as they merely show all the possibilities. The ultimate choice of a particular candidate rests with the modeller or the human decision-maker. Site suitability modelling is underpinned by the Boolean logic of AND because the final outcome must meet all the imposed conditions. Prior to the overlay analysis, it may be necessary to buffer some of the input layers and then recode them. However, it is also possible to simply overlay all the input layers without any processing. The final modelling output can be expressed as continuous, as shown in Figure 6.6a. The output is usually generalised by grouping the modelled suitability. This categorisation into a few classes can also minimise uncertainty in the modelled outcome.

Spatial extent modelling aims to identify the area to be encroached upon from an existing area or point (Figure 6.6b), or the area to be changed in future. It is called modelling of spatial extent because it yields an outcome that illustrates the potentially impacted area of a variable, such as the area subject to the risk of groundwater contamination caused by a leaking pipeline, or one that will be affected by projected sea-level rise, or the spatial extent of a glacier that will be melted by climate warming in the foreseeable future. The modelled outcome is simply binary, with the values of 0 (not affected) and 1 (affected). Spatially, the modelled outcome can be continuous or isolated. In the former case, the affected area represents a continuous extension from (e.g., encroachment of a bushfire upon the nearby forest) or contraction of the current extent (e.g., depletion of ice mass from a glacier tongue), represented by a polygon. Spatially isolated extent is exemplified by urban sprawl, in which the newly urbanised patches may be disconnected from each other or existing urban areas in their spatial distribution.

Potential modelling produces a continuous variation in value at every possible location by accumulating the influence of all variables being considered (Figure 6.6c). The value is estimated using mathematical models or an algebraic combination of various factors represented by spatially referenced data layers. The modelled results are available at every point in space, such as landslide potential and hazard, earthquake damage, and wildfire risk. They are spatially continuous, but may contain holes with a restricted spatial extent. Potential modelling differs from suitability modelling in that all variables are combined arithmetically instead of logically, and they can have varying importance in the modelling, which is reflected in the weights assigned to them. The modelled variable has an attribute of continuous variation that can be grouped into a few categories, just as with the suitability index. However, the differences are reduced if the input layers are not recoded in suitability modelling.

Although predictive in nature, *growth modelling* can be considered as a special case of spatial extent modelling, in which the variable being modelled extends spatially as a result of conversion to it from other cover types. Examples include urban expansion modelling and fire spread modelling. Growth modelling is more complex than spatial extent modelling, in that the newly growing areas can be discontinuous spatially, as exemplified by urban growth in the leapfrog mode. In this mode of growth, newly expanded areas are physically isolated from the existing urban areas. They can be spatially fragmented, or distributed along linear features, such as roads. In case of fire spread modelling, fires can spread from a few ignition points, and the burned areas may not be sufficiently extensive spatially to coalesce into a larger patch. Spatially discontinuous growth is much more difficult to model than spatially continuous extension, as the source of spread has a component of randomness (Table 6.3). Another reason is that the modelled outcome has a component of time (to be discussed in Chapter 7).

Predictive modelling aims to project the spatial behaviour of the dependent variable's attribute in the future under a given set of circumstances (Table 6.4). The difference between static and predictive modelling is the way time is treated. In static modelling, time is regarded as irrelevant as the modelling refers to a particular time or a given time slice. This treatment is considered acceptable if neither of them changes much with time. In predictive modelling, time always refers to moments in the future. Predictive modelling is commonly implemented as spatial simulation (Figure 6.5), in which time is explicitly specified and the simulation is re-run as time increments. Spatial simulation is more dynamic than spatial modelling because of its incorporation of the temporal component. It is good at simulating how a geographic phenomenon varies spatially at different temporal intervals. It is also strong in studying a spatial process with less focus on the outcome than in spatial analysis. Nevertheless, predictive modelling faces a major limitation in validating the

TABLE 6.3
Comparison of Four Objectives of Spatial Modelling and Their Main Features

Nature of modelling	Main features	Examples
Suitability modelling	Narrowing down or grading of choices; provide candidates through exclusion	Where to locate a cell phone transmission tower, habitat suitability modelling
Potential modelling	Derivation of a quantitative index based on all input variables for the whole area	Landslide hazard modelling, earthquake damage modelling, modelling of the quantity of soil eroded from a region
Spatial extent modelling	Prediction of changes to areas, extent of spread into neighbouring areas	Coastal areas to be inundated by the projected sea-level rise, contraction of a glacier tongue size
Growth modelling	Areas to be converted into a type of interest from other covers	Urban expansion modelling, fire spread modelling

TABLE 6.4

Comparison of Main Features of Three Types of Spatial Modelling

Items	Static	Predictive (potential)	Exploratory
Variables	Input and output	State; relationship or rules	Rate; initial conditions; neighbourhood
Treatment of time	Implicit, snapshot	Explicit, time fixed	Time increment
Output/model run	Subject to inputs only, run once	Subject to inputs, run a few times	Time, previous run, initial conditions, multiple runs
Nature of variables	Time-invariant	Temporally fixed	Temporal increment
Assumptions	No assumptions	What happened in the past will happen in the future	Assumptions vital to simplify modelling (e.g., homogeneous space, invariant rates)
Function/best use	Outcome-oriented; assessment of potential	Outcome-oriented; prediction of what is to come	Process-oriented; scenario analysis; sensitivity test
Main features	Generation of an overall value	Generation of spatial patterns for a variable	Exploration of one variable's effects on the modelled outcome
Examples	Landslide potential; vulnerability to groundwater contamination	Distribution of urbanised areas in 20 years	How grazing intensity will affect the recovery of degraded grassland over the next 50 years

modelled outcome, as the truth does not exist or cannot be known. So, in reality, the same simulation must be re-run a number of times to take into account past events against which the modelled outcome can be compared to derive an indication of simulation accuracy. Examples of predictive modelling include urban growth modelling and modelling of meadow degradation under different external influences.

Predictive modelling can be adapted into *exploratory modelling* by withholding all variables except one constant to examine its influence on the dependent variable, which is also known as scenario modelling (Table 6.4). The modelling has to be re-run repeatedly so that multiple outcomes can be generated and compared with each other. Exploratory modelling differs from the other two types of modelling by treating time explicitly. The modelled outcome is a function of time. The main objective of exploratory modelling is to understand spatial processes, particularly how the outcome is affected by individual variables and their changed attribute values.

The three types of spatial modelling differ from each other in the way time and space are treated. In *static modelling*, the dependent variable is implicitly assumed to be time-invariable (Table 6.4). As such, it is ill suited to studying the temporal evolution of the dependent variable and the processes of change, a task that can be achieved through predictive modelling. In contrast, *dynamic modelling* treats time explicitly, either in time slices (e.g., urban expansion in 10 and 20 years) or time increments (e.g., pollutant diffusion every hour). Spatially, static modelling and dynamic modelling also differ from each other in their treatment of the neighbourhood effect. In static modelling, each grid cell is handled individually in isolation from other cells surrounding it. There are no interactions among the neighbouring cells. Thus, the output at a given cell in the spatial layer is unaffected by its location and neighbours, as the neighbourhood is irrelevant. In contrast, in dynamic modelling, the output is subject to the influence of cells in the defined neighbourhood. They interact with each other. The modelled output is subject to not only the neighbourhood size, but also how neighbours are defined. Spatially dynamic modelling is commonly implemented as spatiotemporal simulation (see Chapter 7).

6.2.3 Cartographic Modelling

Both suitability and potential modelling can be generically termed *cartographic modelling*. Cartographic models are temporally static models involving identically geo-referenced layers, each representing a spatially varying factor relevant to the dependent variable. In this kind of static modelling, several inputs are transformed into one spatial output using a set of tools and functions, as well as weights if applicable. It is also possible to introduce mathematical relationships to the modelling process, so it is complex. However, the complexity and the steps involved from the input to the output vary with the nature of the spatial phenomenon being modelled. Conceptually, cartographic modelling can be regarded essentially as Map Algebra of raster layers (Tomlin, 1990), all of which are represented at the same grid cell size and cover the same extent (i.e., the number of cells in both north–south and east–west directions is the same for all layers).

Cartographic modelling is a structured integration of a variety of primitives to create spatial models and for map analysis, in which the sequence of overlaying some layers may or may not make a difference to the outcome, depending on the nature of the modelling (Figure 6.7). If no prior spatial processing is carried out on any input layers, then the order of modelling is inconsequential (Figure 6.7a). However, if some layers must be processed spatially first, then the logical order of modelling must be followed (Figure 6.7b). Since not all the variables in the input are equally important to the output, they should be weighted. Weighting variables is such an important and broad topic that it will be discussed at length in Section 6.4.1. Another issue in cartographic modelling is correlation among the input variables. Whenever two input variables are correlated with each other, their joint influence on the dependent variable should be discounted by assigning a smaller weight to each of them. Although cartographic modelling can be used for dynamic modelling, the modeller has to keep track of time-dependent variables and inputs manually, which is rather burdensome and tedious.

Intrinsically, cartographic modelling is aspatial in nature, since the unknown spatial attribute across space is estimated from its contributing factors. The value at a given location in space is subject to the arithmetical manipulation of all input

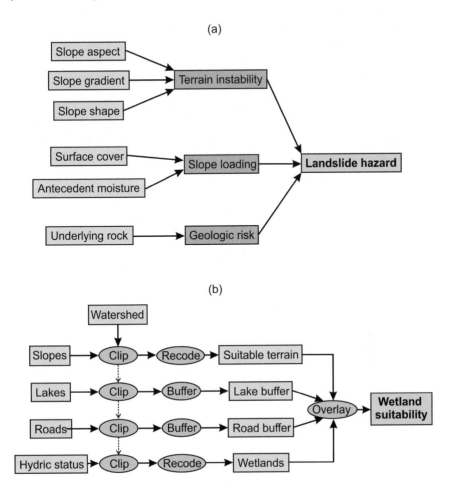

FIGURE 6.7 Two examples of cartographic modelling showing the varying importance of the sequence of analysis: (a) modelling of landslide potential from its important contributors, in which the order of overlay analysis is inconsequential; (b) wetland habitat suitability modelling, in which spatial operations precede overlay analysis, and the sequence of analysis is critical to the modelled outcome, and thus cannot be altered.

variables at the same location. The output at a given grid cell is independent of the influence of other neighbouring cells, without any traces of previous inputs. The modeller does not need to be concerned about where the cell of interest is located, as the same operation is applied to all grid cells in the same layer indiscriminately, regardless of their location. This is permissible as there are no interactions between cells at different locations or within the same neighbourhood. The modelling is considered spatial in the sense that all the input layers are represented as maps spatially and the modelled variable is spatial (e.g., it has a value at any spatial location that varies spatially). The mathematical relationships between the modelled variable and the independent variables are established via regression analysis. Depending on the variable to be modelled and the rules applied to the input variables, the outcome of modelling can be spatially continuous or exclusive. For example, the modelled outcome can be a surface showing vulnerability to coastal inundation in light of anticipated sea-level rise. It can also show a few candidates that meet criteria for being used as a rubbish dump, as in site suitability modelling.

The exact number and type of variables allowable in a cartographic model depend on the nature of the dependent variable and its relationship with the independent variables. In the case of multiple independent variables, they can be combined either additively or factorially, or a combination of both. In the former combination, the influence of all the independent variables is summed up after proper weighting, if necessary. Additive combination should be adopted when the influences of all the variables are independent of each other. For instance, landslide potential is increased under rain, in addition to the risk caused by a steep gradient. Additive combination is exemplified by the Depth to water, net Recharge, Aquifer media, Soil media, Topography, Impact of vadose zone, and hydraulic Conductivity (DRASTIC) model developed by the US Environmental Protection Agency (Aller et al., 1985). The DRASTIC index is calculated by summing all the weighted variables:

$$DRASTIC\ index = DWd + RWr + AWa + SWs + TWt + Iwi + CWc \qquad (6.2)$$

where D = groundwater depth, R = net recharge, A = aquifer media, S = soil media, T = topography, I = vadose zone, and C = hydraulic conductivity.

This linear empirical model is suitable for evaluating the vulnerability of aquifers to groundwater pollution from hydrogeological factors. The weight can be determined using the Delphi technique (see Section 6.4.1). It ranges from 1 to 5, with 5 being the most important and 1 the least important. Sometimes the vulnerability is grouped into classes (Figure 6.8). This grouping is important when the input layers have a higher degree of uncertainty. Without grading, in some cases the modelling outcome can lead to misleading interpretations due to the high level of uncertainty in the input layers, which is amplified in the process of additive combination.

In factorial combination, the influence of all variables is multiplied, causing variable weighting to be redundant. This combination is used when one of the factors can override or amplify the influence of others. For instance, a landslide will not happen on flat terrain no matter how violently the ground shakes in an earthquake. In this case, whether a landslide will eventuate cannot be judged from a single variable in isolation. The answer depends on the complex interactions among multiple variables. A factorial model is exemplified by the Revised Universal Soil Loss Equation (RUSLE) (Kouli et al., 2008):

$$A = R \cdot K \cdot L \cdot S \cdot C \cdot P \qquad (6.3)$$

where R = rainfall-runoff erosivity, K = soil erodibility, L = slope length, S = slope steepness, C = cover and management factor, and P = erosion support practice or land management factor. *A* refers to the average annual soil loss with a unit of

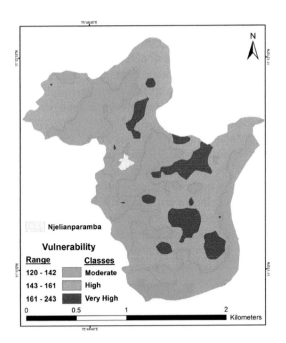

FIGURE 6.8   The vulnerability of an alluvial aquifer modelled using the DRASTIC index-ing method, graded into three classes (source: Jaseela et al., 2016, open access).

tons·ha^{-1}·year^{-1}. R is derived from monthly and annual precipitation data, and K is estimated from a soil map. The RUSLE model requires topographic data in the form of a digital elevation model (DEM) from which L and S are derived, and remotely sensed imagery data from which C can be mapped.

Both additive and factorial models can be implemented easily in the Map Algebra Toolbox in ArcGIS by combining all the weighted layers.

6.2.4   Spatial Dynamic Modelling

Spatial dynamic modelling or spatial simulation is a special type of predictive mod-elling in which time is explicitly involved in the modelling process and the modelled results are time-dependent. At least one of the independent variables must vary with time. If the speed of movement over space is fixed, time is strictly proportional to the distance between observations or the modelled value. In spatial dynamic model-ling, an assumption of temporal sequence is implied. In the raster space, cells further away from the source are affected later than those closer to the source. This can be expressed mathematically using transition rules governing how the state of a cell should change. Common examples of spatial dynamic modelling include surface runoff modelling over a slope and transportation cost (friction) over a surface. In both cases, the dependent variable is spatially varying with the distance of travel.

Rainwater accumulates spatially over a slope as it flows down the slope. The further down the slope it flows, the more rainwater accumulates on the surface, until it reaches a channel. The modelled value or potential at a particular location is affected by a number of variables, including the variable being modelled. Thus, the modelled value is neighbourhood-dependent. In spatial dynamic modelling, the output is time-dependent, and the outcome at time 2 depends on the outcome at time 1, as well as the influence of the neighbours.

In the modelling, the same transition rules are run recursively or iteratively, from one cell to the next in the immediate neighbourhood (Tomlin, 1991). However, a set of initial conditions must be specified to run the model, such as the seed cells from which diffusion takes place, or the ignition cell in bushfire spread modelling. In certain modelling, these seed cells are just selected randomly and the modelling is run multiple times, the results being averaged to eliminate the randomness effect. Thus, a series of predictions are generated from the same transformation rules multiple times. From one run to the next, the transition rules remain unchanged. Only the state of neighbouring cells changes, causing the current cell in question to change its state or attribute value. Whenever the model is re-run in a new session, it must be initialised to the original conditions so that all the modelled outcomes can be benchmarked and compared with each other. This kind of modelling can be terminated at any time or after any number of model runs. It is possible to treat the modelled results as a function of time duration. However, time is not explicitly obvious in the modelling process. Usually, it is automatically incremented by 1 after each iteration. However, one increment can mean a second, a day, a month, or a year, depending on how it is defined by the modeller. The selection of a specific temporal increment pace is subject to the speed at which the phenomenon being modelled changes. Thus, the modelled result is time-dependent and the output must be time-sliced, making spatial dynamic modelling practically synonymous with temporal modelling.

6.3   MODEL DEVELOPMENT AND ACCURACY

No matter how complex a spatial model is, conceptually it can be expressed simply as:

$$\text{Dependent variable} = f\left(\text{var}_1, \text{var}_2, ..., \text{var}_n\right) \qquad (6.4)$$

where f refers to the function in which the exploratory or independent variables var_i ($i = 1, 2, 3, ..., n$) affect the dependent variable. It can take the form of either a linear or factorial equation, or some more complex mathematic equations; var_i ($i = 1, 2, 3, ..., n$) are variables that contribute to the dependent variable (e.g., risk or potential). There is no restriction on the number of variables (n) in the model. More variables will make the model more complex, and likely more accurate, but will demand more data about them. If a large pool of independent variables is available for inclusion in a model, the selection of the best variables to be included in the model can be decided by imposing a threshold, or automatically via machine learning. What variables should be included in a model is ideally addressed via multi-criteria decision analysis. The included variables may be spatial, and their spatial data, usually in the form of geographic information system (GIS) layers in raster or vector format, must be in existence, preferably at the proper spatial scale. They can also be aerial photographs or satellite images, or results derived from them, topographic maps, or spatial data from other sources. Ideally, all the data layers used in the modelling should have a similar currency level, or at least the data variables being portrayed should not have changed much since their collection.

No matter how many variables are included in a model, they must all exert an influence on the dependent variable, even though the exact manner of influence may vary among the variables and the influence may be subject to other factors. For instance, the rate of surface rainwater flow is subject to both slope gradient and surface cover. It is quite possible that the exact way these variables affect the dependent variable is unknown and has to be determined empirically. The development of f involves a number of steps, the most important being feature selection, multi-collinearity testing, model validation and accuracy assessment, and model running (Figure 6.9).

6.3.1 Feature Selection

A very critical consideration in model development is the type and number (n) of independent variables that should be included in the model. They are selected based

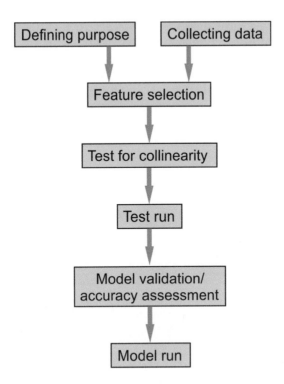

FIGURE 6.9 Major steps in constructing and validating an empirical spatial model.

on specialised knowledge and published literature on the topic. Initially, as many variables as possible should be considered, even though not all of them may affect the dependent variable equally. How to select the most useful independent variables from a large pool of potential variables is an issue known as *feature selection*. It aims to develop a powerful predictive model (e.g., maximal model accuracy) from a reduced number of input variables. Feature selection has two obvious objectives: (1) to keep the model as simple as possible, and (2) to reduce the amount of computation needed. Feature selection can be based on unsupervised or supervised learning (Brownlee, 2019). The former removes redundant variables based on their correlation with the dependent variable via wrappers and filters. Wrappers search for well-performing subsets of variables, while filters select the subset of variables based on their relationship with the dependent variable, using Pearson's correlation coefficient, analysis of variance (ANOVA), and the χ^2 test. Intrinsic selection is performed automatically.

Feature selection can be implemented either statistically, based on feature importance, or automatically during training, using machine learning as part of learning the model (see Section 7.1.4 for more details), such as decision trees and recursive feature elimination (Figure 6.10). Machine learning can determine the importance of the considered variables to the dependent variable being modelled. Many variable selection algorithms based on various evaluation criteria are available for this purpose (Asner and Heidebrecht, 2002). One of them is the CfsSubsetEval algorithm in the Waikato Environment for Knowledge Analysis (WEKA) package by the University of Waikato in Hamilton, New Zealand. WEKA is a freely available data mining and predictive modelling software package based on machine learning. It can search for a subset of features from the available variable pool. The selected features work well jointly, but are barely correlated with each other, though all of

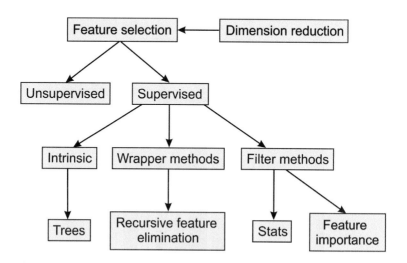

FIGURE 6.10 Techniques commonly used to select variables to be included in a model (source: modified from Brownlee, 2019).

them are highly correlated with the target variable. Apart from quantifying the predictive power of the selected features, WEKA can also yield information on the degree of redundancy between them (Hall, 1998). Among various machine learning algorithms, Random Forest is considered one of the best, and should be used to minimise the inaccuracy of modelling (Pourghasemi et al., 2020). Nevertheless, it is still necessary to account for the joint or overlapping influences among all the selected variables. When multiple independent variables are included in a model, they may be correlated to each other. The co-variance of variables tentatively retained in the model should be analysed via a multi-collinearity test to avoid double counting.

6.3.2   MULTI-COLLINEARITY TEST

The collinearity between multiple independent variables in a model should be tested to see whether they are dependent on each other, an issue commonly resolved via the *multi-collinearity test*. If they are too closely correlated, then they will weaken the prediction power of the regression model (e.g., the p-values become less trustworthy) and affect the proper interpretation of the modelled outcome, because the coefficient of the variables or their weight is highly sensitive to small changes. High multi-collinearity also reduces the precision of the coefficient.

Multi-collinearity is commonly tested via the Variance Inflation Factor (VIF) test. It identifies the strength of correlation between the independent variables and has a value ≥1. A VIF value of 1 indicates the absence of correlation between the tested variables. A value between 1 and 5 indicates a moderate correlation that is not sufficiently severe to warrant corrective efforts. A value over 5 represents critical levels of multi-collinearity, and is indicative of poorly estimated coefficients and questionable p-values. In this case, one of the correlated variables should be dropped from the model.

After all the important variables have been determined and retained in the model, the next step is to assign weights to each of them (if applicable). How to assign a proper weight to a variable to reflect its importance to the dependent variable is such a complex topic that it will be discussed separately in Section 6.4.1. After a model has been constructed, it needs to be tested. A good way of testing is to deliberately add noise to the input variables to examine how the model handles such a situation. However, stringent testing must be based on proper validation.

6.3.3   MODEL VALIDATION

After a model has been properly parameterised, it needs to be validated. *Model validation* is very difficult to achieve, especially in predictive modelling. This problem is dealt with by firstly ensuring the correctness of the model and the logic behind all the variables in it. Next, closely estimated inputs from past events are plugged into the model to see how well the output matches the known results. The degree of agreement between the modelled and observed outcomes indicates model accuracy and the reliability of the modelled results. This method of validation, commonly known as *hindcasting*, is underpinned by an implicit assumption that what has happened in the past will happen exactly the same in the future. Whenever this assumption is violated, the modelled outcome will suffer from inaccuracy. If the validation accuracy is not acceptable, then the model parameters or the transition rules must be modified until the two sets of outcomes converge to a degree considered satisfactory. For instance, certain parameters and/or their weights in the model can be re-parameterised or fine-tuned to reach a closer match between them. Once the model has been calibrated satisfactorily, it can be used to predict the future using the same parameter setting.

A model is never more than an approximation of reality. How good is the approximation? This brings out the issue of model accuracy. *Accuracy* refers to the correctness of the modelled results. If they closely match reality, then they are said to be highly accurate. The difficulty is that future reality is unknown in predictive modelling. In this case, the model can still be validated or calibrated using *bootstrapping*. In bootstrapping, random samples are replaced in each test, in which one of the observations is left out. It has a number of permutations, including Holdout Validation (HV), K-fold Cross-Validation (KCV), and Leave-One-Out Cross-Validation (LOOCV). Each permutation has different requirements and fulfils different needs. HV requires all available samples to be partitioned into two groups at a ratio such as 70%:30%. The second and lower portion, known as the testing dataset, is used to validate the model developed using the remaining 70% of samples, known as the training dataset (Kim, 2009). KCV requires all the available samples to be divided into K equal-sized groups. Each time, only one of them is selected as the testing dataset, while the remaining K–1 groups are reserved for model development. Every group is used in turn just once for validation (Kohavi, 1995). LOOCV is almost identical to KCV except that each sample in the dataset will be used in turn to test the model developed, using the remaining samples minus the one being used for the test each time. The developed model is thus validated n times (n = number of samples) by comparing the observed value with the predicted one (Hoek et al., 2008). The final validation outcome is then calculated by averaging all the individual evaluations. LOOCV overcomes the drawback associated with a small training dataset that cannot be sensibly partitioned into two parts, one for model construction and another for model validation. Instead, all of the samples have to be used in model construction. LOOCV is the most commonly used, even though it is computationally intensive and subject to overfitting, as all the training data except one are fed to the model. All of these validation methods are suitable for a modelled dependent variable whose attribute value can be expressed quantitatively, such as the concentration of air pollutants.

6.3.4 MEASURES OF MODELLING ACCURACY

Accuracy of the modelled outcome can be expressed using a number of measures, each suitable for a unique type of spatial modelling. The goodness of fit between the predicted and observed values can be judged by the coefficients of regression (R^2), root-mean-square error (RMSE), or mean absolute error, all of which are good at indicating the accuracy of general modellings, such as susceptibility to landslides (Table 6.5). These measures are suitable for the model itself, not the modelled

TABLE 6.5

Comparison of Common Measures for Gauging the Accuracy of Spatial Modelling

Measure	Suitability	Main features	Author(s)
R² and RMSE	Model itself, non-spatial	A high R² (up to 1.00) and a low RMSE suggest accurate models	Ma et al. (2020)
AUC	Modelled outcome of a categorical attribute	Accuracy indicated by two parameters: true positive and false positive	Corominas et al. (2013)
Cohen's kappa coefficient	Modelled outcome of a binary nature; measures spatial agreement with chance agreement considered	Value ranging between 0 (no effective agreement) and 1 (perfect agreement); k < 0 means worse than chance agreement	McHugh (2012)
Cohen's kappa simulation coefficient	Similar to kappa, with the agreement adjusted by the proportion of a cover	Identical to kappa; unable to assess the accuracy of fuzzy land cover transition	van Vliet et al. (2011)

outcome, whose accuracy can be measured using the area-under-the-curve (AUC) indicator of the receiver-operating characteristics. Initially developed for military purposes, this measure has found applications in assessing landslide zoning accuracy (Corominas et al., 2013). It is able to reveal the agreement between the observed and modelled landslide potential, expressed as a categorical attribute. The result of comparing the two datasets is represented by a pair of parameters: the true positive rate and the false positive rate. The former refers to the portion of prediction points falling in the correct zone, while the latter indicates the proportion of assessment points falling in the wrong zone. Jointly, these two rates can be used to derive the success rate and the predictive rate. The former indicates the level of agreement between the training dataset and the predicted outcome, and quantitatively indicates the goodness of fit between the two datasets, under the assumption that the model is correct. The prediction rate measures the agreement between the testing data and the predicted outcome (Vakhshoori and Zare, 2018). This measure is suitable for assessing modelled attributes that have been graded into a few categories, such as landslide potential from low to extremely high.

Cohen's *kappa* (*k*) and simulated kappa coefficient are two other indicators of the reliability of the modelled outcome (McHugh, 2012). The kappa statistic measures the agreement between two categorical variables such as land covers, calculated as:

$$k = \frac{p_0 - p_e}{1 - p_e} = 1 - \frac{1 - p_o}{1 - p_e} \tag{6.5}$$

where p_0 stands for the observed agreement (%) between the simulated and observed data, commonly known as the accuracy in percentage, and p_e denotes the probability of chance agreement between them in which the evaluation pixels are randomly selected. If there are N observations that are categorised into k groups, it is calculated as:

$$p_e = \sum_k \hat{p}_{k12} = \sum_k \hat{p}_{k1}\hat{p}_{k2} = \frac{1}{N^2}\sum_k n_{k1}n_{k2} \quad (6.6)$$

$$p_e = \sum_k \quad (6.7)$$

where $\hat{p}_{k12}$ stands for the estimated probability of a land cover modelled as k in both simulations; n_{ki} ($i = 1$ and 2) refer to the number of times the modelled cover i is predicted in category k. Kappa value ranges from 0 to 1, with 0 meaning the absence of effective agreement except that caused by chances, and a value of 1 indicating a perfect agreement. It is possible for kappa to be in negative territory, suggesting that the agreement between the modelled outcome and reality is worse than chance agreement. Kappa is particularly suitable for assessing modelled results that are binary in nature, such as predicted urbanisation, in which there are only three possible types of prediction: hit, miss, and false alarm. The first means that the predicted outcome matches reality and the prediction is correct; the second means that an urbanised cell is not successfully predicted by the model, and the last suggests that the predicted urbanisation does not eventuate. The last two types of prediction are both incorrect.

Kappa is able to indicate the reliability of spatial modelling much more robustly than the simple percentage agreement indicator as it takes into account chance agreement between the validation dataset and the simulated outcome. Nevertheless, it is still flawed, in that not all land covers in the simulation have an equal chance of occurrence. Thus, it has been gradually replaced by *kappa simulation*, which indicates the agreement between the simulated outcome and reality by adjusting the area of change in a land cover instead of its quantity in one of the input datasets (van Vliet et al., 2011). Hence, it more genuinely indicates the reliability of the modelled land cover changes. However, as a cell-by-cell validation method, kappa simulation cannot accommodate fuzzy land cover transitions.

6.4 ISSUES IN SPATIAL MODELLING

6.4.1 ASSIGNMENT OF WEIGHTS

No matter how many variables are included in the spatial model (Equation 6.4), it is essential to quantify their importance, not only to achieve accurate modelling results, but also to identify the variables that affect the accuracy most. Determination of the importance of individual factors in a model is a critical step in model construction and fulfils a vital prerequisite for undertaking spatial modelling. How to weight the variables in a model is the thorniest and most influential issue in additive cartographic modelling, as the weight directly affects the modelled outcome in terms of its validity and accuracy. So far, a number of methods have been devised to accomplish this task. They range from simplistic comparison by experts to sophisticated regression analysis. Other methods may be statistical, such as discriminant analysis, and even machine learning algorithms. Each of them has its own pros and cons, as well as best uses (Table 6.6). No matter which method is used, the weight assigned to each of the input variables should accurately reflect its importance to the dependent variable being modelled.

6.4.1.1 AHP

Devised by Thomas Saaty (1977), the Analytic Hierarchy Process (AHP) is a structured method designed for multiple factor decision-making. In the decision-making process based on the eigenvalue, a complex problem is scaled down to a hierarchy of issues represented as a unidimensional composite priority vector of the alternatives or weights via a pairwise comparison of factors. By simplifying the problem, AHP enables it to be understood more easily. The final decision rests with the weights of the factors and the alternatives. In the pairwise comparison, two factors are compared to each other, and their relative importance to each other is defined by a rank in accordance with the pre-devised system. In the comparison, all variables under consideration have to be paired, and each variable is assigned a pre-determined rank of importance, usually ranging from 1, 3, 5, 7 to 9 (Table 6.7). Since its inception in the 1970s, AHP has been extensively refined to accurately and quantitatively weight the decision criteria, in which experts' opinions are relied on to estimate the relative importance of variables. Usually, a questionnaire is sent to a panel of experts to solicit their judgements and perceptions. They are asked to grade the relative importance of each of the variables being considered, together with a statement of justification (Table 6.7). Each pair is compared and ranked independent of

TABLE 6.6
Pros and Cons of Four Methods Commonly Used to Weight Input Variables in Cartographic Modelling

Method	Pros	Cons	Best uses
AHP	Structured, very simple	Subjective; empirical; discrete; complex if many criteria are considered	When no data about the variables are available; multiple variables to consider
Regression	Objective and reliable	Complex; unable to account for joint effects	When a huge amount of data are available; a large pool of variables to choose from
Weight of evidence	Evidence (weight) based on probability of past events	Not all evidence layers are independent of each other	When historic data of the modelled variable are available to derive the evidence layer
Delphi	Simple ranking by a panel of experts	Imprecise; indicative of rank only; not the actual weight	When it is impossible to attach a weight to a variable quantitatively

TABLE 6.7
Pairwise Comparison of Criteria (Factors) in Deciding Where to Live for a Newly Arrived Resident in a City

Factor 1	Importance	Factor 2	Importance	Justification
Income	3	School	1	Rich suburbs are safer and have better amenities
Income	3	Transport	1	Living in a safe environment is more important than longer commuting
Income	1	Land price	5	Land price dictates real estate price and housing affordability; it is much more important than economic status
Income	5	Ethnicity	1	Income level is much more important than ethnicity; most people are very healthy and safety-conscious
School	3	Transport	1	Children attending a good school is more important than spending a bit more time commuting
School	1	Land price	7	Housing affordability is much more important than whether the school is good or not
School	3	Ethnicity	1	Quality education is important to children; they can get along well with people of any ethnicity
Transport	3	Ethnicity	1	Getting around easily and quickly is more important than who lives in the neighbourhood
Transport	1	Land price	7	Housing affordability is much more important than time spent on commuting
Land price	9	Ethnicity	1	Where one can afford to live is much more important than who the neighbours are

1 = equal importance (both factors play the same role in decision-making); 3 = moderate importance (one factor is moderately more important than the other, based on judgement and experience); 5 = strong importance (one factor is strongly favoured over the other); 7 = very strong importance (one factor overwhelms the other in importance); 9 = extremely strong importance (the importance of one factor over the other is likely in all cases).

other pairs of variables. Of the two variables being considered, one of them always receives a rank of 1. The rank of the other factor is commensurate with its importance. In this way, the comparison simply becomes how many times more important this variable is relative to the other. For instance, school is 3 times more important than transport when a resident decides where to reside in a city. After all pairs have been ranked and graded, a two-dimensional matrix is constructed to calculate the weight (Table 6.8).

TABLE 6.8

Matrix of Pairwise Comparison Outcome Based on the Information Shown in Table 6.7

Criteria	Income	School	Transport	Land price	Ethnicity	Weight*
Income	1 (0.1456)	3 (0.257)	3 (0.209)	1/5 (0.125)	5 (0.238)	0.195
School	1/3 (0.049)	1 (0.086)	3 (0.209)	1/7 (0.090)	3 (0.143)	0.115
Transport	1/3 (0.049)	1/3 (0.029)	1 (0.070)	1/7 (0.090)	3 (0.143)	0.076
Land price	5 (0.728)	7 (0.600)	7 (0.488)	1 (0.626)	9 (0.429)	0.574
Ethnicity	1/5 (0.029)	1/3 (0.029)	1/3 (0.023)	1/9 (0.070)	1 (0.048)	0.040
Column sum	6.866	11.666	14.333	1.597	21.000	1.000

Figures in brackets represent the ratio of the column sum, e.g., 0.1456 = 1/6.866.
* Calculated by dividing the row sum expressed as the portion of the column sum by the number of criteria n (n = 5).

The outcome of the pairwise comparison is represented in a square matrix (Table 6.8) whose dimension equals the number of criteria being considered. The value of all cells ranges from 1/9 to 9. If one variable is 5 times more important than another variable, then its importance value is inversed (e.g., 1/5). In order to calculate the weight, all values must be converted to a ratio to the column sum, expressed as decimal figures, then all the ratios are summed up by row. The weight is the row sum divided by the total number of criteria n (n = 5) (Table 6.8). As shown in Table 6.8, all weights sum up to 1. Since the weights assigned to a variable depend wholly upon the experts' knowledge in the field of study, the assigned weights will inevitably vary from expert to expert. This may be an important source of human bias in the modelled results.

The variability of the weights determined in AHP is measured by the *consistency ratio*. It reflects the fact that different experts will rank the same pair of factors (criteria) differently. This inconsistency issue is likely to worsen when there are a large number of alternatives. The consistency of the matrix shown in Table 6.8 is calculated by multiplying the pairwise comparison matrix by the weight vector:

$$
C = A \cdot w = \begin{bmatrix} 1 & 3 & 3 & \frac{1}{5} & 5 \\ \frac{1}{3} & 1 & 3 & \frac{1}{7} & 3 \\ \frac{1}{3} & \frac{1}{3} & 1 & \frac{1}{7} & 3 \\ 5 & 7 & 7 & 1 & 9 \\ \frac{1}{5} & \frac{1}{3} & \frac{1}{3} & \frac{1}{9} & 1 \end{bmatrix} \cdot \begin{bmatrix} 0.195 \\ 0.115 \\ 0.076 \\ 0.574 \\ 0.040 \end{bmatrix} = \begin{bmatrix} 1.081 \\ 0.609 \\ 0.380 \\ 3.244 \\ 0.206 \end{bmatrix}
$$

$$\text{Consistency index} = \frac{\lambda_{max} - n}{n - 1} \tag{6.8}$$

where n denotes the number of criteria considered and λ_{max} stands for the largest eigenvalue of the weight vector, or:

$$\lambda_{max} = \frac{1}{n}\sum_{i=1}^{n}\frac{c_i}{w_i} = \frac{1}{5}\left(\frac{1.081}{0.195} + \frac{0.609}{0.115} + \frac{0.380}{0.076} + \frac{3.244}{0.574} + \frac{0.206}{0.040}\right) = 5.328$$

$$\text{Consistency index} = \frac{\lambda_{max} - n}{n - 1} = \frac{5.328 - 5}{5 - 1} = 0.082 \tag{6.9}$$

The consistency ratio is calculated by dividing the consistency index (Equation 6.8) by the theoretical random consistency index that has been published in the literature (Table 6.9). It is calculated as:

$$\text{Consistency ratio} = \frac{\text{Consistency index}}{\text{Random index}} = \frac{0.082}{1.12} = 0.073 \tag{6.10}$$

If the weights have been rationally assigned, the consistency ratio should be smaller than 0.1 (e.g., 0.073 in this example). Otherwise, the assigned weights are not reasonable and have to be modified. This deterministic multi-criteria decision-making method relies on expert advice, and the weight may be subjective and empirical. This method of weighting is practical only when a limited number of variables are considered and each does not have any sub-attribute values (e.g., different modes of travel in transport).

6.4.1.2 Regression Analysis

Dissimilar to AHP, regression analysis is a very objective way of determining the importance of independent variables by statistically analysing the strength of their correlation with the dependent variable. Regression analysis can produce highly reliable and precise coefficients for all variables being considered (Table 6.10). If they have been standardised, the larger their coefficients, the more important they are. In spite of this simplicity, it is unlikely that the coefficients of all variables will sum up to 1, due to overlapping influences. Regression analysis is also unable to address the co-variance of all variables included in a model, which can lead to double counting of

---

TABLE 6.9
The Theoretical Random Index (RI) of Consistency Value at Eight Criteria

N	1	2	3	4	5	6	7	8
RI	0.00	0.00	0.58	0.90	1.12	1.24	1.32	1.41

Source: Siddayao et al. (2014).

TABLE 6.10

Variables Selected in a Multi-variate Linear Regression Model with Variable Importance Indicated

Variables in LUR model	Coefficient	Standard error	p-value	VIF
Intercept	19.8151	1.8456	<0.001	—
Main road length_50	4.0958×10^{-2}	0.0051	<0.001	1.0078
Ratio_BH_BN_1000	0.2973	0.0869	<0.001	1.5770
Traffic load_1000	1.0691×10^{-8}	<0.001	<0.001	1.1999
Bus stop nums_100	0.7932	0.2603	0.0030	1.2404
Natural_200	-1.5044×10^{-4}	<0.001	<0.001	1.0693
Elevation	-8.2417×10^{-2}	0.0307	0.0086	1.3165

Source: Ma et al. (2019).

BH = building height; BN = building number; 1,000 = width of buffering analysis.

the same variable. Also unknown is the number of variables that should be included in the model. This question can be decided by imposing a threshold on the p-value of variables. Any variables above the threshold are automatically excluded.

As illustrated in Table 6.10, the number of variables in the land use regression (LUR) model is automatically selected by the software in accordance with the specified selection criteria, together with their coefficients, which range widely in value due to the different units adopted. These coefficients represent the influence of the respective independent variables on the dependent variable (NO_2 concentration), so there is no need to assign a weight to each of them separately. More importantly, this method of weighting can indicate the correlation between them and their contribution to the dependent variable. It is straightforward, but computationally complex. It is also unable to account for the joint effects of the selected variables, such as between main road length and traffic load.

6.4.1.3 Weight of Evidence Model

Weight of evidence is a bivariate statistical analysis method that can be used to attach weights to each independent variable based on past events. It is grounded in a Bayesian rule in a log-linear form using the priori and posterior probabilities. The priori probability (unconditional probability) is the probability of an event that results from the same event in the past. The spatial probability of an event is derived from past events in a given period of time. The posterior probability or conditional probability is the change in probability owing to the additional information given to re-evaluate the priori probability (Samodra et al., 2017; van Westen, 2002). For example, additional use of lithology as an independent prediction variable may change the probability of landslide occurrence based on topographic gradient. However, the weight of evidence in spatial modelling has a prerequisite that evidential layers are approximately and conditionally independent of the target layer, which can be difficult to fulfil in practice (Zhang and Agterberg, 2018).

The spatial association between the dependent variable under study and its determinants or the independent variables determined from historic data is expressed in a pair of weights, W^+ (Equation 6.11) and W^- (Equation 6.12). They can be calculated from the evidence maps, which have to be converted into a binary map of presence/absence of the predictor:

$$W^+ = \ln \frac{P\{F \mid L\}}{P\{F \mid \bar{L}\}} \qquad (6.11)$$

$$W^- = \ln \frac{P\{\bar{F} \mid L\}}{P\{\bar{F} \mid \bar{L}\}} \qquad (6.12)$$

where P stands for the probability, F refers to the presence of the predictive variable, $\bar{F}$ denotes the absence of the predictive variable, L is the presence of the variable to be predicted, and $\bar{L}$ represents the absence of this variable. A positive weight (W^+) suggests the existence of spatial association, and a negative weight (W^-) signifies otherwise. The value of W indicates the strength of the spatial association. A few unique permutations of the weights have special meanings. For instance, if W^+ is positive and W^- is negative, then the variable favours the occurrence of the dependent variable. Conversely, if W^+ is negative and W^- is positive, then the evidence suggests that this factor strongly discourages the eventuation of the dependent variable. If both W^+ and W^- are 0, then the predictive factor is not correlated with the dependent variable.

The difference between W^+ and W^- is termed *weight contrast*. A positive contrast indicates spatial association between the dependent variable and the independent predictive variables (e.g., more events occur in the domain than those caused by chance), and vice versa. The sum of all contrast values forms the weight of the evidence map.

6.4.1.4 Delphi Technique

The *Delphi method* or Delphi technique is a structured method widely used for business forecasting. In spatial modelling, it has been used to determine the relative importance of variables in a manner similar to AHP (Babiker et al., 2005). Instead of pairwise comparisons, Delphi asks the experts to rank all the variables under consideration in a number of rounds. At the end of each round, all the anonymised answers are presented for all the experts to see, and they are encouraged to revise their earlier answers in light of the answers offered by other experts. It is anticipated that the answers of all experts will eventually converge (e.g., every expert will provide the same "correct" answer) after a few rounds. The questioning via a questionnaire is terminated once a pre-determined criterion is reached, such as the number of rounds or the level of disagreement threshold. The end product is the ranked importance of all the variables. It is up to the analyst how this rank will be translated into a numerical weight to be used in the spatial modelling.

6.4.2 SPATIAL MODELLING AND DATA STRUCTURE

Both the feasibility and ease of spatial modelling are subject to the data format. Of the two types of areal data, raster tessellation is much more commonly used in spatial modelling than its vector counterpart. Although it is an inefficient form of feature representation and contains a high degree of data redundancy, raster is popular in grid-based spatial modelling. Although the target of modelling may be the spatial extent, which can be accurately represented in vector format, the modelling is still implemented in raster format, such as modelling snow cover (Cline, 1992). Another example is the modelling of coastal areas to be inundated by anticipated sea-level rise. In both cases, raster format is adopted simply because the required data (e.g., DEM) are recorded and available in this format. Raster is almost the default choice if the other data needed in modelling are collected and stored in this format. In fact, raster is the norm in environmental modellings, such as potential modelling and modelling of vulnerability to pollution and damage caused by earthquakes, for three reasons:

(a) It allows efficient spatial operations with minimal processing required, as all the input layers are made up of regular grid cells of the same size and shape. There is no need to carry out any spatial interpolations as the observed values are available at the desired spatial intervals.

(b) Most geospatial data are suitable for this kind of representation, be they imagery, topographic, light detection and ranging (LiDAR), or thematic (e.g., temperature and pollutant concentration). The wide availability of such environmental and surface cover data enables all kinds of modelling to be implemented in a GIS with great ease.

(c) The same operation can be repeated as many times as needed without modifying the data themselves, as all cells in the input layers are operated on cell by cell.

Admittedly, a large number of intermediate and bulky raster results may be produced in such modelling, but this can be reduced via special scripts.

Vector data such as lines are used when the target of modelling varies along lines, such as traffic volume and channel runoff modelling. The vector data format is an efficient structure, but data may not be available at a desired position. Thus, frequent spatial interpolations have to be evoked to yield the data, slowing down the modelling process considerably. Compared with raster data, vector data are not commonly used in spatial modelling, with only a handful exceptions, such as hydrological modelling involving channel networks, glacier extent modelling, and fire spread modelling. In order to show the direction of channel flow, the Gridded Surface Subsurface Hydrologic Analysis (GSSHA) makes use of vector data for rivers. Similarly, the vector format is used to model glacier extent (Gao, 2004), as the moraine terminus can be expressed accurately by a mathematical formula. The glacier tongue is confined to a narrow valley flanked by deeply incised walls, so is unsuitable to be modelled in raster format as it does not allow the modelled variable to be confined to a spatial extent unless additional layers are involved. In fact, a similar approach based on vector format is adopted for fire spread modelling in which the fire front is treated as composed of a number of line segments represented by vertices (see Section 7.5.1.2). The use of vector data can yield detailed information on the modelled phenomena, such as the precise shape and length of a fire front and its pace of propagation, independent of the raster grid cell size.

6.4.3 PLATFORMS FOR SPATIAL MODELLING

Spatial modelling can be implemented either directly using an existing platform designed specifically for this purpose, or via a specially designed package (Table 6.11). The latter approach is particularly common in novel fields of modelling that cannot be undertaken using generic public domain or commercial platforms. Commercially available packages such as ArcGIS Pro and IDRISI have built-in functions for performing spatial modelling as well as spatial analysis with great ease, each having its own strengths and special fields of applicability. If neither commercial nor public domain packages are available, then spatial modelling has to be implemented using user-written scripts.

6.4.3.1 ArcGIS ModelBuilder

ArcGIS Pro ModelBuilder is an excellent platform for implementing cartographic models with linear and factorial combinations (Equations 6.3 and 6.4). The models can be implemented via stepwise spatial operations. Although simple and easy to understand, this implementation is tedious and slow and creates a large number of intermediate results of no interest. The modelling process can be expedited by creating a model using existing tools such as the ModelBuilder in ArcGIS (Figure 6.11). In addition, models can be built (and have been built) either in ArcMap or ArcCatalog using ArcToolbox. Once built, a model can be saved in Python for further customisation.

ModelBuilder is a visual programming environment that allows the user to create new models and modify existing ones using a chain of spatial operations by picking and mixing functions, specifying the inputs, and displaying the output. In the created workflow, the output of a preceding spatial operation serves as the input to the immediately subsequent operation. The built-in ModelBuilder tool is designed for creating

TABLE 6.11
Comparison of Major Platforms for Implementing Spatial Modelling

Platform/method	Pros	Cons
ArcGIS ModelBuilder/ ERDAS Spatial Modeller	Easy to construct and execute; excellent for cartographic (potential) modelling	Simple, static models only; unable to incorporate external models or logical expressions
ArcGIS Raster Calculator	Able to consider logical and Boolean conditions; flexible	Static and simple models only
IDRISI TerrSet	High degree of automation; good for predictive modelling	Limited domain of applicability (e.g., land use change modelling only)
Scripting	Highly flexible and versatile; fully automated and efficient; good for spatial dynamic modelling; large number of scripts in existence	Programming skills vital; time-consuming to write scripts; hard to detect logical errors in scripts

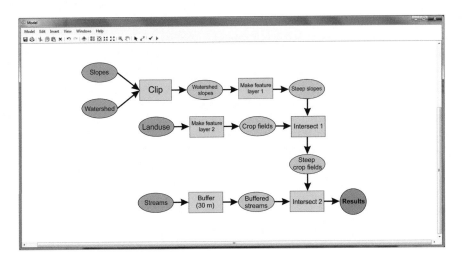

FIGURE 6.11 Identification of cropland located on a steep slope in close proximity to streams using the toolset in ModelBuilder.

reusable and shareable geo-processing workflows to document spatial analysis and modelling automatically. It is also an excellent way of automating cartographic modelling, as well as improving modelling efficiency (Figure 6.11). All such constructed models encompass three elements: variables, tools, and connectors. The necessary variables to be included in a model can be GIS layers, followed by how they are operated on, and multiple spatial operations can be joined together using connectors. The constructed model can be very simple, involving a few steps of operation, or rather complex, involving a long sequence of operations. A graphic model can be created easily by clicking and picking the related files or icons/connectors. Once built, a model can be re-run without changing anything in the model itself except the input and output file names, so it is highly repeatable. Also, such models can be shared with others, from which new models can be created easily via a few modifications.

ModelBuilder has access to all ArcGIS functions, all of which can be embedded into a model. However, looping is not possible at present, but iterative processing of every feature class, raster, file, or table in a workspace is possible. This environment is designed for building static models only. It has no room to incorporate external models. The integration of the built model in ArcGIS with other systems requires scripting.

6.4.3.2 Raster Calculator

ModelBuilder may be easy to use and versatile, but it is impossible to embed logical expressions in the modelling process, as all the input layers are treated universally without imposing any conditions on them. Hence, it is unable to function in case some conditions must be met before modelling can proceed further. For instance, in the RUSLE model (Equation 6.3), the surface may be covered by a lake where no soil is eroded or lost to erosion. In this case, there must be a conditional statement to check whether the surface is covered by water. If so, the output should be defaulted to 0. Such spatial operations involving conditional statements are ideally handled by ArcGIS Raster Calculator. This tool allows flexible conditions to be specified in the modelling process. Three types of options are possible on its dialogue screen: map algebra expression, a panel of calculators and operators, and tools (including mathematical functions). The first allows the selection of layers and variables that can be operated on and the creation and execution of a map algebra expression. It is these operators that distinguish Raster Calculator from ModelBuilder, in that relational and Boolean logic expressions can be embedded in the static models. It offers a quick and easy way to implement simple static modelling. However, Raster Calculator is unable to handle complex spatial modelling involving multiple scenarios.

6.4.3.3 IDRISI TerrSet

This is a desktop raster geographic information and image processing system developed by Clark Labs at Clark University. The latest version, IDRISI TerrSet (Version 18), contains a Land Change Modeller for analysis, modelling, and visualisation of land cover data. It allows land change between two land maps of different times to be analysed, based on which future land cover is predicted via simulation. Apart from rapidly analysing land cover change, this modeller can also identify drivers of the change, and construct empirical relationships between land cover change and the explanatory variables. The simulation of future land change is based on the identified trends of land changes in different scenarios, in which Markov chain simulation is used to estimate the expected quantity of change. The land cover transition potential is expressed as a likelihood determined using several machine learning algorithms, including neural network, logistic regression, and Random Forest, with full information on the explanatory power of every independent variable provided. This modelling platform enables two scenarios of prediction: a hard prediction based on a multi-objective competitive land allocation model, and a soft prediction that outputs vulnerability expressed as a continuous attribute value. This robust platform of land cover simulation takes advantage of IDRISI's suite of remote sensing and GIS functions and analytical tools, accessible as an extension of ArcGIS. It can only run predictive modelling in which time is absent. In addition to the Land Change Modeller, IDRISI TerrSet includes the Habitat and Biodiversity Modeller, a toolset for modelling species distribution based on observed locations and bio-climate variables. As the names suggest, the domain of modelling applications is highly restricted with IDRISI TerrSet.

6.4.3.4 Scripting

When special models need to be incorporated into a GIS to take advantage of its data integration and analytical powers, they have to be scripted, or implemented via scripts that have been written by others. Scripting is particularly crucial in running multi-scale modelling in which only model parameters need to be modified from one form to another. There are a few scripting languages for running spatial modelling, one of which is Python, an object-oriented, high-level programming language with built-in data structures of dynamic syntax. Its attractiveness for spatial analysis and modelling lies in its ability to join existing components. It is very easy to learn scripting with Python, as its syntax is highly readable. More importantly, thousands of third-party Python modules have been written and made available publicly, so the task of creating a fully functioning Python model is eased as some of them can be imported into a script directly. For instance, PyLUR is a Python script that can be used to run LUR modelling (see Section 6.6.2). Similar to ModelBuilder, Python scripts are also a means of streamlining modelling workflow and improving productivity. This powerful scripting language is particularly efficient at processing high-level data structures.

Another resource that has found applications in spatial modelling is ArcScripts, (http://codesharing.arcgis.com/), a website designed for users to share their scripts written specifically for Esri software packages. Of the many online sources in existence, it is by far one of the most comprehensive and well-organised sites. Scripts and models can be searched online by language, software, and keywords, or a combination of any of them, including Python scripts and ModelBuilder models. Some scripts may have been written in Arc Macro Language or Avenue for ArcGIS.

Compared to existing commercial packages, scripting offers a higher degree of flexibility and full automation of repeated modelling. It also enables modelling to be carried out efficiently. The downside is the need to learn the script language and how to detect bugs in the scripts (e.g., debugging). As scripting is gaining more and more popularity and acceptance with spatial modellers, the task of scripting a model executable in a wide range of commercial platforms will become ever easier.

6.5 SPATIAL MODELLING IN GIS PACKAGES

6.5.1 SPATIAL MODELLING AND GIS PACKAGES

There is a close relationship between spatial modelling and GIS technology. Spatial modelling relies on the use of spatially referenced data, spatial analytical functions, and spatial data manipulation operators, all of which can be found in a GIS. A GIS can supply a comprehensive range of spatially referenced data vital for spatial modelling (Figure 6.12). It enables spatial modelling to be carried out easily owing to its powerful analytical and data manipulation functions for preparing inputs to spatial modelling (e.g., coordinate system transformation and standardisation of raster grid cell size). For example, GIS proximity analysis (buffering and intersection) can identify the potential range of influence prior to overlay analysis, and GIS recoding can categorise suitability by excluding irrelevant areas, such as the exclusion of urbanised areas as potential habitats. Certain types of spatial modelling can be undertaken efficiently in a GIS in chronological order of analysis involving variables represented as GIS layers. In particular, GIS modelling is able to quantify, unify, and formulate functions, and integrate data collected by different agencies at multiple levels, ranging from local, regional, and national to international, into a single system (Maidment and Morehouse, 2002). The GIS approach is particularly strong at running cartographic modelling by combining analytical functions. Besides, GIS software offers powerful graphic capabilities excellent for visualising the modelled results almost instantaneously. Although relatively easy to implement,

Land cover

Topography

Transport

Population

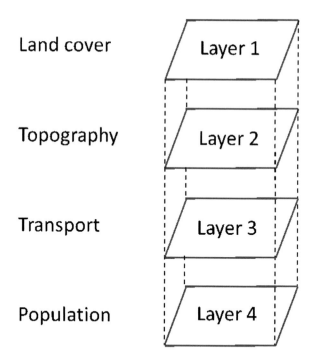

FIGURE 6.12 Conceptualisation of static GIS modelling involving four variables represented by layers that must be registered to the same coordinate system and cover the same spatial extent, and have the same spatial resolution in the case of raster layers.

GIS modelling has limited specialised functionality and applicability, as will be covered in Section 6.5.2.

In spatial modelling, careful considerations need to be given to spatial extent, spatial resolution, and temporal resolution if relevant. Spatial extent is determined by the area under study. All the input layers used in the modelling are required to cover exactly the same spatial extent. This can be achieved by clipping them against a common boundary layer created in accordance with the size and shape of the study area. Spatial resolution is applicable only to raster data. Determination of the most appropriate spatial resolution to use is a thorny issue that requires thoughtful planning. While a fine resolution allows the modelled result to be detailed, it also increases the computation intensity and data volume, which rises exponentially with resolution. At a fine resolution, the data volume will quickly become unmanageable during data analysis and modelling, especially when a larger number of variables are involved in the modelling. To a large degree, the best spatial resolution to adopt is dictated by data availability, especially publicly available imagery and DEM data. If possible, the advice is to adopt the coarsest resolution possible that is still able to meet the modelling needs. Temporal resolution or increment is determined based on the speed at which the dependent variable changes. It can range from seconds in fire spread modelling to decades in urban sprawl modelling.

6.5.2 PROS AND CONS OF GIS MODELLING

Since spatial modelling inevitably involves spatial data that are ideally represented as layers in a GIS, by default such systems are the ideal platform for undertaking a wide range of spatial modelling, for four reasons:

(a) The rich geospatial data stored in a GIS database have the same format residing in the same platform, all of which can be read and manipulated by the GIS. There is no need to change data format or port them from one system to another, or transfer data between different analytical and modelling platforms. This is especially handy when spatial modelling involves data from diverse sources, some of which are in the point format, while others are areal or linear data. All of these data can be easily integrated and saved in the same folder for performing spatial modelling. The ability of a GIS to integrate spatial data from diverse sources, in different formats, structure, or at different spatial resolutions, is a powerful aid to spatially distributed models. A centralised data clearinghouse is easier to upgrade than those widely scattered across different systems.

(b) Some GIS packages have powerful functions and specially designed modules that enable spatially explicit and distributed models to be implemented effortlessly. Certain GIS software may have functions for spatial statistical analysis. Special functions may be available for carrying out a range of standard spatial operations, such as identification of centroid of a polygon, random selection of sample points to create a subset, and so on. Some GIS functions can ease environmental modelling, and support alternative data models, particularly models of continuous spatial variations, and conversions between them using effective methods of spatial interpolation.

(c) The modelled results can be visualised graphically in real time and integrated with data from other sources for further analyses. For instance, after the spatial distribution of air pollutants in a city has been modelled, it can be integrated or overlaid with census data to identify the proportion of elderly residents who are particularly vulnerable to them.

(d) GIS software has more user-friendly interface than other systems. It is equipped with a systematised methodology for modelling, enabling the modelling process to be repeatable and efficiently implemented.

GIS-based modelling eases the task of modelling because the modeller only needs to know how to use the GIS. No developer expertise is necessary whatsoever. This kind of modelling can easily be implemented in a number of platforms, especially ArcGIS Pro and ERDAS Imagine. GIS packages are effective for maintaining and querying a static database for static phenomena. As such, they are good at implementing relatively simple, static modelling if all the model operations are part of the GIS functionality. In fact, all spatial modelling can be implemented entirely within a GIS if no external models are involved. If they must be used, the models need to be mathematically or conceptually simple. Alternatively, they have to be integrated with the GIS software as extensions (see Section 6.5.3.3).

The major limitation of GIS packages in running spatial modelling is their lack of capability in constructing dynamic simulation models because they can only store static information in a discrete-time structure. Thus, they have limited capacity to handle spatial dynamic models and to deal with the temporal dimension of variables in the modelling. In addition, niche or tailor-made models that are required in some specialised domains of spatial modelling may be absent from or impossible to be embedded into existing commercial GIS packages that are designed for generic modelling only. For instance, special ecological and hydrologic models are currently absent from mainstream commercial GIS packages in simulating processes over time. As the field of spatial modelling evolves and develops, more specialist static models can be run from a GIS package in the form of a separate toolbar or an extension. At present, GIS software has limited capabilities in carrying out sophisticated statistical analysis that forms an integral part of certain modelling, such as collinearity testing in screening variables to be included in a model. These deficiencies can be overcome by coupling GIS packages with spatial models.

6.5.3 Coupling of Spatial Models with GIS Software

Occasionally, spatial modelling requires the use of sophisticated and spatially explicit models in certain applications, such as hydrological and air pollution modelling. These models are dynamic, and involve parameters and variables that change with time. They are cumbersome to run in a GIS package because current ones do not explicitly allow for dynamic phenomena to be stored and analysed. Some spatial modelling requires the capability for statistical analyses beyond what is offered by current GIS software. These statistical systems have their own high-level functions for performing data-analytic tasks.

At present, GIS packages are inadequate in incorporating independent models in certain fields into spatial modelling, even though their capabilities have been improved in recent years with more software packages being available. Take ArcGIS Pro as an example. It is a mature and powerful system with its own data model, operating mechanism, and user interface, but it lacks the ability to incorporate complex external models. If the required data analyses go beyond simple regression, more sophisticated external statistical packages such as S-Plus and SAS are needed to perform a variety of stochastic modelling tasks. In order for a spatial model to be used in spatial modelling, it must be integrated with a GIS package. This coupling can take advantage of the strong capabilities of GIS software in handling diverse spatial variables.

A feasible solution to overcoming the lack of dynamic functionality of GIS software is to employ external computer packages to implement the dynamic or complex spatial modelling outside the GIS environment, and then link the output to it. For example, in watershed hydrological modelling, the relationship between surface runoff and rainfall intensity may have to be constructed outside a GIS package, and the established relationship later manually incorporated into the GIS software via special scripts. Although the analogy between the two systems may be valid, there is an important difference between them. The statistical package supports only one basic data model, the table, with one class of records, whereas the GIS package supports a variety of models with many classes of objects and relationships. The frequent transfer of files between a GIS package and an independent package is tiresome, inefficient, and time-consuming. Besides, the data may not be fully compatible in format. Current spatiotemporal models are cumbersome or impractical to run inside a GIS package unless the required data and models can be integrated into one cohesive system. On the other hand, it is vitally important and necessary to integrate GIS software with external models. This is because GIS packages offer a flexible environment for simulation with their standardised array of spatial operators based on mathematical principles that describe the motion, dispersion, transformation, or other meaningful properties of spatially distributed entities.

The integration of specialist modelling systems with GIS software will make the whole of both systems larger than their sum, and can considerably ease spatial modelling and widen its applicability to more fields. The strategies for integrating GIS packages and spatial modelling can be viewed from three perspectives: the technical or programmer's perspective, the functional or user's perspective, and the conceptual perspective. When evaluated from the functional perspective, these systems are shown to support only a restricted form of human–computer interactions. The integration can take place at three levels: loose coupling or supportive but separate, tight coupling or linked, and full integration (Figure 6.13). Their properties are compared in Table 6.12, including typical examples of integration.

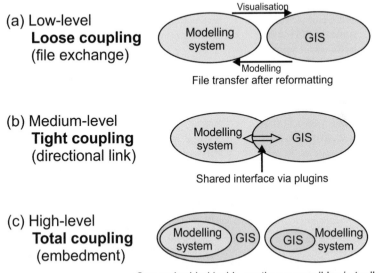

(a) Low-level
Loose coupling
(file exchange)

Visualisation
Modelling system   GIS
Modelling
File transfer after reformatting

(b) Medium-level
Tight coupling
(directional link)

Modelling system   GIS
Shared interface via plugins

(c) High-level
Total coupling
(embedment)

Modelling system GIS   GIS Modelling system
One embedded inside another, accessible via toolbars

FIGURE 6.13   Coupling of GIS and spatial modelling systems at three levels using various strategies (source: modified from Alcaraz et al., 2017).

TABLE 6.12

The Main Features, Pros, and Cons of Coupling Spatial Models with GIS at Three Levels

Level of coupling	Low (loose)	Medium (close and tight)	High (full)
Main features	Separate and distinct systems supportive of each other; separate user interface; frequent file exchange between the two systems; data reformatting vital	Each system performs certain tasks that it is best at; unified user interface; shared data and model base accessible via linkages between two separate systems	Integrated simulation model modification, query, and control; one embedded into another; steerable simulation
Means of coupling	Data reformatted before transfer (e.g., input to and display in GIS)	Linkage; plugin-based toolbars	Scripting of models; one embedded in another; extension
Pros	Less time needed to develop than full integration; use of proven models provides reliable components for spatial modelling; different tools and libraries facilitate independent system development; no need to know both systems well	Automatic data exchange between two systems; inconsistency and data loss avoided; faster modelling; no data redundancy; easier data maintenance	Fully automated; changes easily made to parameters; no need for data conversion; no data redundancy; rapid development of new models; easy maintenance of models
Cons	Frequent change in data format prompted by a new version of GIS software; inefficient for repeated model runs; slow and laborious; error-prone and inflexible	Open GIS structure needed; low-level approach to model formation	Not generically available; costly to develop; lack of specialist knowledge about model may lead to inaccurate outcomes; difficult to alter source codes due to embedment
Examples	AERMOD (Maantay et al., 2009); ArcArAz (Alcaraz et al., 2017); DRASTIC (Jaseela et al., 2016)	PyLUR (Ma et al., 2020); Minitab (Xie et al., 2005); PIHMgis (Bhatt et al., 2014)	AVTOP (Huang & Jiang, 2002); FRAGSTATS (McGarigal et al., 2012); ArcHydro (Maidment and Morehouse, 2002), HEC GeoHMS (US Army Corps of Engineers)

Source: Modified from van Deursen (1995) and Bhatt et al. (2014).

6.5.3.1 Loose Coupling

At this first level of integration, proprietary GIS software is loosely coupled with another package such as statistical software, either proprietary or purpose-built. In this primitive coupling, the modelling and GIS packages are separate but supportive of each other. Loose coupling is necessary when the inputs to the spatial models have a format that is not compatible with or recognised by the GIS (Maantay et al., 2009) and the required inputs are spatial in nature, or when the model is absent from the GIS. This integration is commonly achieved by coupling the input/output routines of the GIS into the model, allowing the model to read and write to the GIS in its native format. In the integration, the GIS can lend support in getting the data into the right shape and format before they are acceptable as the model input. For instance, the input data may be reprojected to another coordinate system, resampled to another cell size, or standardised to the same desirable spatial extent via clipping. Then the special modelling system is activated to run the model, and the modelled results are transferred back to the GIS for visualisation and overlaying with other layers, such as air pollutant distribution with road networks to visualise their association.

A good candidate for loose coupling with GIS software is the external cellular automaton model (Clarke and Gaydos, 1998). Owing to their identical data structures, both systems can interact with the same database. The absence of intermediate conversion steps allows speedy application and the development of interactive applications. A common interface links the GIS software and the modelling process, allowing the transfer of files between the two systems. The modeller has the option of running the analysis and/or modelling in either system.

Loose coupling is advantageous in that the spatial modeller does not have to know anything about the GIS or its data structure and organisation (Figure 6.13a). Loose coupling is quite practical due to the minimal programming required. It offers a fairly high degree of flexibility in processing data of any structure. Since it is not bounded by the GIS, this modelling approach can be adapted for any other systems for further analysis and integration (e.g., with statistical packages) (Table 6.12). The drawback of loose coupling is its inefficiency, as modelling cannot be executed by clicking a few buttons, and every modelling effort requires duplicated fiddling with both systems, so it is not replicable and efficient, especially if the modelling has to be run repeatedly. If a particular spatial model (e.g., a fire spread model) is not available from existing mainstream spatial analysis packages, its incorporation into the modelling process can be achieved only through close coupling or full integration.

6.5.3.2 Tight Coupling

Tight coupling is also known as close coupling or seamless integration, in which the spatial modelling and GIS packages are still separate and distinct, each performing certain specific tasks (Figure 6.13b). Common candidates for coupling are related to statistical packages for spatial logistic regression analysis and identification of spatial predictor drivers, while the GIS package is used to perform spatial analysis on the modelled outcome. Some kind of interface is established between the two systems via a plugin toolbar. A reasonably seamless link between GIS and other packages can be created via special scripts. This linkage avoids direct modification, yet uses operations already present in both systems. It is an appropriate solution for combining complex spatial database and statistical models. It also resolves the data compatibility and user interface issues, so that files can be easily shared between different systems (e.g., all graphic files can be saved as GeoTIFF). However, such linkage does not allow access to fully integrated functionality. This medium level of coupling combines the capabilities of separate modules for GIS functions and environmental simulation. It is effective for pre-processing data or displaying results, but may require investment of substantial resources to build the necessary linkages between the two systems (Steyaert and Goodchild, 1994).

Linked integration, as exemplified by GeoHMS, requires data portability between the two systems, as data structures, modes of operation, and types of user interface in both systems are typically very different and do not lend themselves to comprehensive integration. Data export from and import to either of the systems take place in a way that is transparent to the modeller. Certain processing is done in one system, while others are done in another. For instance, S-Plus builds tree-based models using spatial data in the GIS database, and returns the modelling results to the GIS via the seamlessly communicating Arc Macro Language (AML) program in the GIS (Bao et al., 2000). Another good example of close coupling is PIHMgis integrated with a GIS. PIHMgis's user interface can be accessed from dropdown menus in a plugin-based toolbar (Bhatt et al., 2014). In this integration, PIHMgis supports organisation, development, and assimilation of spatial and temporal data into a hydrologic model, while the GIS provides the platform for display, navigation, and editing of geospatial data layers needed for the modelling. Both systems have a shared user interface, data structure, and method. To a large degree, the hydrologic model is virtually a suite of analytical tools of the GIS. It is not considered full integration, as it is still impossible to modify the modelling parameters inside the GIS, since this is only possible inside PIHMgis.

Tight coupling of GIS and statistical software requires moderate scripting using a programming language that supports modelling within the GIS. Examples include AML, Avenue in ArcView, and more importantly, Python. Such coupling is system-dependent and can accelerate the process of modelling, especially if it has to be run repeatedly.

6.5.3.3   Total Coupling

Also known as complete coupling or high-level integration, full or *total integration* is achievable by embedding the spatial model or statistical analysis system in a GIS, or vice versa (Figure 6.13c). The former is preferred because other GIS functions can be taken advantage of in the modelling. One way of embedding the spatial model to a GIS is to build functionality or extension into it, accessible via a toolbar. Such special extensions can run as a part of the GIS itself. The two systems are used as if they were one, and the modeller is oblivious to their distinction as both are treated as a unit.

This integration is at the highest level of sophistication, because embedded coupling allows running a model directly within a GIS. At present, such embedded coupling into a GIS is practical for finite difference models that can take advantage of the functionality of raster GIS for cell-based modelling, and more spatial analysis functions have been integrated into raster GIS, such as IDRISI, than vector GIS packages. Common examples include AVTOP, FRAGSTATS, ArcHydro, and the Combined Hydrology And Slope Stability Model (CHASM) (Table 6.12). AVTOP represents a GIS realisation of TOPography-based hydrological MODEL (TOPMODEL) using ArcView's macro language, Avenue (Huang and Jiang, 2002).

Initially proposed by Beven and Kirby (1979), TOPMODEL is a physically based watershed model for simulating streamflow based on the variable-source-area concept. Stream discharges are predicted from DEM data and a sequence of rainfall and potential evapotranspiration data. The prediction of distributed hydrological responses is computationally efficient, requiring a relatively simple framework centred around the use of DEM data (Beven, 1997). This full coupling of hydrological models within the GIS environment capitalises on the GIS visualisation and spatial analysis functions, thereby significantly facilitating dynamic hydrological modelling in the entire process of model development, from the initial parameterisation and data transformations to the visualisation of intermediate and final results. This full integration enables the modeller to visually interact with the model and adjust its parameters conveniently in a single platform. The effect of the adjustment on the results can be viewed on-screen instantly, which significantly facilitates exploratory data analysis and decision-making vital for the success of hydrological modelling.

FRAGSTATS is a plugin software package accessible as an ArcGIS extension. It contains an extensive range of routines for spatial analysis. However, the analyst must be very familiar with both the GIS and FRAGSTATS. At one time accessible separately from and outside GIS packages, FRAGSTATS can now be run through GIS look-alike modules customised to descriptive analysis of geospatial data. The input to FRAGSTATS is an output of a GIS analysis (e.g., classified land cover map). The ArcHydro toolbox (toolbar) is an extension of geodatabase models for water resource applications. Its suite of tools supports geodatabase design and basic water resources analyses, such as terrain analysis, watershed delineation and characterisation, and tracing and accumulation through channel networks. CHASM is another example of a fully integrated package for modelling slope hydrology/slope stability. This two-dimensional landslide model can be implemented within a GIS that can supply the required input data, such as geological map, drilling, geophysical investigation, and hydrological monitoring data, all represented as GIS layers in either point or areal format (Thiebes et al., 2013). This Web-based GIS allows the user to carry out limit-equilibrium analyses of slope stability quickly by selecting a variety of input data and model parameters. The simulation results are automatically saved and can be visualised in a GIS for interpretation.

The key to the success of full integration is to embed the mathematical model explicitly into a GIS via purpose-design programs. Programming can be carried out using a variety of scripting languages. In the earlier days, AML was used, later replaced by ArcView Avenue. These days the most popular language is Python, which has found wide applications in implementing spatial models and automating spatial modelling. As more and more models have been developed and incorporated into ArcGIS as extensions, full integration of GIS and spatial modelling will find wider applications with ever-increasing ease, and will play a vital role in certain applications (e.g., specialist numerical modelling). This high level of integration can bring quite a number of benefits, including, but not limited to, accelerated processing and higher efficiency of modelling. Precious time can be saved with less frustration in dealing with multiple systems that may not even be compatible. It provides centralised access to a common, singular platform for both users and maintenance staff.

This coupling allows the use of systematic and standardised methodology and datasets in the modelling, creating a high degree of repeatability. It provides more diverse user interfaces that can simplify modelling. In addition to the common graphical user interface, interactions with the systems can be expanded to include a map-based interface so that any unreasonable outcome can be detected visually early on, to save time. Other generic advantages include:

- Flexible ability to geographically visualise raw and derived data;
- Provision of flexible spatial functions for editing, transformation, aggregation and selection of both raw and derived data;
- Easy access to spatial relationships between entities.

6.6 SPECIAL TYPES OF MODELLING

6.6.1 (SPATIAL) LOGISTIC REGRESSION

Counter-intuitive to its name, *logistic regression* is a statistical approach to describing data and explaining the relationship between a dependent variable and its explanatory variables, whose scale of expression can be nominal, interval, or ratio. Nominal data may be about land covers (e.g., urban, forest, grassland, barren, etc.), categorical data may be related to the modes of transport (e.g., walk, cycle, bus, car), and binary data may depict state (e.g., flooded or intact). All of these attributes must be mutually exclusive. For instance, if an area is covered by forest, it cannot be classed as urban. Urban forest is permissible, but not at the same location as either urban or forest. Logistic regression models are the best at modelling variables that have a linear binomial distribution. As such, they are perfectly suited to modelling land cover change in which there are only two possibilities: change and no change. In its simplest form, a logistic regression model can be expressed as a logit function if the dependent variable is binary. Hence, it is also known as *logit regression* or the auto-logistic model. Logistic regression modelling is highly flexible, in that any number of independent variables can be included in the constructed model as long as they can drive changes of the dependent variable or constrain it (e.g., areas where changes are highly restricted or not permitted).

In logistic regression, the logit function can be expressed as:

$$\text{logit}\left(\overline{p_i}\right) = \log\left(\frac{\overline{p_i}}{1-pi}\right) = \text{logit}\left(\frac{e^y}{1+e^y}\right) \tag{6.13}$$

where $\overline{p_i}$ denotes the expected probability of change at location i and y is a function of the independent variables, expressed as:

$$y = b_0 + b_1 x_1 + b_2 x_2 + \ldots + b_n x_n \tag{6.14}$$

in which x_i represents the ith independent variable, n refers to the number of independent variables being considered, and b_i represent their regression coefficients to be estimated from the observed data that are assumed to be independent of each other. No normal distribution of the data is assumed or required for logistic regression.

In spatial modelling, logistic regression is commonly used to develop the prediction model between the dependent variable (e.g., the spatial extent of snow cover) and its explanatory variables (e.g., unidirectional prevailing wind directions) (Cline, 1992). The developed logistic regression models are non-spatial. They become spatial logistic regression models if the independent variables x_i are spatial in nature, such as land covers represented as a GIS layer. Spatial logistic regression has been commonly used to model rural–urban land conversion in which the outcome has a binomial distribution (Xie et al., 2005). In order for it to function well, the land cover change variable that has been detected from historic land cover maps must be incorporated into the model as one of the independent variables. Logistic regression is particularly suited to modelling land use change because how exactly land use is influenced by its drivers remains mostly unknown, or indescribable mathematically. The best way to approach the influence is via the statistically established driver–change relationship, which can be accomplished through the integration of spatial statistical packages with GIS packages, as discussed in Section 6.5. Independent variables (drivers) commonly considered in logistic regression modelling include topography (slope), soil, geology, roads, current land cover, unit land price, population, and land use policies covering protected ecologically sensitive areas and geohazardous areas. In this kind of modelling, there is a need to consider the spatial dependency of such variables. Spatial dependencies among the considered drivers can be addressed via two methods: incorporation of an autoregressive structure into the model, which inevitably, increases model complexity, or using spatial sampling (Xie et al., 2005). The latter can eliminate spatial auto-correlation to yield more reliable parameter estimation. However, spatial sampling will cause the sample size to drop and lead to the loss of certain information, contravening the large-sample desirability underpinning the maximum likelihood algorithm commonly used to develop a logistic regression model in the first place. The negative impact of a small sample size can be minimised via a reasonably designed spatial sampling scheme. Information on different methods of spatial sampling can be found in Section 3.1.

6.6.2 Land Use Regression Modelling

Similar to logistic regression modelling, *land use regression modelling* is also a statistical approach to estimating the attribute of a given variable in the 2D space, in a manner remarkably reminiscent of spatial interpolation except that the unknown value at the specified location is generated from a pre-established regression model involving the most influential independent variables. Strictly speaking, LUR is a non-spatial modelling approach that is underpinned by a regression relationship between the dependent variable and a plethora of independent variables, all of which are invariably spatial in nature. However, once established, the regression-generated LUR model enables the value of the dependent variable to be estimated at any specified locations or regularly spaced lattices. It is called LUR modelling because all the independent variables in the regressional analysis are related to surface features, such as land covers, road networks, and topography. Since these variables are usually represented as GIS layers and stored in a GIS database, LUR modelling is most effectively implemented in a GIS or in a system closely coupled with it to take advantage of its analytical functions, such as proximity analysis.

In LUR modelling, the relationship between the dependent variable y and a number of exploratory variables x_i is expressed linearly as in Equation (6.14). The construction of this empirical model is based on the observed values of the dependent variable at tens of sites. To yield a reasonable prediction, these sites should be widely scattered spatially throughout the area under study. The observed values at these sites should also be representative of the whole range of the attribute value, otherwise the constructed LUR model's predictive power will be compromised through overestimation of the minimal values and underestimation of the maximal values.

The empirical model in Equation (6.14) is normally constructed using the supervised learning method in five steps:

(1) A total of k regression equations are built by regressing the dependent variable against each of the considered explanatory variables via univariate regression, leading to k regression models. Of these models, the one with the highest R^2 value is deemed the potential "seed model". This status is confirmed if the included explanatory variable in the model has the same effect as its pre-defined criteria.

(2) Each time one of the variables from the pool of variables is added, in turn, to the "seed model" with the R^2 of the newly constructed model noted. This process is iterated until all the variables in the pool have been added to the newly built models except the one that has been included in it already.

(3) The R^2 values of all the updated models is compared, and the one whose R^2 value increased the most over its previous counterpart is considered as the final "new model" if it meets two criteria: (a) the increase in R^2 value exceeds the pre-defined threshold (e.g., >2%), and (b) the direction of the effect of every explanatory variable is consistent with that of its pre-defined criteria. Otherwise, the explanatory variable under consideration will be removed from the new model.

(4) Steps (2) and (3) are iterated for all the remaining variables one at a time. Namely, they are sequentially inserted into the current model if their inclusion in it causes its R^2 to rise by more than the pre-defined threshold (e.g., >1%). If not, then they will be skipped.

(5) All the explanatory variables that have been retained in the current model are scrutinised for their p-value. Any variables whose p-value exceeds the pre-defined threshold (e.g., >0.10) are sequentially removed from the current model, starting from the variable with the highest p-value. This elimination process continues until all the variables retained in the current model are statistically significant, and the newly created model is regarded as the final model.

In the case of a large variable pool, some of the variables can be eliminated outright to expedite model construction via pre-processing, such as calculation of correlation.

A simple criterion based on the correlation coefficient between the dependent and each of the independent (predictor) variables can be applied to screening the explanatory variables. If the correlation coefficient falls below a certain threshold, then this predictor variable can be safely excluded from consideration. After the potential LUR model is finalised, it may still be necessary to subject it to various checks for collinearity among the retained variables and their spatial auto-correlation. The topic of the multi-collinearity test was covered in Section 6.3.2. Any predictor variables with a VIF above the defined threshold are deemed unacceptable, and should be removed from the model. Spatial auto-correlation (normally $p > 0.05$) can be checked against Moran's I (see Section 4.5.2).

After the finalised LUR model has been successfully validated, it can be applied to predicting the value of the dependent variable at any points of interest. This is when LUR modelling becomes spatial, as the modelled output is extended to the spatial domain. A spatial distribution map of the dependent variable can be produced over an area if the predictions are made at regular lattices (Ma et al., 2020) (Figure 6.14). There is no limit on the spatial resolution of the output that affects only the data volume and speed of processing, especially for a large study area. A finer resolution output is feasible only when the study area is small (e.g., tens of km²). In implementing LUR modelling, it may be necessary to determine the sphere of influence of a potential variable through proximity analysis (see Table 6.10). In spite of the gradual attenuation of the influence of a predictor variable (e.g., degradation of noise level with distance), such analysis has to be performed at a certain number of representative, discrete buffers to keep the analysis manageable (Tables 6.10 and 6.13).

Originally, LUR modelling was developed to assess exposure to vehicle-originated air pollution, usually in densely populated urban areas, but has been expanded to study air pollution epidemiology (Ryan and LeMasters, 2007). LUR is particularly

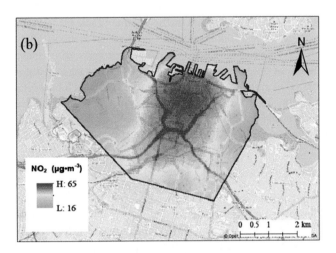

FIGURE 6.14 Spatial distribution of vehicle-originated NO_2 in central Auckland, New Zealand, predicted from LUR modelling at 50 m resolution (source: Ma et al., 2020, with permission).

TABLE 6.13
Predictor Variables Commonly Considered in LUR Modelling of Air Pollutants and Their Attributes

Predictor variables	Specific attributes
Land use	Residential, industrial, and urban greenfield
Urban configuration	Buildings (quantity, height, area and volume); sky view factor
Transport infrastructure	Distance to bus stop, train station, ferry terminal, and the coast; number of car parks
Population*	Height of residence; number of households and inhabitants
Road network	Number of road intersections; distance to the nearest road intersection/main road; length of all main (major) roads; total (heavy-duty) traffic load of major roads
Traffic load and volume*	Total (heavy-duty) traffic load of all roads; length of all roads; heavy-duty traffic volume on the nearest (major) road; distance to the nearest (major) road; traffic volume on the nearest (major) road

*Apart from population, traffic load, and volume, buffer analysis of various radii is needed to identify the sphere of influence for all other linear variables (e.g., distance to the nearest major road).

useful in assessing the spatial distribution of traffic-induced pollutants at a local scale. In LUR modelling, the spatial distribution of the air pollutant under study is confined to the Earth's surface, even though it is intrinsically 3D in reality. This is inconsequential, as human exposure to pollutants takes place as people traverse the surface of the Earth. The variables commonly considered in LUR modelling fall into a number of categories, such as land use, urban architecture configuration, buildings, transport facilities, road networks, and traffic volume (Table 6.13). Some of them are related to the source of the pollutants, while others affect their spatial dispersion and diffusion in the 2D space. All of them have different units, directions, and spheres of influence. Some of the variables in Table 6.13 will not be relevant if the air pollutant originates from a plant or exposure is studied at a much broader scale beyond the urban landscape. In this case, the process governing the spatial dispersion of pollutants in the air must be taken into consideration, such as the predominant wind direction and speed, as well as atmospheric turbulence.

6.6.3 HYDROLOGICAL MODELLING

Hydrology is the field studying both surface water and ground water, of which surface water is much easier to study. Surface water hydrology is concerned with how rainwater is distributed spatially over a hill slope and in a catchment, and how much rainwater associated with a rainfall event eventually ends up in a channel. Both tasks are intrinsically spatial and highly complex, as they involve a large number of factors related to water, land, air, and the biosphere. It is hence highly challenging to carry out hydrological modelling, because it requires consideration of a number of processes that may be interacting with each other. On the other hand, hydrological modelling can yield information that is vitally important in mitigating the risks of flooding. Hydrological modelling is significant as it can identify areas that are highly vulnerable to potential flooding so that flood-prone areas can be protected from urban development and road infrastructure can be better designed to minimise flood damage (e.g., bridges must be constructed at certain heights above the highest water level). Hydrological modelling can predict the quantity of rainwater generated from a watershed following a rainfall event of a certain intensity (e.g., a 50-year rainfall). Dissimilar to hydrological modelling, *hydraulic modelling* aims to predict the amount of surface runoff that ends up in a river channel, the speed of channel flow, depth of flow water, areas to be inundated by the rainwater, and the peak flow and its arrival time. Such information is crucial for emergency evacuation and disaster prevention.

In hydrological modelling, the explanatory variables are related to climate (rainfall and evaporation), soil, topography, and land cover. Soil is important to consider because it affects the rate of water infiltration. The antecedent soil moisture governs how much the rainwater can be absorbed by the ground and when its moisture content reaches saturation. Represented as a DEM, topography dictates the velocity of water flow over a surface, flow direction and length, and where the surface runoff converges and is discharged to (i.e., within the catchment or outside it). Land cover data that can be derived from satellite images are needed to determine how much the rainwater is intercepted by the vegetation canopy and the proportion of rainwater that percolates down to the ground surface. Since such data are commonly stored in a GIS database, naturally, GIS packages provide the default platform for hydrological modelling, even though they have a limited but expanding range of hydrological models for this purpose, such as those depicting the relationship between rainfall and runoff.

The modelling of the quantity of rainwater ending up in a channel is essentially to budget water in a catchment involving input and output. The former takes the form of mostly rainwater as precipitation (and occasionally snowmelt). Water can be lost from a catchment via infiltration, evapotranspiration, surface and subsurface flows, and outflow to external areas. Each process must be considered carefully to yield a viable prediction. Essential in the modelling is the reliance on physically based hydrological models. They are grounded on scientific knowledge about energy and water fluxes. In physically based hydrologic modelling, the hydrologic process of water movement is modelled by approximating the partial differential equation representing the mass, momentum, and energy balance, or via empirical equations (Abbott et al., 1986). For instance, the relationship between streamflow and basin characteristics can be established statistically via regression analysis.

Surface hydrological modelling is usually preceded by catchment analysis, which can yield vital information needed to run surface hydrological modelling in a number of steps (Figure 6.15). Catchment analysis is usually performed on a DEM, in which special geomorphological features such as pits are identified. Surface analysis may also include the determination of slope aspect, gradient, flow direction and length, drainage area, and delineation of the catchment boundary. Flow accumulation can

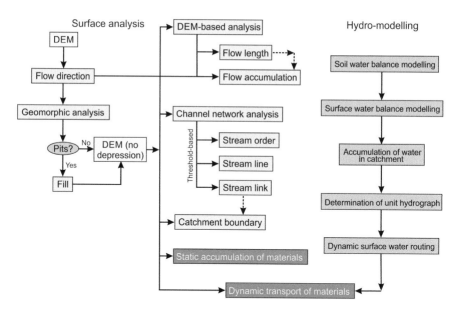

FIGURE 6.15 Main components and procedure of catchment-level surface analysis (blue boxes) and hydrological modelling (green boxes) using DEM. Dark blue boxes show modelling of silts and sediments in a channel flow not part of hydrological modelling (sources: modified from van Deursen (1995) and Djokic, https://proceedings.esri.com/library/userconf /proc15/tech-workshops/tw_382-228.pdf).

be derived from flow length. All of these analyses can be easily implemented using currently available GIS tools. Another major component of surface analysis is the detection of channel network and ordering, which is indispensable for flow routing. The last two analyses in Figure 6.15 (dark blue boxes) are not related to hydrological modelling *per se*. Instead, they are designed to model materials carried by river water, such as sediments.

After this preparatory procedure that addresses how the rainwater flows from the slope to the channel, it is time to run surface hydrological modelling. It comprises a few types of modelling, including soil water balance modelling, surface water balance modelling, and accumulation of water in catchment modelling, all of which require the use of mathematical models. Peak flows can be modelled from dynamic surface water routing and transport of materials (e.g., sediments) on the basis of the calculated unit hydrograph.

So far, many models of vastly varying complexity have been developed, each addressing one or more components of the hydrological cycle. The earliest system is PCRaster by van Deursen (1995), with ArcHydro commonly used at present (Table 6.14). PCRaster can be used to model the effects of climate change on river discharge (including sediments), but it is difficult to use, as documentation is sketchy and the system is not well supported by the user community. The Watershed Modeling System by AQUAVeo is a modelling environment designed to perform

TABLE 6.14
Computer Packages/Systems for Undertaking Hydrological Modelling and Their Main Features

Water type	Name	Main features	Author(s)
Surface	PCRaster	Able to model the effects of climate change on river runoff; supports transport process modelling	van Duersen (1993)
Surface and sub-surface	GSSHA	Able to model surface water and groundwater hydrology, erosion and sediment transport; diverse inputs (meteorological, surface energy balance, seasonality in evapotranspiration); channel in vector format	Downer et al. (2006)
Surface and ground	Watershed Modelling System	Mostly for watershed analysis; floodplain modelling, storm drain modelling, and groundwater-surface water interaction modelling; flood forecasting	www.aquaveo.com
Surface and ground	ArcHydro	Modelling of precipitation-runoff responses; space and time connected; surface and ground water integrated	Maidment and Morehouse (2002)

a variety of analytical functions in implementing existing hydrological models, such as automated basin delineation, geometric parameter calculations, GIS overlay computations, and cross-section extraction of terrain data (Table 6.14). It supports hydrologic modelling and facilitates hydrologic model development using a suite of tools for watershed-level spatial analysis. The modelled output is needed to run other hydrological modellings, such as surface water and groundwater modelling using the software produced by the same company. ArcHydro contains a suite of tools for modelling runoff response to precipitation. It connects space and time via time-series HydroFeatures data, and integrates surface water with groundwater data (e.g., the relationship between surface water features such as streams and groundwater features such as aquifers) (Maidment and Morehouse, 2002). Core to it are two modules, the Hydrologic Engineering Centre (HEC) Hydrological Modelling System (HMS) for modelling the precipitation-runoff processes, and HEC GeoHMS. The latter is a GIS pre-processor for HMS in which drainage paths and watershed boundaries derived from a DEM are transformed into the hydrological data structure in preparation for the precipitation-runoff response modelling.

The modelling of channel hydraulics and groundwater interaction can be achieved using the Gridded Surface Subsurface Hydrologic Analysis (GSSHA) system. This grid-based 2D physical watershed hydrologic model was developed by the Engineer Research and Development Centre of the US Army Corps of Engineers (Downer et al., 2006). It is able to model surface water and groundwater hydrology, erosion, and sediment transport. A large number of variables are considered by the system, including factors related to meteorology, surface energy balance, and even seasonality in evapotranspiration. Although surface hydrological modelling is implemented in raster format, channels are represented in vector format in modelling channel discharge.

6.7 THREE CASES OF SPATIAL MODELLING

6.7.1 EARTHQUAKE DAMAGE MODELLING

Earthquakes are a kind of destructive natural hazard that can do tremendous damage to infrastructure and properties, and can cause a heavy loss of lives within minutes. In order to mitigate earthquake damage and minimise the losses, it is crucial to model the spatial distribution of potential damage so as to plan urban development accordingly. Earthquake damage can be caused by liquefaction and ground shaking. The former refers to the flow of moisture-saturated soil to the surface because of the sudden loss of soil strength in response to rapid shaking (Figure 6.16). It usually takes place in urban areas built on floodplains or former river beds. During an earthquake, the vibration and swing of the ground can also topple buildings and rupture the ground surface, doing further damage to bridges and roads and making them unnavigable, which hampers rapid aftershock emergency response and evacuation. Tearing apart of land and infrastructure can also burst underground pipelines and topple utility poles, further triggering fires.

The severity of earthquake damage to properties is dependent on several factors, including distance from and depth of the epicentre, the magnitude of shaking, building height and density, population distribution, and infrastructure layout. The exact location and the depth of the epicentre usually remain unknown prior to an earthquake, and is very difficult to predict. It is commonly handled via scenario analysis. Also unknown beforehand is the magnitude of shaking produced by an earthquake.

FIGURE 6.16   Kilmore Street in Christchurch, New Zealand, damaged by liquefaction triggered by a 5.8-magnitude earthquake in 2011 (source: Woodford 2011).

It is dealt with in the same way as the epicentre in modelling. The magnitude of ground shaking is measured at various scales, one of which is the Modified Mercalli (MM) Intensity Scale. Without any mathematical foundation, it is just a subjective scale of ten levels, ranging from 0 (not felt) to 10 (extreme) based on the observed effects of shaking. It is commonly thought that noticeable damage occurs starting from level 5 – moderate shaking (i.e., felt by everyone). The relationship between shaking intensity and the portion of the damage is well established in the literature, and the proportion (%) increases exponentially with the MM intensity after level 6 (Figure 6.17a). However, the damage curve varies with building type (assuming all are equally vulnerable). In general, much more loss can incur to industrial/commercial buildings than residential buildings at the same MM intensity (Figure 6.17b). This relationship between the expected loss and shaking intensity can be used to translate the damage ratio to the expected loss as a percentage value.

Reliable modelling of earthquake damage requires information on geology, especially the location of fault lines, along which an earthquake is likely to occur. A geotechnical map of the area under study (e.g., zones prone to liquefaction) is also needed. Since the damage at a location is dependent on building height and density, population/infrastructure distribution, and proximity to the epicentre, the scenario-based MM intensity must be adjusted by distance. The expected damage is then converted to a percentage of loss or replacement value based on property valuation and the number of building floors. The former can be obtained from the local council, as it must be estimated for taxation purposes. The latter can be acquired from LiDAR data. If modelled in the raster environment, the value of the loss is aggregated to grid cells (Figure 6.18). In the modelled output based on scenarios of both ground shaking and liquefaction, central Wellington and Lower Hutt in lower North Island, New Zealand will incur the highest damage due to the higher concentration of commercial/industrial buildings in these areas. The damage level gradually decreases away from these central areas towards the outer suburbs. Nevertheless, there is no way the modelled value can be substantiated or verified, even if an earthquake does eventuate. This is because the epicentre and the magnitude of shaking used in the scenario

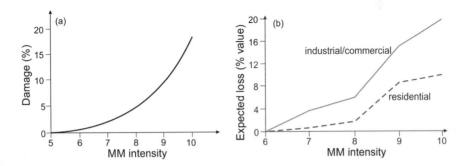

FIGURE 6.17 Conversion of damage ratio to expected loss (% value) based on the relationship between expected loss and shaking intensity (a), and the vulnerability curve for building content (b) (source: modified from Aggett, 1994).

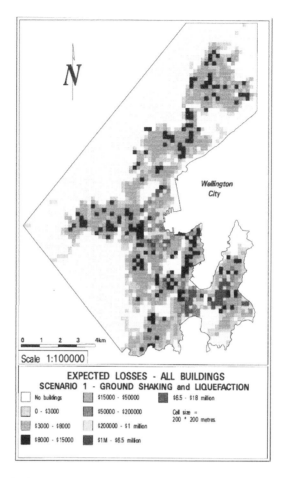

FIGURE 6.18 Distribution of expected losses to all buildings in a scenario of both ground shaking and liquefaction (source: Aggett, 1994).

analysis may not match reality closely. Since the absolute value of properties changes with time, the modelled loss in terms of monetary value may become obsolete a few years after the modelling. However, the expected loss expressed as a percentage will not change, and the modelled results can be updated easily using the latest valuation data instead of re-running the modelling from scratch.

6.7.2 Landslide Susceptibility Modelling

Landslides are a natural hazard that can destroy properties and infrastructure, and result in mortality. Each year, landslides cause thousands of deaths and tens times more injuries around the world (CRED, 2018). Such destruction can be mitigated by modelling the spatial distribution of landslide susceptibility so that precautions can be taken to avoid the most vulnerable areas before disaster strikes. Authentic modelling

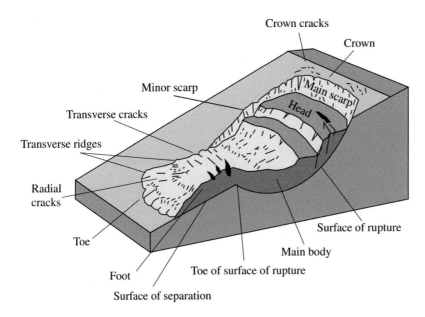

FIGURE 6.19   An idealised slump-earth flow showing common nomenclature for describing the various components of a landslide (source: USGS, 2004). Roughly, the zone of depletion lies above the surface of rupture, below which lies the zone of accumulation. In modelling landslide hazard, the runout zone or the zone of accumulation has to be modelled.

of landslide susceptibility requires the dissection of a typical landslide path into three zones: source, transportation, and deposition or runout (Figure 6.19). The *source zone* refers to the area where the mobilised materials originate from, with a semi-circular shape at the crown. Its topography is usually steep, and its surface denudated after the earth materials have been displaced downslope. This zone can be delineated from large-scale aerial photographs based on the discontinuity in surface morphology, gradient, and the abrupt change in land cover, facilitated by a DEM. The transportation or *transitional zone* is usually elongated in shape and lies between the source zone and the depositional zone. The runout or *depositional zone* has a round lobe whose slope is rather steep, and whose surface may still be covered by the original vegetation, especially if the landslide debris is displaced laterally with little rotational movement. Nevertheless, not all three zones are present or distinctive in all landslide chutes. For instance, the transitional zone can be absent if the debris does not flow much downslope. The depositional zone may become indistinctive or disappear completely after the debris has been removed by erosion several years after the landsliding event.

Landslides are commonly triggered as a consequence of the complex interactions between a plethora of climatic, topographic, hydrologic, land use, and environmental/ geological variables. Some of them can be further divided into more sub-factors, such as slope gradient and orientation under topography (Table 6.15). Occasionally, the surface equilibrium may be tipped by external events, such as earthquakes and storms, both of which can trigger landslides. Modelling of landslide susceptibility

TABLE 6.15

Variables Commonly Considered in Modelling Landslide Susceptibility/ Hazard

Category	Variables	Significance
Topography	Slope	Gravity, shear stress, acceleration of surface runoff
	Aspect	Antecedent soil moisture, exposure to solar insolation and evapotranspiration
	Curvature	Slope loading and stability
Hydrology	Stream power index	Strength of stream power and potential bank erosion
	Topographic wetness index	Soil water content/slope loading
	Distance to rivers	Lateral stream erosion
Geology	Lithology	Shear strength, permeability, and susceptibility to weathering
Soil	Soil types	Geotechnical properties
Land cover	Land cover types	Location of at-risk elements
Transport	Distance to roads	Artificial slope modification, traffic-induced instability (e.g., ground vibration)

involves an implicit assumption that those areas that have suffered from landslide events in the past are vulnerable and prone to future landslides. Those areas distributed with a higher density of landslides hence face a higher risk of future landslides. Other areas within a certain geographic proximity that share a similar environmental setting to these vulnerable areas are also classed as landslide-prone. While the spatial distribution of landslides can be easily mapped from aerial photographs or using LiDAR data, the modelling of landslide susceptibility is not so straightforward.

In modelling landslide susceptibility, the source zone should be treated differently from the other two zones because it directly governs the volume of the mobilised debris and the damage to the local area from where the debris materials are mobilised. In comparison, more damage may occur in the transportation and runout zones if the mobilised materials are displaced further downslope at a higher velocity. How far the landslide debris slides down the slope is dependent on its gradient and the momentum of the debris materials themselves. Since it is difficult to accurately delineate each zone from remote sensing imagery or in the field, the runout zone is commonly determined via runout distance modelling, prior to the modelling of the overall landslide susceptibility.

Runout distance modelling requires two inputs, the landslide source zone and a DEM. In the modelling, the area affected by a landslide is subject to the propagation of the mobilised debris (e.g., a combination of flow direction and runout distance). The former is relatively easy to determine, as the flow always follows the terrain downhill in the steepest direction, which can be determined from the DEM. Runout distance is related to the flow path of the displaced material (Horton et al., 2008). Both flow direction and runout distance can be calculated using the multiple flow direction algorithm D-infinity (Tarboton, 1997) and the D-infinity Avalanche Runout in TauDEM, respectively. The termination of the mobilised debris flow is decided based on some assumptions. For instance, it will stop once the topographic surface is too level, such as a slope gradient <5°.

The overall landslide susceptibility can be derived from the integrated slope failure susceptibility at the source of slides and the modelled landslide runout path (Figure 6.20), categorised at five levels – very low, low, medium, high, and very high

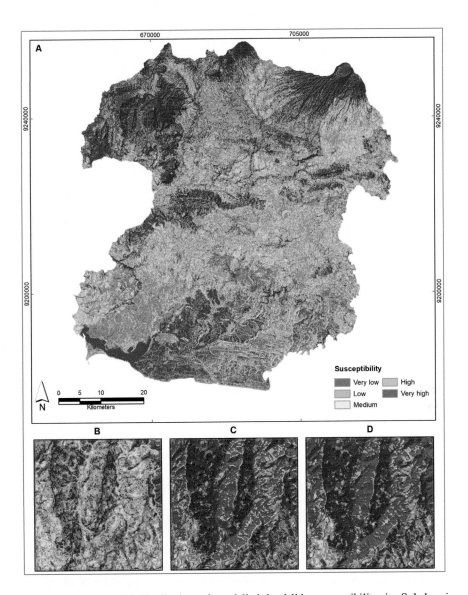

FIGURE 6.20 Spatial distribution of modelled landslide susceptibility in Sukabumi, Indonesia (A), and comparison among slope failure susceptibility (B), runout zone (C), and final landslide susceptibility (D) in 2017 (source: Wiguna, 2019).

risk – using natural breaks in ArcGIS (Corominas et al., 2013). Such a grouping is helpful to reveal the spatial pattern of susceptibility. It is also a means of suppressing uncertainties associated with the input variables and those caused by the aforementioned assumptions that may not be totally valid. In the modelling, if a grid cell has more than one susceptibility value due to the overlapping of two contributors, then it is assigned the higher susceptibility. For example, if a cell's slope failure susceptibility is very low, but the cell falls into the runout path in the moderate and low susceptible zones, then it receives an overall level of moderate susceptibility.

As shown in Figure 6.20A, in general, areas having a (very) high susceptibility level are located in the north of Sukabumi, Indonesia while (very) low susceptibility levels are found in the south. The influence of the runout zone on susceptibility zonation is shown in Figure 6.20B. The terrain that is highly susceptible to landslides is located on the upper part of a hill. In comparison, the runout zone of very high susceptibility (Figure 6.20C) shows that the areas potentially devastated by landslides are not confined to the top of the hills. Landslides may also affect a downslope region with lower susceptibility. When these two maps are combined, the final susceptibility may have a larger area than that shown on the slope failure susceptibility map (Figure 6.20D). The accuracy of the modelled landslide susceptibility is assessed using the area under the curve (AUC) Receiver Operating Characteristics (ROC) method. Comparison of the training and testing sample data with the modelled landslide susceptibility map yields a similar success and prediction rate of 0.89 and 0.88, respectively. These accuracy levels are highly comparable to those reported by Althuwaynee and Pradhan (2016) in a previous study.

6.7.3 GLACIER EXTENT MODELLING

The spatial extent of a glacier, especially its tongue, is highly sensitive to climate change, even though it may respond to climatic and atmospheric conditions in a complex way. In order to predict how the ice extent of a glacier will change in the future as a result of climate change, it is imperative to know its behaviour in the past. The past spatial extent of a glacier, especially its terminus position, can be mapped reliably from historical aerial photographs or large-scale satellite images. In order to avoid seasonal change, the terminus position should be surveyed at the same time of each year, preferably at the end of the summer season. However, this requirement may be difficult to fulfil, as the terminus position depends on the timing of aerial photography if mapped from aerial photographs.

Once the spatial extent of a glacier has been mapped from time-series aerial photographs, it is possible to detect the temporal changes in its tongue area from one year to the next using a GIS. Since the source of a glacier (the cirque) is full of compacted snow perennially, its spatial extent barely changes over a short period of time (e.g., decades). Thus, it can be excluded from consideration. Instead, attention can be concentrated exclusively on its tongue, where the snow will eventually end up. Under the effect of gravity, the accumulated snow creeps downslope in a confined, deeply incised valley whose walls hardly shift with time. Thus, only the tongue terminus varies with the amount of snowfall, and the modelling of the spatial extent of a glacier can be reduced to the modelling of its terminus position after the glacier terminus is approximated by a parabolic curve. The modelling can be accomplished by determining the apex of the parabolic curve from an arbitrarily defined reference point within the cirque, or the medial tongue length. This refers to the horizontal Mahalanobis distance between the two points. If the tongue does not follow a straight course due to a bend in the glacier valley, it needs to be measured by splitting the whole medial line into a number of straight-line segments. All the individual line segment lengths are then summed to derive the total length, the dependent variable in the modelling. Naturally, the glacier tongue area can also serve as the dependent variable. As shown in Equation (6.15), the model accuracy varies only slightly (a difference of only 0.74% in R²) between the two variables, but tongue length is much easier to measure than tongue area, and hence should be used.

$$\begin{cases} \text{Length} = 4100.779 - 1.2346 \cdot PPT + 91.9162 \cdot T - 29.30 \cdot \beta (R^2 = 80.77\%) & (1) \\ \text{Area} = 308.3207 - 0.0342 \cdot PPT + 3.3100 \cdot T - 0.9893 \cdot \beta (R^2 = 81.51\%) & (2) \end{cases} \quad (6.15)$$

The environmental variables critically affecting glacier extent include temperature, precipitation, and valley floor gradient. Temperature and precipitation have differing impacts on glacier tongue area in terms of timing. Of the two, current-year summer temperature has an immediate impact on tongue size. Summer temperature (December–February in the southern hemisphere) dictates the speed at which the frozen ice is depleted from the tongue via thawing and melting. A high temperature accelerates the melting of the accumulated and compacted snow, ablates the tongue size, and speeds up the downhill creeping of the snow. The exact relationship between the summer temperature of the current year and glacier tongue length is complicated by the slope gradient of the underlying topography. On a gentle slope, a slightly warmer temperature will lead to a tangible retreat of the terminus position. Conversely, the same rise in temperature may not translate into a perceptible retreat if the underlying topography is rather steep and a huge mass of ice has accumulated over it, even though the same quantity of snow is melted in both cases. This relationship is confirmed by Figure 6.21, which illustrates an inverse correlation between glacier tongue size and the gradient of the underlying topography at the glacier terminus. The only outlier of the relationship is the gentlest gradient. Its obvious deviation from the general trend may be attributed to the enormous uncertainty in its derivation caused by a coarse DEM grid size of 30 m. Slope gradient may not contribute to the ice mass of a glacier, but this scatterplot unambiguously illustrates that it exerts a deterministic effect on the spatial extent of the redistributed frozen snow in the glacier tongue, in that a gentle terrain is conducive to forming a long tongue, while a steep one confines the snow mass to a smaller area.

In terms of seasonal precipitation, glacier extent is critically affected by winter precipitation only, usually in the form of snow. It governs the quantity of compacted snow accumulated in the cirque, taking into account its tendency to creep gradually downhill towards the low-lying tongue. Thus, precipitation has a delayed effect on the size of a glacier tongue. The time it takes for snow to reach the terminus from the cirque varies with the valley floor gradient. The approximate time of travel can

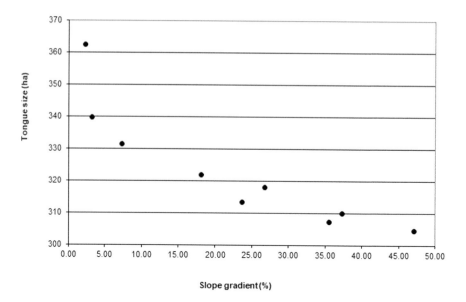

FIGURE 6.21   Scatterplots of glacier tongue area (ha) versus slope gradient at the glacier terminus (source: Gao, 2004, with permission).

be determined via correlation analysis of the cumulative snow of preceding winters with the current-year glacier tongue area. Thus, multi-linear regression models need to be established between winter precipitation of previous years and the current year's glacier tongue area or the medial length. The model with the maximum R^2 value can reveal the rough time it takes for snow to reach the terminus.

The change in glacier tongue area/medial length can be predicted from both climate variables through regression analysis (Equation 6.15). The nature of the regression equation (e.g., linear, non-linear, or exponential) requires plotting of the two sets of variables as a scatterplot, from which the relationship is assessed visually. Only the variables achieving a certain level of R^2 value are retained in the final prediction model. The tongue area or medial length of the Franz Josef glacier in the South Island of New Zealand can be reliably predicted from the summer temperature of the current year and the 3-year mean winter precipitation of the preceding 5 years at an R^2 over 80% (Equation 6.15), even without consideration of the underlying topographic gradient. These empirical relationships are accurate if established from a sufficient number of observations spanning a long period of time (e.g., decades).

In this example, the modelling was implemented in vector format in ArcMap. In the modelling, the terminus apex was represented by a pair of horizontal coordinates, through which the terminal moraine, approximated as a parabolic curve, passes. The whole extent is determined by overlaying the predicted terminal moraine with the unchanged outline of the valley walls (tongue sides) against the backdrop of contours (Figure 6.22). A new prediction can be made by plugging the anticipated rainfall and temperature into Equation (6.15), and the predicted glacier extent is then displayed on-screen.

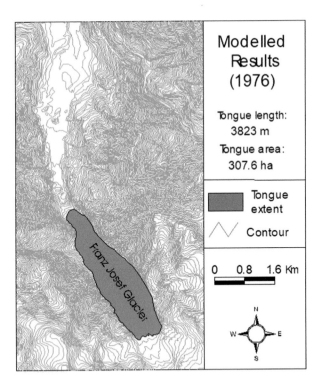

FIGURE 6.22   The spatial extent of Franz Josef Glacier in the South Island of New Zealand, modelled in vector format, in which the terminus position is approximated as a parabolic curve, and modelled from current-year summer temperature and 3-year mean winter precipitation over the preceding 5 years, against the backdrop of contours (source: Gao, 2004, with permission).

The empirical models of glacier extent (Equation 6.15) ignore the effect of the underlying topographic gradient. Thus, the same change in temperature or precipitation will cause the glacier terminus position to shift by a variable horizontal distance, leading to a variable relationship between the change in the climate variables and the pace of retreat. A better alternative is to make use of ice volume, which can be estimated by taking into account the ice pack depth determined using ground penetration radar. It is expected that the depletion of ice mass from the glacier can be predicted from climate variables at a higher accuracy than the models shown in Equation (6.15). Also, the climate data used should be recorded locally. Since this was not possible, the temperature data recorded at the nearest weather station were used. They may differ from the air temperature near the glacier tongue. Furthermore, temperature itself is elevation-dependent. The temperature at the cirque is lower than at the low-lying glacier terminus. The discrepancy between temperatures observed at the nearby weather station and those near the terminus may decrease the accuracy of the empirical models, and should be avoided if possible.

REVIEW QUESTIONS

1. What are the common criteria used to classify spatial models? Which of them critically affect the manner in which spatial modelling is implemented?
2. Compare and contrast spatially explicit with spatially implicit, time-explicit with time-implicit models.
3. In which ways does spatial modelling differ from spatial analysis? What is their relationship?
4. What variables are commonly considered in spatial modelling? How are they affected by the dimensionality of spatial modelling?
5. It can be argued that the three types of static modelling (suitability, extent and probability) are virtually the same. To what degree do you agree or disagree with this statement? Explain.
6. Compare and contrast predictive spatial modelling with diagnostic (i.e., identification of areas vulnerable to floods) spatial modelling.
7. What is cartographic modelling? Use an example each to illustrate that the sequence of modelling is significant, and irrelevant.
8. Compare and contrast spatial dynamic modelling with spatial static modelling.
9. What are the main steps you should follow to develop a robust model? How is the quality of the model judged?
10. Compare and contrast the major criteria that have been adopted to evaluate the accuracy of spatial modelling/simulation.
11. How are the variables considered in spatial modelling commonly weighted? Compare and contrast the strengths and limitations of common variable weighting strategies.
12. The common environments for implementing spatial modelling include ArcGIS ModelBuilder, Raster Calculator, IDRISI TerrSet, and scripting. Use an example each to justify why you would select each of them to perform your modelling task.
13. Why is spatial modelling closely associated with GIS? What limitations does GIS face in some spatial modelling applications?
14. How should spatial models be coupled with GIS to enhance spatial modelling? What are the pros and cons of loose, tight and total coupling?
15. Compare and contrast logistic regression and land use regression in spatial modelling.
16. It can be argued that hydrological modelling is non-spatial. To what extent do you agree or disagree with this claim?

REFERENCES

Abbott, M. B., Bathurst, J. C., Cunge, J. A., O'Connell, P. E., and Rasmussen, J. (1986) An introduction to the European Hydrological System—Systeme Hydrologique Europeen 'SHE'. 2: Structure of a physically based, distributed modelling system. *Journal of Hydrology*, 87: 61–77.

Achinstein, P. (1965) Theoretical models. *The British Journal for the Philosophy of Science*, 16(62): 102–120, http://www.jstor.org/stable/686152.

Aggett, G. (1994) A GIS-based assessment of seismic risk for Wellington City. M.Sc. thesis, University of Auckland, p. 121.

Alcaraz, M., Vázquez-Suñé, E., Velasco, V., and Manjarrez, R. A. C. (2017) A loosely coupled GIS and hydrogeological modeling framework. *Environmental Earth Sciences*, 76: 382, doi: 10.1007/s12665-017-6709-3.

Aller, L., Bennett, T., Lehr, J., and Petty, R. (1985) *DRASTIC: A Standardized System for Evaluating Ground Water Pollution Potential Using Hydrogeologic Settings.* Washington, DC: U.S. Environmental Protection Agency, EPA/600/2-85/018.

Althuwaynee, O. F., and Pradhan, B. (2016) Semi-quantitative landslide risk assessment using GIS-based exposure analysis in Kuala Lumpur City. *Geomatics, Natural Hazards and Risk*, 8(2): 706–732, doi: 10.1080/19475705.2016.1255670.

Asner, G. P., and Heidebrecht, K. B. (2002) Spectral unmixing of vegetation, soil and dry carbon cover in arid regions: Comparing multispectral and hyperspectral observations. *International Journal of Remote Sensing*, 23: 3939–58.

Babiker, I. S., Mohamed, A. A., Hiyama, T., and Kato, K. A. (2005) GIS-based DRACTIC model for assessing aquifer vulnerability in Kakamigahara Heights, Gifu Prefecture, central Japan. *Science of the Total Environment*, 345: 127–40.

Bao, S., Anselin, L., Martin, D., and Stralberg, D. (2000) Seamless integration of spatial statistics and GIS: The S-Plus for ArcView and the S+Grassland links. *Journal of Geograph System*, 2: 287–306, doi: 10.1007/PL00011459.

Beven, K. (1997) TOPMODEL: A critique. *Hydrological Processes*, 11(9): 1069–1086.

Beven, K. L., and Kirby, M. J. (1979) A physically based variable contributing area model of basin hydrology. *Hydrological Science Bulletin*, 24: 43–69.

Bhatt, G., Kumar, M., and Duffy, C. J. (2014) A tightly coupled GIS and distributed hydrologic modeling framework. *Environmental Modelling & Software*, 62: 70–84, doi: 10.1016/j.envsoft.2014.08.003.

Brownlee, J. (2019) How to choose a feature selection method for machine learning. In *Data Preparation*, retrieved from https://machinelearningmastery.com/feature-selection-with-real-and-categorical-data/.

Clarke, K. C., and Gaydos, L. J. (1998) Loose-coupling a cellular automaton model and GIS: Long-term urban growth prediction for San Francisco and Washington/Baltimore. *International Journal of Geographical Information Science*, 12(7): 699–714.

Cline, D. W. (1992) Modeling the redistribution of snow in alpine areas using geographic information processing techniques. *Proceeding 49th Eastern Snow Conference*, Oswego, New York, 1992, pp. 13–24.

CRED - Centre for Research on Epidemiology of Disasters (2018) *Natural Disasters 2018.* Brussels, retrieved from https://cred.be/sites/default/files/adsr_2018.pdf.

Corominas, J., van Westen, C., Frattini, P., Cascini, L., Malet, J. P., Fotopoulou, S., . . . and Smith, J. T. (2013) Recommendations for the quantitative analysis of landslide risk. *Bulletin of Engineering Geology and the Environment*, doi: 10.1007/s10064-013-0538-8.

Cova, T. J., and Church, R. L. (1997) Modelling community evacuation vulnerability using GIS. *International Journal of Geographical Information Science*, 11(8): 763–84.

Dorigon, L. P., de Costa, M.C., and Amorim, T. (2019) Spatial modeling of an urban Brazilian heat island in a tropical continental climate. *Urban Climate*, 28, doi: 10.1016/j.uclim.2019.100461.

Downer, C. W., Ogden, F. L., Niedzialek, J. M., and Liu, S. (2006) Gridded surface/subsurface hydrologic analysis (GSSHA) model: A model for simulating diverse streamflow producing processes. In *Watershed Models*, V. P. Singh, and D. Frevert (eds.). Taylor and Francis Group, CRC Press, pp. 131–59.

Gao, J. (2004) Modelling the spatial extent of Franz Josef Glacier, New Zealand from environmental variables using remote sensing and GIS. *GeoCarto International*, 19(1): 19–27.

Hall, M. A. (1998) Correlation-based feature subset selection for machine learning. Ph.D. thesis, University of Waikato, p. 178.

Hoek, G., Beelen, R., De Hoogh, K., Vienneau, D., Gulliver, J., Fischer, P., and Briggs, D. (2008) A review of land-use regression models to assess spatial variation of outdoor air pollution. *Atmospheric Environment*, 42(33): 7561–78.

Horton, P., Jaboyedoff, M., and Bardou, E. (2008) Debris flow susceptibility mapping at a regional scale. In *Proceedings of the 4th Canadian Conference on Geohazards: From Causes to Management*, J. Locat, D. Perret, D. Turmel, D. Demers, and S. Leroueil (eds.), Québec: Presse de l'Université Laval, p. 594.

Huang, B., and Jiang, B. (2002) AVTOP: A full integration of TOPMODEL into GIS. *Environmental Modelling & Software*, 17(3): 261–8, doi: 10.1016/S1364-8152(01) 00073-1.

Jaseela, C., Prabhakar, K., and Harikumar, P. S. P. (2016) Application of GIS and DRASTIC modeling for evaluation of groundwater vulnerability near a solid waste disposal site. *International Journal of Geosciences*, 7: 558–71, doi: 10.4236/ijg.2016.74043.

Kim, J. H. (2009) Estimating classification error rate: Repeated cross-validation, repeated hold-out and bootstrap. *Computational Statistics & Data Analysis*, 53(11): 3735–45.

Kohavi, R. (1995) A study of cross-validation and bootstrap for accuracy estimation and model selection. *International Joint Conference on Artificial Intelligence*, 14(2): 1137–45.

Kouli, M., Soupios, P., and Vallianatos, F. (2008) Soil erosion prediction using the Revised Universal Soil Loss Equation (RUSLE) in a GIS framework, Chania, Northwestern Crete, Greece. *Environmental Geology*, 57: 483–97, doi: 10.007/s00254-008-1318-9.

Lim, C. H., Kim, Y. S., Won, M., Kim, S. J., and Lee, W.-K. (2019) Can satellite-based data substitute for surveyed data to predict the spatial probability of forest fire? A geostatistical approach to forest fire in the Republic of Korea. *Geomatics, Natural Hazards and Risk*, 10(1): 719–739, doi: 10.1080/19475705.2018.1543210.

Ma, X., Longley, I., Gao, J. Kachhara, A., and Salmond, J. (2019) A site-optimised multi-scale GIS based land use regression model for simulating local scale patterns in air pollution. *Science of the Total Environment*, 685: 134–49.

Ma, X., Longley, I., Salmond, J., and Gao, J. (2020) PyLUR: Efficient software for land use regression modelling the spatial distribution of air pollutants using GDAL/OGR library in Python. *Frontiers in Environment Science Engineering*, 14(3): 44, doi: 10.1007/ s11783-020-1221-5.

Maantay, J. A., Tu, J., and Maroko, A. R. (2009) Loose-coupling an air dispersion model and a geographic information system (GIS) for studying air pollution and asthma in the Bronx, New York City. *International Journal of Environmental Health Research*, 19(1): 59–79, doi: 10.1080/09603120802392868.

Maidment, D. R., and Morehouse, S. (2002) *Arc Hydro: GIS for Water Resources*. ESRI Inc., p. 203.

McGarigal, K., Cushman, S. A., and Ene, E. (2012) *FRAGSTATS v4: Spatial Pattern Analysis Program for Categorical and Continuous Maps*. Computer software program produced by the authors at the University of Massachusetts, Amherst, received from http://www. umass.edu/landeco/research/fragstats/fragstats.html.

McHugh, M. L. (2012). Interrater reliability: The kappa statistic. *Biochemia Medica*, 22(3): 276–82.

Pourghasemi, H. R., Sadhasivam, N., Kariminejad, N., and Collins, A. L. (2020) Gully erosion spatial modelling: Role of machine learning algorithms in selection of the best controlling factors and modelling process. *Geoscience Frontiers*, 11(6): 2207–19, doi: 10.1016/j.gsf.2020.03.005.

Ryan, P. H., and LeMasters, G. K. (2007) A review of land-use regression models for characterizing intraurban air pollution exposure. *Inhalation Toxicology*, 19(Suppl 1): 127–33, doi: 10.1080/08958370701495998.

Saaty, T. L. (1977) A scaling method for priorities in hierarchical structures. *Journal of Mathematical Psychology*, 15(3): 234–81.

Samodra, G., Chen, G., Sartohadi, J., and Kasama, K. (2017) Comparing data-driven landslide susceptibility models based on participatory landslide inventory mapping in Purwosari area, Yogyakarta, Java. *Environmental Earth Sciences*, 76(4), doi: 10.1007/s12665-017-6475-2.

Siddayao, G. P., Valdez, S.E., and Fernandez, P. L. (2014) Analytic Hierarchy Process (AHP) in spatial modeling for floodplain risk assessment. *International Journal of Machine Learning and Computing*, 4(5): 450–7, doi: 10.7763/IJMLC.2014.V4.453.

Simmonds, D., and Feldman, O. (2011) Alternative approaches to spatial modelling. *Research in Transportation Economics*, 31(1): 2–11.

Smith, L., Beckman, R., Baggerly, K., Anson, D., and Williams, M. (1995) TRANSIMS: Transportation analysis and simulation system. Technical Report, doi: 10.2172/88648.

Steyaert, L. T., and Goodchild, M. F. (1994) Integrating geographic information systems and environmental simulation models: A status review. In *Environmental Information Management and Analysis - Ecosystem to Global Scales*, W. K. Michener, J. W. Brunt, and S. G. Stafford, (eds.), London: Taylor & Francis, p. 333–55.

Store, R., and Jokimäki, J. (2003) A GIS-based multi-scale approach to habitat suitability modeling. *Ecological Modelling*, 169(1): 1–15, doi: 10.1016/S0304-3800(03)00203-5.

Tarboton, D. G. (1997) A new method for the determination of flow directions and upslope areas in grid digital elevation models. *Water Resources Research*, 33(2): 309–19, doi: 10.1029/96wr03137.

Thiebes, B., Bell, R., Glade, T., Jager, S., Anderson, M. G., and Holcombe, E. A. (2013) A WebGIS decision-support system for slope stability based on limit-equilibrium modelling. *Quarterly Journal of Engineering Geology*, 158: 109–118.

Tomlin, D. C. (1990) *GIS and Cartographic Modeling*. New Jersey: Prentice Hall.

Tomlin, C. D. (1991) *Geographic Information Systems and Cartographic Modelling*. Prentice Hall, p. 249.

U.S.G.S. (U.S. Geological Survey) (2004). Landslide types and processes. U.S. Department of the Interior; U.S.G.S. Science for a changing world, https://pubs.usgs.gov/fs/2004/3072/pdf/fs2004-3072.pdf.

Vakhshoori, V., and Zare, M. (2018). Is the ROC curve a reliable tool to compare the validity of landslide susceptibility maps? *Geomatics, Natural Hazards and Risk*, 9(1): 249–66.

van Deursen, W. P. A. (1995) Geographical information systems and dynamic models development and application of a prototype spatial modelling language, Ph.D. thesis. The Netherlands: University of Utrecht.

van Vliet, J., Bregt, A. K., and Hagen-Zanker, A. (2011) Revisiting Kappa to account for change in the accuracy assessment of land-use change models. *Ecological Modelling*, 222(8): 1367–75.

van Westen, C. (2002). *Use of Weights of Evidence Modeling for Landslide Susceptibility Mapping*. Enschede, the Netherlands: International Institute for Geoinformation Science and Earth Observation.

Wiguna, S. (2019) Modelling of future landslide exposure in Sukabumi, Indonesia in two scenarios of land cover changes. M.Sc. thesis, University of Auckland, p. 165.

Woodford, K (2011) Understanding the Christchurch earthquake: Building damage, https://keithwoodford.wordpress.com/2011/02/27/understanding-the-christchurch-earthquake-building-damage/, accessed on 14 May 2021.

Xie, C, Huang, B., Claramunt, C., and Chandramouli, M. (2005). Spatial logistic regression and GIS to model rural-urban land conversion. *Paper presented at PROCESSUS - Second International Colloquium on the Behavioural Foundations of Integrated Land-use and Transportation Models: Frameworks, Models and Applications, 2005/6/12.*

Zhang, D., and Agterberg, F. (2018) Modified weights-of-evidence modeling with example of missing geochemical data. *Complexity*, 2018, doi: 10.1155/2018/7945960.

7 Spatial Simulation

7.1 INTRODUCTION

7.1.1 Spatial Simulation

Spatial simulation is defined as a computer-based operation that approximately imitates the functioning of real-world temporal processes. It is an approximation because the actual process is so complex that it must be simplified to understand how it operates in relation to a few vital independent variables. Spatial simulation requires building models and condensing data to provide parameters for a detailed model of a specific situation. It seeks to predict the future states over the simulation span of space and time based on two assumptions:

(1) The state of every system at every moment in every place can be quantified.
(2) Changes in a system can be described mathematically or logically.

In spatial simulation, the value of the attribute under study at a particular location is derived or estimated from that of adjacent locations based on mathematical models or equations. It is a window-based operation in which the location of a spatial entity in question and other observations within the defined neighbourhood also exert an influence on the simulated outcome. Apart from such spatial dependency, there is a variation in time. Temporarily, not all operations are taking place simultaneously. It is implicitly assumed that some operations must follow others, and all operations are repeated as many times as necessary. Simulation is accomplished via running the same function (mathematical model) as an increment of a fixed duration, each operation representing a unique time. The same model rules are applied with only the initial conditions, and the neighbours are changed from one iteration to the next. By incorporating the implicit temporal component, spatial simulation is able to show how the process under study (e.g., grassland degradation) evolves both spatially and temporally.

Unlike static modelling, covered in Chapter 6, spatial simulation involves variables that can be mobile agents. They roam the simulated space and behave as in the real world. Through these agents, it is possible to take into account the mutual interactions between adjacent spatial entities. For instance, if a grassland patch has been grazed to such a degree as to become barren, the livestock will move on to an immediately neighbouring patch of healthy meadow to graze. Thus, it is more dynamic. A *spatial dynamic simulation model* is a combination of a model with simulation. This computer-based mathematical model can realistically simulate spatially distributed, time-dependent processes. Simulation has the most significant advantage of being able to assess the influence of a given variable by controlling other variables unchanged, which is valuable as in reality, a number of variables may be at play

DOI: 10.1201/9781003220527-7

simultaneously. Thus it is difficult to isolate the influence of one variable from that of others. Through spatial simulation, it is also possible to carry out sensitivity analysis by changing a variable's value over a range to examine whether it is appropriate to include this variable in the simulation or whether it has been properly parameterised.

7.1.2 SPATIOTEMPORAL DYNAMIC SIMULATION

Spatiotemporal dynamic simulation models can be stochastic without parameters, known as "non-parametric". These models should be parameterised to make a quantitative prediction. They can be further grouped as spatially and temporally dynamic and explicit models, and are used chiefly for simulating a process to see how a variable changes over time spatially. Spatially and temporally dynamic and explicit models are able to handle variations in time and space. The value at a particular location is affected by the conditions at nearby locations and other constraints, all of which are a function of time. Time is implicitly incorporated in the simulation via iteration, with the temporal interval set by the modeller. Whatever time increment is chosen, it cannot be altered during simulation.

The independent variables in all spatiotemporal simulation models fall into three groups: state variables, driving variables, and rate variables. The state variables simply describe the nature or condition of the driving (independent) variables, such as population, biomass, and topography. They usually have a limited number of permutations. The driving variables or forcing functions characterise the external influence that is not subject to the dynamic process itself. For instance, the rate of bushfire spread in the landscape may be accelerated by strong winds in the right direction, but not subject to its location. These variables affect how (fast) the dependent variable spreads out, be it dispersed or diffused. The rate variables stipulate the speed at which the state variables change. Rates may not be constant throughout the simulation process, and can vary with time. For example, the amount of fresh biomass a sheep devours daily probably does not vary much from day to day. However, it will vary considerably with age. The quantity is lower at a young age, but higher in adulthood. Thus, the rate of consumption is not constant over the lifespan of livestock. The exact value of consumption depends upon the state and driving variables according to the established rules. Model performance is highly dependent on transition rules and proper parameterisation of rate variables, without which no realistic results can be expected.

Spatiotemporal modelling offers the following advantages:

- A means for describing and exploring the interactions between natural and human factors if both play a role in the process;
- Evaluation of spatial patterns;
- Exploration of the implications of interactions between the modelled variable and environmental variables;
- Prediction and simulation of spatial and temporal evolution;
- Ease of sensitivity analysis.

Spatiotemporal dynamic simulation has found applications in a variety of disciplines. Common examples include modelling of oil slick spread in environmental science, fire spread and species dispersion in ecology, disease diffusion in epidemiology, hydro-chemical contaminant and spread in heterogeneous basins in hydrology, simulation of urban population growth and population density in urban areas, urban sprawl, and dynamic population modelling in geography. Frequently, spatiotemporal dynamic modelling is able to accomplish a number of fundamental functions, such as:

(a) To quantitatively describe and evaluate the spatial pattern generated via simulation;
(b) To predict the temporal evolution of the attribute being modelled at different time increments;
(c) To identify the hot spots of frequent changes;
(d) To integrate spatial and temporal scales;
(e) To carry out sensitivity tests.

These functions can be accomplished in two environments, *cellular automata* (CA) and *agent-based modelling* (ABM), or via a combination of both using advanced machine learning algorithms. These will be discussed in detail in Sections 7.2 and 7.3, respectively. The implementation in both environments requires spatially explicit models.

7.1.3 SPATIALLY EXPLICIT SIMULATION MODELS

Spatially explicit simulation models are concerned with spatially dependent processes that can be simulated with incredible precision at a fine scale owing to the advent of simulation platforms and advanced computing technologies. The shift from spatially implicit modelling to spatially explicit modelling has triggered a transformation from capturing fine-scale details to improved understanding of ecosystem mechanisms (De Angelis and Yurek, 2017).

Space in spatially explicit simulation models is without exception 2D. This flat space is represented by a lattice of interconnecting squares or hexagonal cells. All cells have exactly the same size and orientation, resulting in the simulated area always having a regular shape. This size may be different from the actual size of a study area with an irregular boundary. Since simulation attempts to examine the influence of certain variables in theory, the simulated results can be applied to any geographic areas as long as the rules and assumptions are not violated. The exact bound of this geographic area is irrelevant. The spatial layout is explicitly incorporated into the simulation, in which location (site) and spatial extent of the variable imply dynamic changes from one iteration to the next. The temporal increment in the simulation can vary widely, depending on how it is set by the simulator. Both space and time are considered discrete. Together with other variables in the simulation model, they change incrementally from one iteration to the next.

There are two classes of spatially explicit models, CA and *percolation models*. The main difference between them lies in the nature of the state variables. In CA models, space, time, and state variables are all discrete. Rooted in percolation theory, which deals with fluid flow or a similar process in random media, percolation models also treat space and time as discrete, but treat state as continuous. Percolation models are also extremely flexible, and allow spatial heterogeneity and stochasticity to be incorporated into the simulation. In both CA and percolation models, the simulation space is the same, comprising a lattice of cells. Since percolation models are rarely used in spatial modelling, they will not be covered further. Instead, CA will be introduced in detail in Section 7.2.

The continuous advancement in the capacity and speed of computers allows the analysis of large datasets and makes it practical to develop spatially explicit simulation models. Their development also benefits from remote sensing and geographic information system (GIS) tools. The former is able to produce and supply spatially referenced data that serve as essential inputs to running the models successfully. The latter are able to supply spatial data readily available in a format that is directly transformable into a 2D lattice. Without the data or output from them, it is difficult, if not impossible, to develop spatially explicit models. Furthermore, the models can be validated against the ground truth obtained from remotely sensed images or other datasets already stored in a GIS database.

Spatially explicit models have several advantages over other models. For instance, emerging phenomena can be visualised as an array of cells, which is not possible using other models. They are rule-based, with impossible mechanisms ruled out. Such lattice models may be used for parameter testing and quick identification of qualitative changes in system behaviour over a range of parameter values. They are easy to implement, and provide rapid feedback, tremendous flexibility, and relative ease of simulation.

7.1.4 Spatial Simulation and Machine Learning

Machine learning is a generic term describing a set of methods and tools used by computers to learn and adapt on their own in solving a problem. As a subset of artificial intelligence, it is able to create algorithms via self-modification without human intervention. The desired output is produced through training or learning from the samples fed to the machine. Recently, machine learning has found increasing applications in spatial simulation because of its ability to capture non-linear relationships between phenomena under modelling and their drivers or influencers in an effort to improve the reliability of the simulation outcome. Apart from assigning weights to variables, it can also be used for their selection (Brownlee, 2019). There are three broad types of learning paradigms: supervised, unsupervised, and reinforcement. Both unsupervised and reinforcement learning do not require inputs, and as such are ill-suited to modelling spatial changes that are usually predicted from what has taken place in the past. Supervised learning is hence the only choice left, and it requires feeding the machine with a training dataset, from which a function is generated to map the inputs to the desired outputs. This process continues until the model reaches an accuracy level deemed acceptable. Common supervised learning algorithms include (logistic) regression, decision tree, support vector machine, and Random Forest (Du et al., 2018). The support vector machine algorithm is popular in image classification, but has seldom been used for spatial simulation.

Machine learning algorithms such as Artificial Neural Networks (ANNs) have been widely exploited in spatial simulation because of their ability to self-adapt, self-organise, and self-learn. One of the most critical strengths of machine learning algorithms is their independence of particular functions, avoiding the need to make any assumptions about the data distribution. ANNs can more reliably handle different types of data, regardless of data redundancy and noise (Du et al., 2018). So far, ANNs such as multilayer perceptron, general regression neural networks, and self-organising maps are used chiefly for spatial data analysis. Examples of applications include the classification of remote sensing images, spatial prediction/mapping, non-linear dimensionality reduction, and visualisation of high-dimensional multivariate socioeconomic data. ANNs are better suited for spatiotemporal simulation than other statistical methods as the exact relationship between the independent factors and the phenomenon being modelled remains largely unknown, whether the relationship is linear or non-linear. Of the various types of ANNs, the feed-forward, error Back-Propagation Multi-Layer Perceptron ANN has been commonly adopted to simulate spatial change owing to its simplicity, ease of training, efficiency, and its abilities of reasonable associative memory and prediction.

In spatiotemporal modelling, in recent years machine learning has been used chiefly to automatically derive decision rules and to construct non-linear relationships between independent and dependent variables. Such non-spatial models are applied to mapping out the spatial behaviour of the dependent variable in a manner resembling land use regression (LUR) modelling. In simulating urban expansion, an ANN is able to generate, from a plethora of factors, the urban suitability index (USI), which is vital in simulating urban expansion (Xu et al., 2019). Tree-based ensemble machine learning offers a new avenue for exploring causal relationships between the dependent variable and its explanatory variables (Knudby et al., 2010). It must be cautioned that the use of ANNs in developing decision rules faces the risk of overfitting to the training data. This is because at a given location, the dependent variable's attribute may be under the influence of neighbouring cells, which is not represented in the training data. Overfitting can also occur owing to misinterpreting variables (e.g., land covers are actually related to elevation).

The effectiveness of three machine learning algorithms in simulating urban expansion is compared in Table 7.1. In this comparison, all three were integrated with CA–Markov chain (MC) models. They enabled the modelling of newly urbanised areas that bear a close resemblance to the reference data (kappa > 91%). Of these algorithms, logistic regression is the worst performer, achieving the lowest kappa simulation and fuzzy kappa simulation values. The deterministic Analytic Hierarchy Process (AHP) algorithm outperforms logistic regression with a (fuzzy) kappa simulation > 0.452, but its performance is still noticeably inferior to ANN's. Integrated with CA–Markov chain models, ANNs achieve the best predictions of

TABLE 7.1

Comparison of Three Machine Learning Algorithms in Simulating Urban Expansion

Model	Kappa	Kappa simulation	Fuzzy kappa simulation
ANN	0.941	0.547	0.572
AHP	0.929	0.452	0.475
Logistic regression	0.918	0.369	0.386

Source: Xu et al. (2019), with permission.

urban expansion. Their kappa simulation (0.547) and fuzzy kappa simulation (0.572) are nearly 20% higher than those of logistic regression and 10% higher than those of AHP, thus they should be adopted widely.

7.2   CELLULAR AUTOMATA SIMULATION

As a raster way of looking at the planar world, cellular automata partition it into an array of regular grids, usually square in shape, implemented in a lattice environment. The CA environment comprises a lattice of interconnecting square, triangular, or hexagonal grid cells, in one, two, or higher dimensions. Each cell has a specified state that changes over discrete time and space from one iteration to the next at the pre-defined interval of increment (e.g., hourly, daily, annual, or monthly), according to the local and global cell state configurations. This spatial layout is explicitly incorporated into spatial simulation in which location and spatial extent of the dependent variable imply dynamic changes. CA is a simple way of demonstrating how a complex system behaves. It is a processing mechanism for information fed to it from the surroundings with temporally varying characteristics. There are two classes of automata tools for spatial modelling, CA and ABM. The former is covered in this section, and the latter in Section 7.3.

7.2.1   Automata

The CA modelling and simulation environment utilises cell state, neighbourhood relationships, and transition rules to create discrete dynamic systems over time and space. Decisions are made based on internal settings, rules, and external inputs. Rules stipulate the reaction of a cell's state to the inputs. The reaction leads to two types of decisions for the cell's state: transition (change) and no transition (no change). Whether a cell's state is changed or not depends on the transition rules governing the conditions required for the cell state to change, subject to the state of cells in close proximity to it within the defined neighbourhood (Figure 7.1). As a "bottom-up", neighbourhood-based model, with the neighbourhood size stipulated by the adjacency criterion, CA may be described as "something which has the power

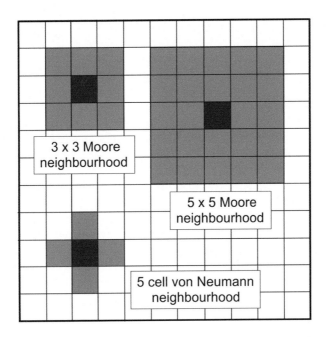

FIGURE 7.1 Two types of neighbourhood and neighbourhood size commonly adopted in spatial dynamic simulation. Central cell (dark blue) = automata.

of spontaneous movement or self-motion". A macro-level spatial process or change is reduced down to micro-level changes in cell state. CA consist of:

- A cellular space normally composed of 2D grid cells extending infinitely in every direction;
- A defined neighbourhood of a cell;
- A set of possible cell states;
- A set of transition rules.

The above key components of CA or the Wolfram formal framework can be expressed as:

{Cells, States, Initial conditions, Neighbourhood, Rules, Time}.

Each cell in the 2D space has a set of possible states associated with it. The state of a cell is not fixed. Instead, it is likely to transit among all the possible states under a set of transition rules from one iteration to the next. They govern how the state is changed over time. Initial conditions refer to the state prior to running CA. Every time a model is run, the current state must be re-initialised or set to the default state. Sometimes the cells with a given initial state may be unknown (e.g., where a bushfire is ignited). In this case, they have to be allocated spatially in a random manner (randomness is cancelled out by averaging results from multiple repeated runs).

Of the six elements of CA, neighbourhood is the simplest to define and address. It refers to the spatial extent of localised areas comprising several CA neighbouring the target automaton. It has two connotations. First, it refers to the size of a window defining the adjacent cells. Commonly adopted neighbourhood size ranges from 3 by 3 to 5 by 5 (Figure 7.1). The window size must be an odd number, with the central cell being the target of interest or automaton. In the case of a 3 by 3 window, each cell in the neighbourhood is surrounded by eight cells. Only those cells within the neighbourhood are considered in simulation, and only they influence the cell in question and interact with each other. The second connotation is the number of connections between neighbouring cells. There are two types of connectivity: four neighbours and eight neighbours. The former connectivity is called the von Neumann neighbourhood, in which only the immediately adjoining cells are considered neighbours. The latter connectivity is known as the Moore neighbourhood, in which all the cells in touch with the automaton are considered as neighbours, even if they do not share any common boundary with the central cell. CA simulation is a window-based operation during which the central cell is the focus, while taking into consideration neighbourhood effects and the transition rules (Figure 7.1).

7.2.2 ENVIRONMENT

Environment refers to the space or virtual world in which cells interact with each other, and with the agent in the case of agent-based models. The cell in question interacts only with its immediate neighbours. All cells outside the neighbourhood (e.g., beyond the window size) exert no influence on the central cell. The "backcloth" upon which the dynamics of the automaton unfold is a cellular grid of some kind, usually a rectangular partitioning of R^n. Two-dimensional CA for which the grids form the state-space provide the grid layout. The cellular state space is the equivalent to an individual pixel of a raster layer. This environment can be discrete or continuous, or even a network, depending on the nature of the problem being modelled. It can be the geographic space of modelling, or some physical features in this space. This environment can also impose constraints on the movement of agents, such as a barrier. In spatial simulation, this environment is always spatially explicit. For instance, the spatial extent of simulation and cell size must be specified beforehand. Sometimes, the location of an agent is tracked during modelling to examine the spatial extent of its roaming, whether it has acquired resources, and whether it has encountered other agents (e.g., whether a sheep has encountered a wolf).

The CA environment possesses the following three unique traits:

(a) *Finite states*: Each cell of the state-space can assume only a finite number of k values. If the CA lattice has N cells, then the total number of states is also finite ($=k^N$). All the possible number of states a cell can have must be pre-defined. At any given moment, however, each cell can have only one of these possible states. Since the modelled phenomenon is in constant change, it is likely that its state will change with time or from one iteration to the next during simulation. The decision about change from the current state to another state for a cell is made on joint consideration of its current state, the state of some of its neighbouring cells, the neighbourhood effect, and the transition rules.

(b) *Homogeneity*: Each cell in the CA environment is exactly the same as other cells, in that all of them can take on exactly the same set of k possible states at any one moment and the same transition rules are applicable to all the cells within the defined neighbourhood. Longer temporal ranges for the modelling of space-time diffusive processes are likely to require "evolving" sets of k values.

(c) *Locality*: The state transitions are local in space and time (local in the sense of the "range" of possible transition rules). The next state of a given cell depends on the state of the cells in the neighbourhood of the previous iteration. What are called non-local interactions in "traditional CA" can be dealt with in a number of ways. For example, neighbourhood can be expanded or contracted, dependent on the interaction of model elements.

7.2.3 RULES

CA are fundamentally rule-based. The simplest class of CA has rules which depend only on the sum of the states of the neighbourhood. More complicated systems can involve directional rules. Both the nature and the number of rules in a CA simulation vary with the domain of applications. A rule can be a simple logical comparison or a highly complex mathematical operation, or a combination of both, without considering other drivers or hampers. It may be expanded to include stochastic functions. More complex rules can be constructed quantitatively or via spatiotemporal methods using machine learning. A set of rules can be combined in reaching a decision. Some of them may prescribe the relationship between cells determined by cell state. All the rules in a CA fall into three types: transition rules, neighbourhood (including movement) rules, and constraint rules (Table 7.2). Of these three types of rules, transition rules are the most important as they govern whether and how the current

TABLE 7.2

Comparison of Three Types of Rules in CA Simulation

Nature of rules	Purpose	Main features
Transition	Change in cell state from one iteration to the next	Can be complex; multiple rules possible
Neighbourhood	Define the influence of neighbouring cells	Simple; two types (neighbourhood size and connectivity)
Constraint	Define the space of simulation; exclude areas from simulation; impose resistance	Additional inputs needed; a simple check or multiplication of a resistance value

state of a cell S^t at time t should be changed at time $t+1$ based on current input I^t. Conceptually, a transition rule can be expressed as:

$$A \sim \left(S, N, R_s, I\right); R_s : \left(I^t, S^t\right) \rightarrow S^{t+1} \qquad (7.1)$$

where A is a basic automaton, S stands for the state variable that describes the condition of A, S^{t+1} denotes the cell state at time $t+1$, N represents the effect of a cell's neighbourhood, and R is the transition rule. The rule of transition for a cell is a deterministic function of its current state and the states of the neighbouring cells, as shown in Equation (7.1). A transition rule articulates how one state is transitioned to the next, and prescribes state transition at each time interval.

The neighbourhood rules spell out how neighbouring cells affect each other, and define the cell "neighbourhood' within which each cell operates or responds to other cell states. These rules may also prescribe the movement, which falls into two types:

(a) Immediate movement rule – only the automaton next to the one in question exerts some influence;
(b) Nearby movement rule – similar to above, except the neighbourhood size is larger.

The constraint rules are enacted for accommodating additional conditions imposed on the 2D space to exclude certain areas from consideration and to confine simulation within the specified boundary. Typical constraints may include physical barriers (rivers, geological hazards, and reserves) and the boundary of the study area. These constraints can be embodied as an extra layer of data in the input. For instance, any water bodies will constrain urban development. Prior to running a CA-based simulation, it is imperative to check where the cell under consideration coincides with a lake spatially or falls into some protective zones where urban development is forbidden in land cover change simulation.

The success or failure of CA simulation depends mainly on the comprehensiveness and authenticity of the transition rules, which ultimately affect the accuracy or even validity of the simulation outcome. It is not an exaggeration to claim that the development of transition rules is the bottleneck in CA simulation. How to define rules and articulate them clearly is a crucial step in CA simulation that will likely take up a large amount of time and effort. After development, the validity of all rules needs to be checked prior to implementation in a simulation. Naturally, any imprecise rules will lead to inaccurate or unreliable simulation results.

7.2.4 CA AND SPATIAL SIMULATION

CA are fully compatible with raster data, as their 2D space is analogous to the tessellation model of raster GIS packages. CA simulation can be easily implemented in a raster GIS due to its open structure. CA are gradually becoming a significant avenue for raster GIS modelling, as both possess compatible qualities in structure and support temporal behaviour. A finite set of grid layers of data offers the structure on which neighbourhood operations are carried out. In CA-based spatial simulation, the state of a given cell in the lattice space is simulated individually, taking into account the neighbourhood effect. Each cell's value is decided in isolation, and may be subject to that of its neighbours (e.g., a cell cannot be burned unless its neighbouring cell is on fire) and the transition rules. Thus, all the cells vary spatially in their state and change according to the behaviour of the phenomenon being simulated (e.g., fire spread in the landscape). With CA, it is possible to increase the spatial resolution of dynamic models by several orders of magnitude in comparison with what is possible with traditional techniques, which permits the models to address the issues of spatial complexity directly. CA can provide an essential theoretical approach in which temporal processes can be modelled within a GIS, creating a spatiotemporal simulation model with transparent assumptions. In particular, CA are good at simulating the interactive relationships between drivers of a process and the environment. CA simulation outcome depends on the inter-relationship between the spatial factors under consideration and the environment.

As a grid-based and spatially explicit model, CA can make use of both qualitative and quantitative knowledge in the simulation as long as they can be represented as rules or mathematical equations. In fact, 2D CA have become a powerful method in simulating a number of diffusion processes, such as modelling of land use change and urban sprawl, and forest fire propagation modelling, all of which require space-time diffusion determined by certain rules or constraints essential in growth simulation. All such dynamic and complex spatial simulations share the same trait, in that the relationship between the spatial process and its drivers cannot be articulated precisely. CA are particularly powerful at and suited to modelling the dynamics of spatially complex phenomena, such as ecological modelling, in which paucity of data is the norm. More importantly, such models can be easily coupled with raster GIS, which supplies a variety of auxiliary data essential for the simulation (e.g., digital elevation models [DEMs]).

7.2.5 Two Examples of CA Models

One of the most commonly used CA models is called Slope, Land Use, Exclusion, Urban extent over time, Transportation, and Hillshading (SLEUTH). The Hillshaded backdrop is designed only for visualising the modelled outcome (in contrast to no-data cells), and plays no role in modelling itself. SLEUTH is a spatially explicit model whose name indicates all the input data layers required for simulating land use change in general, and urban expansion in particular. The Slope layer contains information on topography, with rugged terrain being less favourable for urban development than flat land. The Land layer shows current land use, especially urban areas, and hence potential areas that can be converted to urban use. The Land Use layers illustrate the spatial distribution of all possible types of land covers in the simulation space. Such layers must be available at a minimum of four times, from which information about how the urban areas have changed in the past in relation to other factors considered in the model can be acquired. These layers can reveal land cover dynamics in a given time interval based on historical land covers. Also, one of the recent ones can be used to validate the model's accuracy, or calibrate the model if the accuracy is not satisfactory. The Exclusion layer imposes constraints on urban growth. It dictates where urbanisation can take place, such as outside protected ecologically sensitive and valuable zones or geologically hazardous areas. Alternatively, this layer can be weighted to indicate the resistance to urban growth (e.g., a slower growth rate due to unfavourable terrain or lack of infrastructure or amenities). The Urban layer is indispensable in the model, as it shows the urban extent from simulation.

SLEUTH comprises two subcomponents within the framework of the model: the urban growth model, and the Deltatron land use change dynamics model, which can be used to simulate urban growth and land use change, including urban land use dynamics. These two models are tightly coupled (Dietzel and Clarke, 2007; Clarke, 2008). In the first model, whether a cell is transitioned to urban from non-urban is evaluated individually based on weighted maps of the study area. The diffuse, outward growth of the urban area is governed by five critical parameters: (1) terrain (slope gradient), (2) distance from existing urban areas, (3) proximity to transportation routes, (4) zoning regulations, and (5) mode of diffusion (Figure 7.2).

The implementation of SLEUTH requires the input of initial conditions and transition rules. The former specify the seed cells of urban areas from which growth and change take place one cell at a time. The latter stipulate whether a cell is converted from non-urban to urban, and they are applicable to all cells, subject to the spatial properties of the neighbouring cells and the influences of other factors under consideration within the neighbourhood. The next group of model inputs is related to diffusion, or how dispersive the spread of urban areas is overall. This is determined by four coefficients:

(1) The breed coefficient specifies the chance of a newly urbanised detached settlement to grow on its own.
(2) The spread coefficient denotes the degree of contagion diffusion radiating from existent urban cells.
(3) The slope coefficient indicates the level of resistance imposed by steep terrain.

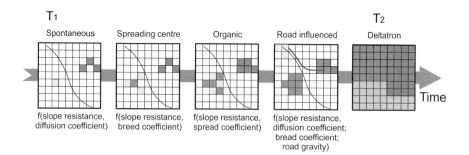

FIGURE 7.2   Cell state transition from T_1 to T_2 in urban expansion simulation using SLEUTH CA, in which different expansion coefficients are used, subject to the mode of expansion (source: modified from Clarke, https://slideplayer.com/slide/6990931/).

(4) The road gravity coefficient identifies the attractiveness of new development next to roads.

Jointly, these four coefficients stipulate how the growth rules are applied after they have been calibrated by comparing the simulated land cover change with historical data. These coefficients, in turn, are self-modified in response to the change in the growth rates. They can decrease or increase, depending on whether the growth is rapid or constrained. The growth rules also lay out initial conditions and a set of decision rules governing how land cover changes under the influences of different factors.

In order to produce reliable simulation outcomes, SLEUTH must be carefully calibrated. One straightforward method of calibration is to predict the present from the past, then compare the simulated with the observed. Alternatively, calibration can be based on Monte Carlo simulation to produce a range of possible outcomes and the probabilities of their occurrence (see Section 6.3.3). The goodness of fit indicates the effectiveness of model calibration. In order to run SLEUTH successfully, the modeller must supply three sets of input data represented as rudimentary layers in GIF format, as well as a scenario file that contains most variables and settings to run the model, particularly the coefficients. This file also instructs the computer how to execute the model and where to save the output both graphically and statistically (see Figure 7.3).

The second example is the TRansportation ANalysis and SImulation System (TRANSIMS) model. This integrated regional transportation system analysis package

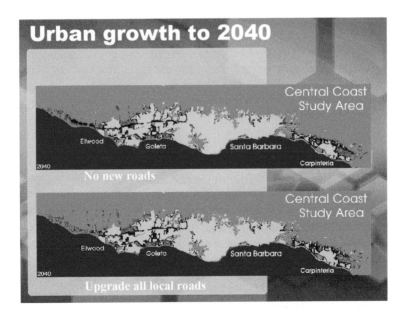

FIGURE 7.3   Urban growth of Santa Barbara by 2040 simulated using the SLEUTH land use change model (source: https://slideplayer.com/slide/6990931/, copyright © Keith Clarke, with permission).

is developed for mutually supporting realistic simulations, models, and databases using advanced computational and analytic methods. It can yield detailed simulation results that enhance our understanding of complex issues in transportation planning, such as environmental pollution, energy consumption, traffic congestion, land use planning, vehicle efficacies, and the effect of transportation infrastructure on quality of life, productivity, and the economy. It is composed of four modules: household and commercial activity disaggregation, intermodal route planner, travel microsimulation, and environmental models and simulations (Smith et al., 1995). Comprising two sub-modules (synthetic populations, and activity demand and behaviour), the first module extracts population data from census and other sources. The travel microsimulation module mimics the movement and interactions of travellers throughout the transportation system of a metropolis. It is designed to simulate the location of every vehicle in the transport network at a given moment. The environmental models and simulations translate traveller behaviour into resultant air quality, energy consumption, and carbon dioxide emissions by taking into account the microsimulation results. In turn, they feed information on fog to the microsimulator to yield more reliable predictions. It is the travel microsimulation module that makes use of CA to microsimulate traffic in the transportation network. In the simulation, the road network is divided into a finite number of cells whose size roughly matches the length of a vehicle. At each temporal increment, a cell is checked for vehicle occupancy. If a vehicle has occupied it already, then it moves on to the next cell in accordance with the pre-set rules. Despite its low fidelity, this CA micro-simulation approach is a viable means of simulating a large number of vehicles quickly, and it is possible to increase the fidelity by reducing the cell size and by taking into account vehicle attributes with more sophisticated rules at the expense of a slower computational speed. This microsimulator outputs the location of each vehicle at any moment, from which additional information such as travel speed and average travel time can be derived. Since TRANSIMS is an open-source software platform, anyone who wants to make improvements to it can do so freely. Inputs from a multitude of users can improve and perfect its functionality rapidly.

7.2.6 INTEGRATION WITH OTHER MODELS

CA transit cell states based on consideration of only the neighbourhood effect, without quantitatively considering the role of driving variables in the transition. As a micro-level model, CA do not possess a global perspective on where the newly transitioned cells should be spatially allocated. Another critical limitation of CA is their inability to determine the total amount of cells that should change their state irrespective of the neighbourhood effect or transition rules, or time (fixed). In other words, traditional CA are unable to gauge the actual quantity of expected transitions or changes. To a large degree, these two limitations can be circumvented by integrating CA with other models. Of all the possible candidates for integration, Markov chain is the most suitable. This stochastic method models a series of sequential events whose occurrence depends solely on the state of a preceding event. Whether a subsequent event results or not is subject to the transition rules based on probabilities.

The coupling of CA with MC, also called locally interacting MC or probabilistic CA, allows the state of an array of cells to be updated in accordance with some simple universal rules, either in parallel or synchronously, and is thus particularly suited to GIS-based simulation. The integration of CA with MC is commonly known as the CA-MC model. In this integrated simulation model, MC determines the probability and the amount of the expected transition between different cell states, usually through analysis of historic multi-temporal data and detection of changes between any two states in the interim. Such time-series data can be easily sourced from classified remote sensing images and data stored in a GIS database. On the basis of the detected changes, it is possible to calculate the transition probability for a particular state. The integration of CA with MC in a GIS allows expected urban growth to be optimally allocated within the study area, which is perfectly suited to simulating dynamic urban expansion.

This coupling of CA with MC can bring a number of benefits, such as the ability to take into account the effect of external drivers of change. This ability is particularly significant in modelling urban expansion in which historic land cover layers are used to calculate the non-urban to urban transition possibilities and spatially allocate newly urbanised cells. In this integration, MC determines the quantity of cells whose state will be transitioned from time 1 to time 2, and CA predicts the cell state at time 2 based on its current state and that of all other cells in its neighbourhood.

The third critical limitation of CA models is that all cells in the simulation process are spatially immobile, which cannot meet the requirements of spatiotemporal dynamic modelling involving mobile agents. For instance, a grassland patch cannot be grazed if it has become denuded, so the livestock will move on to the next patch. Similarly, if a person infected with a communicable virus moves around to be in close proximity to other people, they will become infected as well. The modelling of grassland grazing and the spread of a communicable virus in the community require the ability to address the impact of mobile agents. To overcome this limitation, CA must be integrated with agent-based simulation, or ABM. This coupling allows an agent to roam into every potential neighbouring cell to determine whether its state should be transitioned to another during one iteration. Another suitable candidate for integration is the ANN approach, which can be used to establish transition rules for a CA model in simulating human decisions (Zhao and Peng, 2014).

7.3   AGENT-BASED SIMULATION

7.3.1   Agents and Geographic Agents

There is no universal agreement on the precise definition of *agent*, a subject of much discussion and occasional contention (Macal and North, 2009). In general, it suffices to say that an agent is an independent computational component that mimics the behaviour of a living being in automatically yielding relevant results in a specific domain of applications under a given circumstance. These results can be highly accurate owing to the agent's capability of learning from the environment and dynamically changing its behaviours according to past experience. The nature and type of agents in different simulation applications vary with the issue to be simulated. In spatial simulation, agents can be any variable that exerts an impact on the modelled variable. They can be geographic, environmental, and even socioeconomic and political. Geographic agents in agent-based modelling are multiple in quantity. They can be lifeless or live agents, such as trees and land covers. They can also be residents, households, employers, developers, planners, and policy-makers in the case of urban sprawl simulation (Torrens, 2006). Agents can have different hierarchies. A macro agent may be the decision-maker controlling where new urban development is allowed to take place (e.g., not in flood-prone and ecologically sensitive areas). Micro agents may be residents who decide where to live in a newly urbanised area of a city. They can be mobile, such as livestock, and "roam" the modelling space at will. How an agent roams the space is dictated by how it reacts to the environment and the action of other agents.

Typically, agents are situated in space and time and reside in lattice-like neighbourhoods, able to make their own decisions independently. The number of agents and their variety in a simulation model vary with the nature of the simulation problem. It is permissible for multiple agents to coexist in the same environment and interact with each other at the same level of hierarchy or across the hierarchy. There is no limit to the number of agents a model can have, though some systems may restrict it to a fixed number due to the constraints of computing capability, or to simplify a highly complex modelling issue to a manageable degree. Some simulators may impose a constraint on the number of agents a cell can have at a time. In most cases, only one agent is allowed to occupy one cell at a time. For instance, if a patch is grazed by a sheep, it cannot be grazed by another sheep simultaneously. No matter how many agents are involved in a simulation, they must all be defined, specifying the information agents need in terms of data structures and mechanisms for how they manipulate information. They must be coded in the model and are time-invariant. In other words, their reaction to the same situation will always remain the same regardless of iteration (or time increment).

From the practical perspective of simulation, an agent should possess four properties or attributes:

(1) *Autonomy*: An agent can function independently in its neighbourhood and freely interact with other agents, generally within the specified neighbourhood. Thus, there is little or no centralised control over agent behaviour. All agents are at liberty to make their own decisions irrespective of other agents unless instructed otherwise.

(2) *Interactions*: Agents interact with other agents and with the environment in accordance with the pre-determined protocols or mechanisms of behaviours (Figure 7.4). The interaction protocols can range from competition for space, collision avoidance, and mutual influence to information sharing.

(3) *Mobility*: Agents may be mobile and able to roam the environment so that their geographic location is not fixed during the simulation (Figure 7.5). The direction and possible range of movement are dictated by the pre-determined rules, such as one cell per iteration. Although an agent can move to

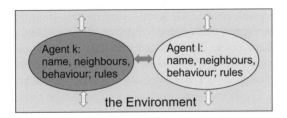

FIGURE 7.4 Interactions (orange arrows) between agent *k* and agent *l*, and between them and the external environment (yellow arrows).

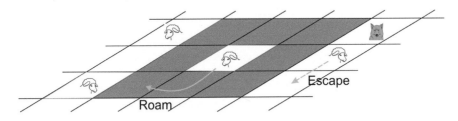

FIGURE 7.5 Mobility of agents over the simulation space in two contexts: interactions with the environment (solid arrow) because the forage has been devoured, and with another agent (dashed arrow) to escape from imminent danger.

multiple places, it is not possible for the same agent to appear in more than one cell simultaneously.

(4) *Behaviour*: Agent behaviour falls into two types, responsive and adaptive. Responsive behaviour is simplistic and reactive, in that a given condition will trigger one of the pre-determined responses. It can be expressed as a conditional statement in the form of "if …, then …". In addition, an agent's behaviour can be more complex, determined via machine learning. For instance, whether a vacant land plot will be urbanised can only be determined through analysis of events in the past under a variety of contributing or influential variables. Adaptive behaviours are modified changes based on past behaviours or experience. For instance, an adult yak may consume more forage than a juvenile one. Its grazing behaviour change in relation to age should be embodied in the decision rules.

7.3.2 Agent-based Modelling

Although spatial modelling and spatial simulation are used interchangeably by some authors (e.g., Smith et al., 1995), agent-based simulation is used exclusively here instead of modelling because the outcome is dynamic (e.g., time-dependent). *Agent-based models* are computational models for simulating the dynamic interactions of autonomous, rule-based agents and assessing their effects on the environment. They represent a combination of game theory, complex systems, and Monte Carlo simulation for randomness that is indispensable for initialising the placement of mobile agents in the simulation space. All agent-based models comprise three essential ingredients: agents, agents' relationships and ways of interacting, and agents' environment. All of them except agent interactions have already been covered earlier in this chapter. Interactions can take place between agents and between agents and the environment (Figure 7.5). The former is exemplified by the predator–prey relationship when two agents encounter each other in landscape ecological simulation. During the interactions, it is possible for agents to pass informational messages to each other and act on what has been learned from them. For instance, a ewe will give birth to offspring at a certain age and die at an old age. As an agent, the same sheep can react to the environment. For instance, if a healthy meadow is covered by grass, the livestock will graze it. After the forage has been devoured completely, the livestock will move on to the next cell, or they will die if there is no grass left for grazing.

It must be noted that not all agents interact directly with other agents all the time. For instance, if a wolf is spatially distant from a sheep or it is not hungry, it will not prey on the sheep. They interact with each other only when they are within the predefined physical distance from each other. Modelling agent relationships and interactions is as important as modelling agent behaviours. Modelling agent behaviours or the nature of interactions requires careful considerations of a number of issues, such as possible connections between agents and the mechanisms of their dynamic interactions. Once these issues are codified into rules, ABM can be implemented as continuous and repeated executions of agents' behaviour and interactions iteratively, with each run representing one temporal increment, in a fashion identical to that of CA simulation. The model run can be halted through external control (interruption) or end naturally. In ABM, spatially dynamic processes are simulated through self-activities and interactions between various agents. Reliable simulation outcomes are possible only if there is a sufficient number of accurate data layers and all model parameters have been properly configured.

ABM differs from cartographic modelling in 12 respects (Table 7.3). The most significant difference between them is their focus. ABM focuses on the dynamic processes of agent interactions with each other and with the environment that are simulated repeatedly over time. This kind of simulation is process-oriented, time-stepped, discrete, and dynamic. ABM is able to produce multiple outcomes, each corresponding to a unique time increment. Through sensitivity analysis, it is possible to account for the influence of a given variable or its value range. In contrast, cartographic modelling ignores time (everything is treated as time-invariant), and the modelled results are static without any rules involved in the decision-making. In this outcome-oriented approach, the number of variables involved is limited, and their role in the outcome remains mostly unaccounted. ABM is much more diverse and robust than cartographic modelling, in that the simulated results can yield insights into a spatial process that are not obtainable otherwise.

7.3.3 RULES

ABM rules are the same as the CA rules presented in Section 7.2.3, except that more rules are needed about agents and their interactions. Also known as assumptions, rules in ABM are statements about agent behaviours, such as their movements

TABLE 7.3
Comparison of Traditional Cartographic Modelling with Agent-based Simulation

Aspect	Cartographic modelling	Agent-based simulation
Nature	Deterministic (one outcome)	Stochastic (multiple outcomes)
Philosophy	Allocative (top-down)	Aggregative (bottom-up)
Focus	Outcome	Spatial process
Implementation	Equation-based formulas	Adaptive behaviours
Ability to explain	No explanations given	Powerful explanations given
No. of parameters	Few, similar in type (e.g., all spatial layers)	Many, diversely natured (rates, neighbourhood, and rules)
Spatiality	Spatially implicit	Spatially explicit
Treatment of time	Time-invariant	Time-dependent (increment)
Cell state	Static	Dynamic
Interactions	None	Between agents and between agents and the environment
Rules	Irrelevant (weights)	Essential, diverse types
Initial conditions	Non-existent	Must be specified, even unknowns

and interactions, to simplify the simulation issue. Rules are needed to spell out how agents move and turn around in the environment and interact with it and other agents. The development of rules is a critical component of all competent ABM and the bottleneck of successful ABM. They can be empirical, based on knowledge gained from the published literature. They can also be tested and supported by results from field experiments. In the field, attention must be paid to the representativeness of the collected samples. The more representative the field samples, the more widely applicable the established rules based on the field data, and the more authentic the simulation outcome. Rules can be expressed logically or mathematically. Logical rules are simple, qualitative, and can be implemented easily as "if ..., then ..." statements. Mathematical rules, on the other hand, are much more complex and may involve multiple agents. They are particularly applicable to adaptive behaviours in which certain conditions are spelled out. The number of rules needed for a simulation and their nature depend totally on the spatial phenomenon being simulated. In the simulation of sustainable grassland grazing, rules may be concerned with the pace of livestock movement, life expectancy, life cycle, and the quantity of forage consumption by livestock, as well as the rate at which new biomass is yielded in a grassland. Needless to say, the authenticity and reliability of rules directly affect the quality of the ABM outcome.

7.3.4 CA or ABM?

CA and ABM are two general approaches to spatial simulation. Both are complex systems built on micro-scale actions. They are good at studying the aggregate behaviour of variables, which is not achievable by other forms of modelling. In reality, the two are frequently used jointly, as in the TRANSIMS model. They share some common traits, but have vastly differing strengths. CA are strong in spatial modelling compared to other types of analysis or modelling. These non-linear dynamic mathematical systems are the most advantageous in modelling and exploring natural processes if coupled with GIS packages. One of the strengths of CA is the simplicity of the rule-based algorithms in investigating long-term "evolutionary" behaviour. As stated previously, macro-scale spatial changes can be made via micro-level transitions in the cell state. Thus, "local" effects across space and time can be investigated at the "global" scale, which is significant in studying complex systems and processes. CA serve as an alternative to more traditional analytical methods for understanding system structures in which it is vital to have a good comprehension of the hierarchical organisation of the system, such as the significance of the elements in relation to one another and their effect on the outcome of a spatiotemporal process. As an analytical tool, CA alleviate this "need" because behaviour is revealed through time and space as the simulation progresses through each iteration, driven by a simple rule base.

CA have been frequently confused with multi-agent systems, even though geographically, they are quite distinctive from each other in their spatial attributes (Torrens, 2006). CA cells are static (e.g., the automata locations are fixed) and immobile, while agents are mobile. In terms of spatial interactions, information in CA is transmitted via diffusion over the neighbourhood (Table 7.4). In ABM, agents transmit information by themselves and can move at any distance from the current position to a new one, subject to the defined neighbourhood size. In CA, each cell has only a mono-attribute (one cell, one attribute), with the input fixed. In contrast, agents are dynamic. They can have a life span (birth and death) and may produce offspring at the proper age. Their initialisation is usually random. In terms of focus, CA are excellent for studying the emerging properties from interactions (outcome-oriented).

TABLE 7.4
Comparison of Main Features of CA with ABM in Spatial Simulation

Feature	CA	ABM
Space	Finite, discrete, fixed number of cells, state changes	Continuous, cardinality is infinite
Neighbourhood	Fixed	Variable
Rules	Simple, flexible	Sophisticated, diverse
Function/capability	Limited, static	Powerful, adaptable
Drivers of process	Unable to consider	Multi-drivers permissible
Agents	Limited, immobile	Multiple, unlimited, mobile
Cell properties	Immobile, mono-state,	Agents roaming among cells, dynamic (death and birth)
Focus	Outcome-oriented	Process-oriented
Main applications	Limited, simplified outcome	Simulation, realistic outcome

In contrast, ABM simulates complex situations with a focus on the process itself, while the outcome is not the primary concern.

Also known as the individual-based model, ABM can be off-lattice, or on-lattice, in which space may be absent, such as tracking individuals over time. Non-spatial ABM is commonly used in social sciences and economics. The CA model is a specific case or subset of agent-based models in which space must be represented as a lattice of grids. If applied to the spatial domain, ABM's nearest neighbours vary with time as there are no restrictions on the movement and interactions of agents with each other and with the environment (Table 7.4). In comparison, the neighbourhoods of CA are always fixed. ABM is more powerful in that agents' behaviour can be modelled through machine learning (e.g., artificial intelligence). Rules in CA are simple and flexible, and allow other variables to be easily incorporated into the models. They just need proper parameterisation.

7.3.5 DESIGNING AND DEVELOPING ABMs

The development of spatially explicit models is made easier by the continuous advancements in computing technology. The capacity and speed of present-day computers allow the analysis of large datasets and make it practical to develop spatially explicit simulation models, facilitated by the rapid advances in percolation theory, non-linear dynamic, and CA theories developed over the last couple of decades. In particular, remote sensing and GIS packages supply spatially referenced data readily available in a directly transformable format into a two-dimensional lattice. Nevertheless, designing a reasonably functioning ABM requires careful thought and planning. The best place to start is to ask questions about the model. Consideration should be given to the problem to be addressed, the kind and nature of agents to be included in the model, the agents' environment and mobility, agents' behaviours, such as the mechanisms of their decision-making and actions, data sources, and model validation.

The six critical considerations in developing an ABM can be summarised as follows (DeMers, 1997):

(1) *Model objectives*: A model is developed with a unique purpose or domain of application in mind. In order for the model to work properly, it is crucial to define its purpose and understand its limitations. The modeller needs to ask various questions about it. For instance, what is the problem to be addressed by the model? What is the purpose of developing the model? Is it for explaining a complex phenomenon, for predicting relationships or consequences, or for evaluating situations for resource use? What questions can be answered by the model output? What is it about the ABM approach that is essential so that it cannot be replaced by other modelling approaches?

(2) *Components*: Which model to use? How to apply existing tools to appropriately develop meaningful models? What agents should be included in the model? Which entities in the model have behaviours (e.g., range of roaming)? If so, how do they work? Do agents have a lifespan? What agent

attributes must be specified beforehand, and what attributes can be calculated endogenously by the model and updated in the process of modelling? How should their behaviour be quantified: using experimental field results, or based on findings in the literature? Are there any rates involved in the model? If so, how should they be determined?

(3) *Specification of connectedness*: What should the neighbourhood size be? How large is the extent of influence?

(4) *Mechanism of interactions*: How do agents affect one another? How do they interact with the environment? What is their life cycle?

(5) *Implementation environment*: How large is the simulation space? What is a suitable lattice size? What is the range of mobility of agents in the environment? Is the environment homogeneous or heterogeneous? Are there any constraints in the environment?

(6) *Data needs*: What data are essential to run the model? Where will they come from? Some of them may be spatial in nature. If they are raster data, what will be the spatial resolution?

The development of an ABM can follow a standard procedure consisting of a number of sequential stages (Figure 7.6). Most of them have been covered already, except model calibration. This refers to re-parameterising model variables so that the modelled outcome closely resembles reality. Of the seven identified stages, the most difficult is stage 4: articulating the full range of the dynamics of agents and their possible interactions with each other and with the environment. A sketch flow diagram will be helpful to set out all the possible types of interactions (Figure 7.7). This diagram will become rather complex and messy when multiple agents are involved and each has various paths of variation and interactions with other agents and the environment. Unless such interactions are clearly visualised in a sketch and coded in the model, the simulation will not produce authentic outcomes.

After this crucial step, model implementation becomes a matter of simple codification of the identified relationships among agents and their interactions, and model calibration. The coding requirements and peculiarities vary with the system or toolkit (see Section 7.3.6). A major issue in model implementation is how to parameterise

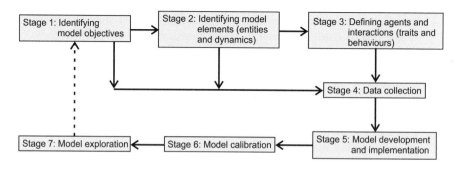

FIGURE 7.6   A typical procedure of and main stages in developing an ABM.

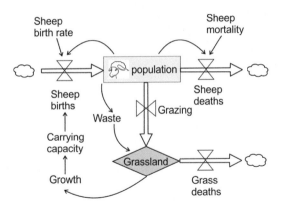

FIGURE 7.7   A schematic model representing the interactions between agents (sheep) and the environment (grassland). Arrows show the relationship between variables; clouds show everything outside the ecosystem; arrows with "vale" symbols show flows; grey boxes show agents; the green diamond shows the environment.

model variables, especially the rate variables, and set the initial conditions. Proper parameterisation of a model demands considerable effort and time. Model parameterisation can be based on the literature or results from field experiments. The former is easier, but the parameterisation established for one geographical area may not be appropriate for other regions. The alternative is to make use of experimental field data, which are essential in setting up the appropriate initial values for some parameters. Field experiments are costly and time-consuming, and may take a long time to generate reliable results. A short experiment may not gather reliable data, which in turn affects the success or failure of the ABM simulation. The more authentically parameterised the model inputs, the more realistic the model outputs. The appropriateness of model parameterisation can be checked by experimentally running the model, then comparing the modelled outcome with the literature or reality to see whether they match closely. If not, some of the parameters need to be modified one at a time until a close match is reached. This is known as the model calibration stage, when the value of some parameters may be fine-tuned to see how the simulated outcome responds to the changes. Once the model is deemed reasonable, then it can be run to explore scenarios.

7.3.6   TOOLKITS FOR IMPLEMENTING ABM

ABM can be implemented using a number of toolkits (Table 7.5). They differ widely in their ability to work with files in different formats and in their treatment of spatial proximity, each having its own unique features and requirements. The ability of six toolkits to work with geospatial data is compared in Table 7.5. Of these toolkits, NetLogo will be discussed separately in Section 7.4, so it is not included in the comparison. Featured prominently in the comparison is the toolkits' ability to represent neighbourhood and to place agents into cells. Cormas can accept data stored in the

TABLE 7.5

Comparison of Main Features of Toolkits for ABM with Geospatial Data

Toolkit	Space	Agents	Neighbourhood		Placement	
			Agent	Cell	Cell→Agent	Agent→Cell
Cormas	Raster or vector	*	Network functions	Euclidean	Manipulate, create	*
GRSP	Vector	Vector	*	Euclidean	Manipulate, create	*
NetLogo	Raster	*	*	Euclidean	Manipulate	*
OBEUS	Vector	Vector	Transition	Complex operators	Manipulate, create	Manipulate, create
Repast	Raster or vector	Vector data	Network functions	Complex operators	Manipulate, create	*
TerraME	Cellular data	*	*	Generalised proximity matrices	*	Manipulate, create

Source: De Andrade (2010).

*Up to the modeller. Time is not included in the comparison as it is up to the modeller as well.

MID/MIF and ENV formats. Repast, OBEUS and GRSP all treat space and agents as geospatial entities, but they differ from each other in their ability to handle data in different formats. Repast is highly compatible with ArcGIS by recognising shapefiles for vector data, but only ASCII for raster data. OBEUS is a GIS package in which both the properties and locations of objects can change with time. GRSP is the most restrictive, as it makes use of PostgreSQL databases. TerraME is a kind of CA, but the properties of regular grids are stored in TerraLib databases. More critically, it is unable to represent agents with geospatial data (De Andrade, 2010).

The placement of agents into cells (an explicit 1 agent → 1 cell representation) and neighbourhood of cells is based on Euclidean proximities in all toolkits except OBEUS, Repast, and TerraME. Complex vector operations, such as point-in-polygon, buffering, and intersection to determine proximity between cells, are possible with OBEUS and Repast. OBEUS takes advantage of the leader and follower approach to transit agent to agent from other relations. The general concept of non-proximal spaces is used in TerraME, which calculates proximity from complex neighbourhoods created by two objects, as well as additional layers such as a transport network. However, TerraME cannot represent agents as geospatial entities, so it cannot create a neighbourhood for them. It enables dynamic spatial modelling that requires the cellular space to be linked to geospatial databases for data storage and retrieval. Since its initial development, new methodologies have been proposed to increase its functions. Now, general types can be defined to meet the needs of spatially explicit dynamic modelling in studying human–environment interactions (De Andrade, 2010).

7.4 NETLOGO FOR DYNAMIC SIMULATION

7.4.1 GENERAL FEATURES

The NetLogo scripting language has an easy-to-learn, user-friendly syntax and structure. One of its strengths is that each line (command) in the source code is executed immediately, so any logical errors in it can be detected in a timely fashion without the need to search for the wrong expressions throughout the script, which accelerates the process of scripting a grammatically correct model. Another strength of NetLogo is its modular structure. All the modules written by different modellers can be assembled easily in the control program by simply spelling out the modules' names without any knowledge or concern about their internal structure and the variables local to them. Such a modular structure offers a high degree of autonomy and flexibility in undertaking a vast and complex simulation project by multiple scripters at the same time without the need to communicate with each other.

NetLogo has an easy-to-use graphical user interface (GUI). The user can interface with the scripted model by setting up buttons on the screen to control its execution and to change the values of certain variables. The attribute values of certain variables can be changed on-screen independent of the script itself, which provides tremendous ease and flexibility in running simulation models and in testing the sensitivity of certain variables. On the graphical interface screen, various buttons and sliders can be set up to initialise the simulation model. All the parameters are set to the default values or generated in a spatially random manner (e.g., spatially randomised) prior to running the model with a simple click on "setup" (Figure 7.8). The slider buttons allow the user to change the value of rate variables. The "go" button controls the execution of the script. A single-click activates the execution, and a double-click terminates it. The speed slider controls the running speed and parameter values so that the modeller can see exactly what is happening on-screen. This is handy as there are no written records of all the intermediate results. It is also possible to turn on/off a variable in the simulation model via its "switch" button. This is useful when running scenarios in which some variables may be absent. It is also possible to display the values of important attributes during different model runs (iterations) on-screen. They are usually displayed as time-dependent line charts, together with the spatial output (Figure 7.8).

In order to understand NetLogo and make use of it to run simulation models, it is essential to know some of the jargon unique to it. The NetLogo world is made up of *patches*, *turtles*, *links*, and an *observer*, all being NetLogo agents. *Patches* refer to grids in the environment representing the simulation space – the ground over which turtles move. All patches are autonomous agents with an independent state and behaviour. *Turtles* are mobile agents that can roam the simulation space

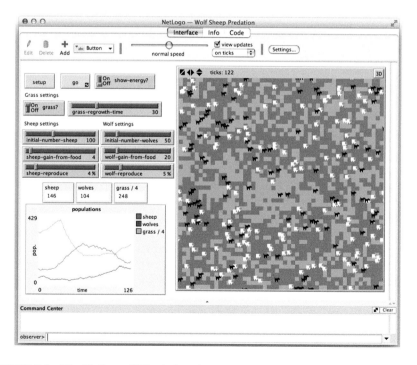

FIGURE 7.8 The NetLogo GUI window that allows the value of model parameters to be changed on-screen outside the script. The modelled results of sheep–wolf ABM are displayed in real time both graphically and as a line chart.

as interacting entities. The direction and pace of movement are governed by the rules set in the model. Turtles are associated with certain primitives, such as *die* (kills a turtle), *hatch* (creates a given number of offspring at the exact location as the mother), (move) *forward*, (turn) *left*, and (turn) *right*. These are pre-defined functions that can be used directly in a NetLogo script. *Links* are the connections between turtles. The *observer* oversees everything that is going on and controls and monitors the progress of simulation from outside the environment. Commands to all agents can be issued via the Command Center at the bottom of the screen. Variables can be associated with a turtle, a patch, or a link, so there are turtle variables, such as life expectancy, growth, reproduction rate, and mortality. Variables can also be global or local, depending on where they are defined (e.g., inside a module or at the beginning of the script). A global variable can have only one value (e.g., it does not change with time) and is valid throughout all modules. In contrast, a local variable defined inside a module is applicable to it alone. It is not recognised beyond this module. NetLogo also has its own built-in variables, such as *xcor*, *ycor*, etc. It is important to understand these terms as some actions can only be initiated by certain types of agents. A permissible action for turtles may not be valid for a patch. To some degree, they represent different classes of features in the script.

NetLogo has the ability to save all scripts automatically, so there is no need to type all the codes that have already been entered into the system in a previous session, which considerably boosts the efficiency of programming as some pre-existing functions can be imported into a script easily. More importantly, a large number of NetLogo scripts have already been written by other modellers, and they are freely shared among the spatiotemporal modelling community. They can be tailored to suit a particular simulation application via some modifications. In addition, NetLogo's model library contains an extensive collection of pre-written simulation scripts in a diverse range of disciplines, ranging from traffic jam modelling in transportation to fire modelling in ecology, all of which can be modified to suit a particular need.

7.4.2 ANATOMY OF A NETLOGO MODEL

All NetLogo programs follow the same standard structure with a few essential ingredients. First of all, the variables of the model must be defined. They can be global variables that are applicable to all the functions and subroutines, or turtles' own variables that are applicable to only turtles, so have no effects on patches. Next, the model needs initialisation to set it up. It is important to initialise the model after every run, as the value of certain variables may have changed during a previous run. This setup procedure may involve a degree of randomness (e.g., initial values randomly assigned to certain variables) because the true value is unknown. Another essential item is to define agent procedures for both turtles and patches. Here is a sample script:

```
Defining the model's variables (global …, turtles-own …);
Model initialization (to setup);
Agent procedures (to turtle-procedure, to patches procedure);
Simulation part (to go)
Plot or output.
```

The second part of a NetLogo script contains the actual simulation module starting with **to go**. All the mathematical calculations or logical reasoning are implemented in this part. The last component of a NetLogo program is the output module. It **plots** out the simulated outcomes instantly on-screen as the model run progresses, either spatially or as time-dependent curves. It is also possible to insert a pause button in the script to stop the plotting temporarily to examine the output value in detail:

```
to go
    move-turtles
    eat-grass
    reproduce
    check-death
    regrow-grass
    tick
end

to reproduce
    ask turtles [
        if energy > 50 [
            set energy energy - 50
            hatch 1 [ set energy 50 ]
        ]
]
    end ;; this is a procedure called reproduce
```

The above codes represent two **procedures** in which a series of actions are lumped together. Each procedure must start with **to** and end with **end**. The word immediately after **to** represents the procedure name. Each procedure may contain more procedures (the first one) or a series of commands (the second one). These commands should be indented to maximise coding clarity so that any missing brackets can be detected easily. In the above codes, **if** is a conditional statement. The condition after it is assessed. If it is true, the action immediately following it is executed. Sometimes the alternative action (if present) is executed if the condition is not true. The expressions **set** and **hatch** are NetLogo primitives. The expression **set energy energy - 50** means 50 is deducted from energy after this operation, whereas **set energy 50** just initialises energy to a value of 50.

7.4.3 Sensitivity Analysis

Sensitivity analysis aims to ascertain the effect of the varying values of a group of independent variables on the dependent variable under certain specific conditions. It is especially valuable in analysing a "black box process" in which how the independent variables affect the dependent variable individually remains unknown or not well understood, and hence cannot be described mathematically. Sensitivity analysis plays a vital role in model development and evaluation by assessing the effects of variability on model parameter values and inputs on model outputs. In order to test the sensitivity of a given input parameter and generate a meaningful output, ideally sensitivity analysis should be carried out by holding all variables' attributes value unchanged apart from one variable whose value is allowed to change so as to quantify the response of the dependent variable to the change (Babiker et al., 2005). As such, sensitivity analysis can identify the level of dependency of the output on a particular input value. It is also useful in examining how the uncertainty of the input variables affects the modelled outputs by changing their values slightly within a small range. Through sensitivity analysis, it is possible to reduce the uncertainty of parameters in the simulation model and increase our overall confidence in the model outputs. It can also facilitate the search for mistakes in the model. Sensitivity analysis also enables the identification of the most important and influential input variables among all the considered ones. These are the inputs that must be correct. It is thus possible to identify a subset of variables that dictates the output variability. Sensitivity analysis is also valuable for predicting an outcome under hypothetical circumstances or for what-if scenario analysis. It is particularly essential to carry out a sensitivity analysis in ecological modelling involving a wide range of factors (see Section 7.5.3). Finally, sensitivity analysis is able to reveal how errors propagate in the model by examining the impacts of input errors on outputs, mostly via simulation. The input variable's value is deliberately altered, while all other parameter settings remain unchanged. Such an analysis can shed light on the risk of a strategy.

The most fundamental way of undertaking a sensitivity analysis is one at a time: only one parameter value is allowed to vary. The derived sensitivity is usually expressed as a ratio of the change (%) in input to that in output.

7.4.4 NETLOGO VERSUS SPATIOTEMPORAL SIMULATIONS

NetLogo is the ideal platform for running highly complex agent-based spatiotemporal simulations. It is relatively flexible because some variables or agents can be turned on or off at will without modifying the source code. It is also good at analysing scenarios by simply modifying the values of a given parameter on the interface screen and carrying out sensitivity analysis (see Section 7.3). Nevertheless, caution must be exercised in interpreting the simulation outcome, as model performance is only as good as the rules themselves in the script. The modeller can give instructions to hundreds or thousands of "agents", all operating independently of each other. In using NetLogo for simulation, the same run must be repeated owing to spatial randomness in the initial setting of certain variables. In order to avoid any potential bias caused by the randomness assumption, the same model has to be run tens or even hundreds of times repeatedly using exactly the same settings or parameterisation. Only averaging of all these results can lead to a genuine outcome of simulation. Apart from sensitivity analysis, calibration and statistical validation may also need to be carried out.

In spite of the relative ease in running spatiotemporal simulation using NetLogo, its applications in simulating some ecological processes are not without controversies. NetLogo-based spatiotemporal simulation faces three common limitations related to assumptions, external influences, and scaling. The assumption about the simulation space being invariably homogeneous and spatially continuous is not always valid.

There is no room to accommodate spatial heterogeneity. The space of some ecological processes is neither homogeneous nor continuous due to the existence of ecological barriers. Spatial heterogeneity can easily be caused by topography. For instance, soil moisture, nutrients, and even temperature all vary with topography. Unless the surface is perfectly flat, which is rather rare at a broad scale, these environmental variables are going to vary spatially with slope gradient and aspect. More realistic simulation must take such spatial heterogeneity into consideration by incorporating an additional raster layer (e.g., a DEM) into the simulation process, against which the same functions can be modified for steep cells or cells lying in the shade.

The second limitation of NetLogo-based ecological simulation is the utter disregard for external changes. In most modelling applications, attention is devoted only to internal changes. Thus, the same iteration is repeated from one run to the next as if the world never changes no matter how long the simulation period spans. This indiscriminate treatment of time increment ignores possible changes in the outer environment. It may not change much over a short span, such as months or even years. However, environmental variables, especially climate variables such as temperature, rainfall, and sunshine hours, may change considerably over decades or even longer temporal scales. Thus, the same process may be subject to the influence of more variables, and the same assumptions about the same variable may not be valid any more. Such annual or decadal variations must not be ignored in the simulation in order to generate reliable predictions.

Finally, the simulation of processes at a micro-scale may not be truly reflective of the processes operating at a much broader spatial scale. No matter how large the lattice space that is adopted in a NetLogo simulation, it is always square in shape and partitioned into a uniform cell size. In order to keep the simulation to a manageable level, the simulation space is restricted to tens of square kilometres at most. This may appear to be rather large, but can never match the catchment or watershed size at which some ecological processes are operating. Also, the local-scale processes and dynamics of agent behaviour and interactions may not be exactly replicable at the regional scale and beyond. Thus, caution needs to be exercised in upscaling the results obtained from the local-scale simulation to the landscape-scale and regional-scale processes.

7.5 SPECIAL CASES OF SPATIAL SIMULATION

7.5.1 WILDFIRE SIMULATION

Simulation of wildfire or wildland fire spread refers to numerical modelling of fire propagation in the wildness at the landscape scale in an attempt to understand and predict the behaviour of fires, such as the direction and speed of spread, the proportion of fuel burned, and the heat generated. Such information plays a vital role in extinguishing a fire, mitigating fire damage, and increasing the safety of firefighters and the public. *Wildfire simulation* also plays an essential role in protecting ecosystems and watersheds and in improving air quality. Wildfire modelling can also assess the ecological and hydrological effects, tree mortality resulting from a fire, and the quantity of smoke produced by a fire.

At the core of wildfire simulation are models of fire propagation, fuel distribution, and combustion. Fires commonly propagate in two modes, network and cellular (from cell to cell), resulting in fires being modelled in two different environments, vector and raster. Fuel models prescribe the type of fuel in the fuel bed that is burned in a fire. Commonly used fuel models include the Albini (1976) models and the Anderson (1982) models, each having its own assumptions and applicable scales. Wildfire fuel can be broadly classed as grass, dead wood, and live plants, and their height. Small, dry twigs close to the ground burn much faster than large trunks close to the canopy. However, models of combustion may not be needed in modelling surface fires.

In addition to models about fires themselves, successful and realistic simulation of wildfire spread must take into consideration environmental factors. These fall into two categories: terrain and weather. Terrain can be easily considered via a DEM, from which elevation and aspect can be derived. Of all the weather variables, the two most important in wildfire simulation are wind (speed and direction) and antecedent moisture, which is related to rainfall, evaporation, temperature, and humidity. Wind is the most influential to the direction and velocity of fire spread, but is the most difficult to predict. Related to moisture is temperature. At a higher temperature, the same fuel will burn much faster. Although rainfall can suppress and even extinguish a fire directly, it is seldom considered in wildfire simulation as it is unpredictable and rarely coincides with a fire event. Other unusual events, such as cold fronts, thunderstorms, sea and land breezes, as well as diurnal slope winds, may also be considered in a sound wildfire simulation.

7.5.1.1 Wildfire Models

All wildfire simulation models fall into three categories: physical, empirical, and stochastic (Table 7.6). Physical models are based on the physical and chemical mechanisms of fire spread processes in the combustion of biomass fuel and its interaction with the atmosphere. These models require an understanding of several complex processes, such as convection, radiation, and turbulence. They make use of mathematical equations to predict the rate of fire spread, defined as the steady propagation of the flame front through the fuel layer (Koo et al., 2005). The shape of the flame, defined by its length and angle, determines the spread rate. Physical models are rather computationally demanding and data-intensive, requiring consideration of a large number of parameters and boundary conditions, all of which are difficult to parameterise. Hence, they are frequently simplified in practice, such as simplified chemistry, averaging time, and turbulence modelling. Even with these simplifications, physical and quasi-physical models take a long time to run. They are suitable for studying fires in a laboratory, where the weather conditions can be strictly controlled. In reality, physical models have found few applications to the heterogeneous meso-scale landscape due to the difficulty of model parameterisation. These challenges can be circumvented by semi-physical models that are still physics-based, but involve assumptions about a number of processes and fuels (Drissi, 2015). Semi-physical models may be created by combining a network model with a quasi-physical model of the interaction between burning and non-burning cells. It is physical in the

TABLE 7.6

Three Categories of Surface Wildfire Models and Typical Examples in Each Category, and Their Main Properties

Category	Model name	Author(s)	Main features
Physical	Simple	Koo et al. (2005)	Prediction of flame spread (one-dimensional steady-state contiguous spread) rate based on energy conservation and heat transfer mechanism; non-spatial; time-irrelevant
	Semi-physical	Drissi (2015)	Combination of a network model with a quasi-physical model to simulate fire patterns in heterogeneous landscapes; time sequence results
(Semi-)empirical	Rothermel	Rothermel (1972)	Consideration of fire environment factors such as wind velocity, slope, propagating flux, and fuel characteristics to predict the rate of fire spread
Stochastic	Shortest path	Hajian et al. (2016)	Wind speed treated as unpredictable; landscape represented as a network; Monte Carlo simulation for determining fire travel time distribution; spatially explicit
	Cellular automata	Freire & DaCamara (2019), Braun & Woolford (2013)	Possible to couple with existing fire spread models; simulation of burn probability

sense that radiation and convection from the flaming zone and radiative heat loss to the ambient are all considered in the model. The alternative to complex physical models is empirical models.

Empirical models are based on historical wildfires, experiments, and observations. One of the best-known empirical models is the Rothermel (1972) model for simulating the rate of surface fire spread and burning intensity. Empirical and quasi-empirical models may involve two components, a fire behaviour model and a fire spread model (Ghisu et al., 2015). Fire behaviour is commonly modelled from a number of variables, including wind speed, slope gradient, fuel moisture, fuel load, and density, all of which affect the rate of fire spread and burn intensity. The fire spread model spells out how the fire perimeter evolves across the landscape, based on local settings of fuel, weather, and topography. Apart from the rate of spread, empirical models cannot generate information on fire size, heat flux, or fuel temperature. Although much simpler than physical models, empirical models are highly limited in their applicability. They must be calibrated before being applied to an area that is different from where they were originally developed. In contrast, semi-empirical models include at least one physical law. As such, they are more adaptable from one scale to another due to the consideration of fire physics.

Neither physical nor empirical models can address the uncertainties associated with the weather, especially wind. The best way of dealing with such uncertainties is to use stochastic models that have an element of randomness, and the simulated outcome may be more authentic due to the redundancy of ad hoc fine-tuning of the model parameters, such as the shortest path of fire propagation (Hajian et al., 2016). Stochastic models are commonly implemented as simulation models in the GIS environment, in either vector or raster format, especially CA (Freire and Dacamara, 2019). Stochastic models are effective at addressing the effect of weather, which is critical to fire behaviour, but has a degree of variation that is unable to be captured by the weather data.

7.5.1.2 Raster or Vector Simulation Models

All the simulation models in Table 7.6 are implemented in one of three approaches in fire spread modelling: shape (pre-determined fire perimeters and areas), lattice (spread of fire from one cell to another), and vector (expanding fire polygon). In terms of data format, they can be grouped as vector and raster simulations. Vector-based simulation treats fire as a polygon, and the fire perimeter is represented as a closed, discrete curve, usually elliptical in shape, defined by a number of vertices. These vertices expand as the fire spreads out according to the local conditions. This format is the default choice if fire propagation is thought to take place along a network. The simulation of fire growth becomes the determination of the perimeter of the fire front. The outer shape of all individual fires constitutes the new perimeter, which is discretised as the fire expands. The main drawback of the vector implementation is the high intensity of computation in generating the convex hull fire-spread perimeter at every temporal increment. The computation is even more intensive in light of fire crossovers and unburned islands (Ghisu et al., 2015). Another limitation of vector implementation is the inability to represent the spatial heterogeneity of fuel and terrain, which can be avoided by using the raster format.

Lattice-based fire simulation is fully compatible with CA modelling. As a fire grows in size, it may spread out in multiple directions at various speeds. It is this kind of growth that can be simulated easily using CA. CA-based fire spread simulation assumes fuel and terrain conditions within each cell to be homogeneous, and fire propagates cell-wise. How the fire spreads from one cell to the next depends on a set of transition rules, as well as the defined adjacency or spread template (Tymastra et al., 2010). The transition rules can be based on wave propagation, percolation theory, and stochastics. In wave propagation, every point on a wave front of light is considered as a source of individual wavelets. A fire is represented as a polygon, and its front comprises a string of straight-line segments forming a closed path. The vertices defining a line segment propagate in accordance with the partial differential equations as the fire expands. In percolation theory, fire spread is simulated as a diffusion process, and the transition rules are based on probabilities of fire propagation from burning cells to unburned cells. This allows for modelling of probability processes, but cannot yield realistic information on fire perimeters.

In fire simulation, the spread template is subject to the direction and shape of the flames. In addition to the already-mentioned 8-cell Moore neighbourhood template (Figure 7.1), 16- and 32-cell neighbourhoods have also found uses in fire modelling (Figure 7.9a and b). In comparison with these templates, the 8-cell spread template is inferior as it produces unsatisfactory shapes of flames, a problem not found with the 32- and 64-cell spread templates (O'Regan et al., 1976). The first two adjacency templates in Figure 7.9 are symmetrical, which may lead to a large error in simulating wind-driven fire spread from an ignition point. Symmetrical adjacency templates are suitable for wildfires that produce regular shapes such as ellipses, double slips, and ovoids under homogeneous conditions (e.g., constant fuels, weather, and topography). The use of symmetrical templates is prone to large errors in case of heterogeneity, which can be reduced via the use of asymmetrical templates (Figure 7.9c). Instead of a square neighbourhood, it can be elongated towards the dominant wind direction. Another way of remedying the errors is to apply a correction factor to either the maximum rate of spread, the ellipse eccentricity, the dependence of the fire spread rate on the angle from the effective wind direction, or the rear propagation speed (Ghisu et al., 2015). It is even possible to modify the fire spread rules to non-adjacent cells, depending on wind speed for large fires at the explosive stage of combustion (Freire and DaCamara, 2019). In the simulation, each grid cell has one of three possible states – unburned fuel, burning fuel, and burnt out – of which the burning fuel cell is the point of ignition.

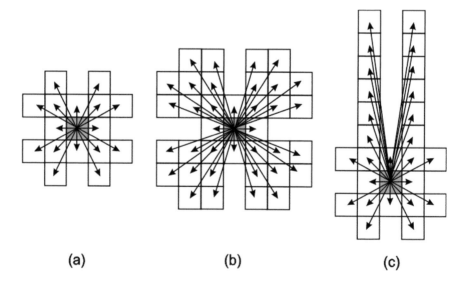

(a)   **(b)**   **(c)**

FIGURE 7.9   Three types of adjacency templates commonly adopted in CA-based simulation of fire spread: (a) 16-cell; (b) 32-cell; and (c) asymmetric adjacency template suitable for simulating wind-driven fire spread. Blue cell = point of ignition (source: modified from Tymastra et al., 2010).

The transition between these states is unidirectional, always from unburned fuel to burning fuel, and eventually to burnt out.

Raster-based fire simulation is spatiotemporal modelling, in that a spatial output can be generated at the user-specified time interval. Raster-based CA simulation of wildfire spread is computationally efficient, but the simulated outcome suffers from distorted fire shapes caused by inconstant winds. Another major drawback of CA-based fire simulation is the treatment of fuel and terrain as a constant within a cell. Also, the fire spread is confined to only eight possible directions. These deficiencies can be overcome by redefining the spread velocity by modifying the equations commonly used in vector-based simulation (Ghisu et al., 2015). Since neither vector nor raster simulation are perfect, a natural solution is to combine them: fire perimeter is simulated in vector format, and fuel and terrain are represented in raster format to take into account their spatial heterogeneity.

7.5.1.3 Common Wildfire Simulation Models

So far, several wildfire simulation models have been developed for different countries to achieve different tasks, based on different fire behaviour models and for different kinds of local environments (Table 7.7). Of these models, FARSITE is the most mature and widely used. It is designed to simulate fire spread under realistic slope and wind conditions by incorporating a number of existing models covering surface fire, crown fire, point-source fire acceleration, spotting, and fuel moisture (Finney, 2004). It is able to produce a 2D fire perimeter in vector format at specified time intervals. The required inputs include elevation, slope, aspect, canopy cover (%), crown height, crown base height, crown bulk density, weather and wind, and fuel moisture. It is able to handle interactions between multiple fires. Initially developed for the US, it may need calibration for local fuel settings if applied elsewhere. In addition, site-specific fuel models may be used to replace the standard fuel models by sampling the main natural vegetation type complexes (Jahdi et al., 2015).

Another very successful model is Prometheus, the Canadian Wildland Fire Growth Simulation Model (Tymstra et al., 2010). It produces the fire front perimeter at user-specified temporal intervals (Figure 7.10). This vector-based deterministic model is underpinned by the principle of wave propagation. Its required inputs include topography (slope, aspect, and elevation), fuel types, and weather. It is flexible and able to yield diverse outputs, including the rate of fire spread, detailed information on fire perimeter and intensity, as well as the fuel burned. Its major limitation is that it can only run scenario-based simulations. It is also rather complex and unable to take into consideration fire interactions in cases of multiple fires.

Developed for France, ForeFire is a simplified physical model involving ten assumptions about flame shape, heat convection ahead of the flame front, flame velocity and tilt angle, surface fuel distribution, mass loss rate, radiative tangent plane, radiative factors, and so on (Balbi et al., 2009). It can be regarded as a reduced 2D physical model that is based on a full physical 3D model, thereby giving it wide applicability. This model assumes the rate of surface fire spread to be a function of wind, fuel, and topography. It is able to generate realistic and detailed flame geometry (height, depth, and tilt angle) and thermodynamic quantities (temperature,

TABLE 7.7

Comparison of Major Fire Simulation Models, Their Best Uses, Strengths, and Limitations

Model/country	Author(s)	Main features	Pros	Cons
FARSITE/US	Finney (2004); Jahdi et al. (2015)	Spatially and temporally explicit; semi-empirical; required inputs: topography, surface fuel models, canopy characteristics, and fuel bed, fuel moisture, and climate data	Able to simulate the rate of spread and flame length; both spatial and temporal growth of fire possible; results realistic	Effect of wind on crown fire not authentic; shape of fire assumed to be elliptical; more fire spread models need to be considered
Prometheus/Canada	Tymstra et al. (2010)	Vector model based on the principle of wave propagation; required inputs: topography (slope, aspect, and elevation), fuel types, and weather	Flexible, diverse outputs (rate of spread, detailed data on fire perimeter and intensity, fuel burned)	Scenario-based simulation; complex; unable to consider fire interactions
ForeFire/France	Balbi et al. (2009)	Reduced physical model based on the full physical 3D model via assumptions; rate of surface fire spread treated as a function of wind, slope, and vegetation	Realistic and detailed simulation of flame geometry and thermodynamic quantities; faster than real-time; easy to use; broad applicability	Ten assumptions may not be entirely valid
SiroFire/Australia	Coleman and Sullivan (1996)	Vector fire spread prediction system incorporating fire behaviour models for fire spread (raster used for terrain and fuel conditions)	Simplistic; real-time; five spread models to choose from; reasonable performance	Grass and forest fuel only; rate of fire spread only; no full fire perimeter; not all critical factors considered
Burn-P3/Canada	Parisien et al. (2005)	Landscape-level Monte Carlo simulation based on hybridised local-scale deterministic modelling with large-scale probabilistic fire ignition, spread, and weather	Diverse output of rate of spread, fire intensity, and crown fraction burned; burn probability	Data-demanding; computationally intensive; empirical ignition patterns and burning conditions

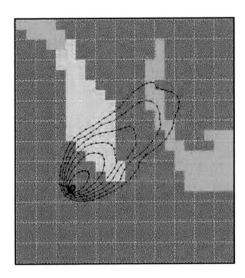

FIGURE 7.10 Spread of a wildfire from the ignition point (red circle) during the first 35 minutes, simulated using the Prometheus model. The black dots represent the individual vertices along the fire perimeters. Cell size is 25 m; temporal interval is 5 minutes. Colours show fuel types (blue is C-1, grey is C-2, and beige is O-1a) (source: Tymstra et al., 2010).

radiant flux, and fire front intensity). The computation is faster than real-time, making it rather valuable for firefighting and containing a fire. However, the ten assumptions may not be valid in all cases. The system is easy to use as there are only three parameters to configure: wind, terrain, and vegetation.

Developed for Australia, the SiroFire model incorporates five fire spread models to suit users familiar with different models. Only two major fuel types (grass and forest) commonly found in Australia are considered in the model. The potential spread of fire across the landscape is calculated from GIS-derived maps and DEMs (Coleman and Sullivan, 1996). It makes use of both vector and raster data. Fire spread is simulated in vector format, with fuel and terrain conditions represented in raster format to take into account their spatial heterogeneity. However, the model has found few applications outside Australia.

All the aforementioned models are suitable for simulating fire propagation by yielding information on the rate of fire spread, flame geometry and thermal dynamic quantities for specific parts of the world. As such, they are unable to model burn probability, which can be achieved using the Burn-P3 (probability, prediction, planning) simulator (Parisien et al., 2005). It is able to simulate landscape-level burn probability at annual intervals based on scenarios using Monte Carlo simulation, in which local-scale deterministic modelling is hybridised with large-scale probabilistic fire ignition, historic fire spread events, and the weather conditions. In addition to burn probability, it is able to output diverse simulation outcomes, including the rate of spread, fire intensity, and crown fraction burned. However, it is data-demanding and computationally intensive. Also, both ignition patterns and burning conditions are empirical, which may restrict its applicability to certain areas. Apart from Burn-P3, burn probability can also be generated by repeatedly running a stochastic lattice-spread model and calculating the time required for individual grid cells to be burned (Braun and Woolford, 2013). Such a stochastic fire spread simulation is superior to Burn-P3 in that it avoids the uncertainties intrinsic to fire behaviour. The stochastic simulation also offers another advantage by holding all variables constant except one. In this way, the effect of one variable such as scale can be determined precisely.

Some of the aforementioned fire models can be implemented in BehavePlus. This Windows-based system contains a total of nine modules for modelling both surface and crown fires. Its GUI page layout is intuitive and allows the user to enter the necessary parameters. Users unfamiliar with the system are guided by worksheets to enter the appropriate parameters. Results can be generated quickly using the default settings (Heinsch and Andrews, 2010). The output layout page displays results that encompass surface and crown fire spread rate and intensity, probability of ignition, fire size, spotting distance, and tree mortality in tabular and graphic formats. These outputs can be used to cater for a host of fire management needs, such as projecting the behaviour of an ongoing fire, planning prescribed fires, and assessing fire hazards.

7.5.2 Urban Expansion Simulation

At present, approximately half of the world's population live in urban areas, and this proportion is projected to surpass 60% by 2030 (Hertel, 2017). Consequently, the total urban area of the world is expected to grow rapidly to reach 1.25 million km^2 by 2030 (Angel et al., 2011). Without proper planning, rapid urban growth will cause grave problems, such as traffic congestion, air pollution, environmental degradation, and loss of amenity. In order to avoid these problems, there is a need to understand the spatiotemporal behaviour of urban growth and identify the importance of different socioeconomic, cultural, physical, and environmental factors in the process of urbanisation through simulation. Urban expansion simulation can also yield vital information for properly planning suitable and sustainable urban developments. Specifically, urban growth simulation can help us:

(1) Identify high potential growth areas and make preparatory plans for the anticipated growth prior to the advent of problems associated with chaotic urban development;
(2) Minimise the negative impacts of the anticipated urban growth through better planning;
(3) Explore alternatives in urban design to make urban development sustainable and residents- and environment-friendly.

All simulations and predictions of urban growth are implicitly underpinned by the assumption that "the historical urban areas will affect the future expansion patterns through interactions among different land use types" (Clarke et al., 1997). Whenever this assumption is violated, inaccurate simulation will occur. In urban expansion simulation, historic land covers must be available, from which past behaviour of certain land covers, particularly changes in the targeted land covers, can be learned. Usually, at least three time maps must be used. Analysis of the first two time maps can produce training samples of both transformed and untransformed cells in the raster modelling environment. The second change map produced from the last two time maps can be used to validate the simulation model and indicate its accuracy. Finally, reliable simulation of urban expansion requires an understanding of how urban areas expand spatially or the modes of urban expansion.

7.5.2.1 Modes of Urban Expansion

So far, a number of urban expansion modes have been identified, such as infilling, edge expansion, road-like (linear), and leapfrog (outlying or spontaneous) or sprawling (Figure 7.11). Infilling expansion takes place within the existing urban limit, within which non-urban areas are converted to urban uses. It affects the total urban area, but does not affect the outer urban boundary. Edge expansion or concentric growth refers to the physical extension of the current urban limit as new urban areas emerge right next to existing ones. This expansion bears little relationship to transport. In contrast, linear development or sectoral expansion takes place in linear form along artery roads or in the least resistant direction from the city centre. This mode is also known as ribbon or strip growth. Sprawling or leapfrog expansion means that newly urbanised areas are physically detached from existing urban areas, with the distance between existing urban and newly developed areas varying. These growths may form multi-nucleus functional zones or a satellite city of a metropolis. To some degree, road-like expansion is a special case of leapfrog growth, in that the new urban areas expand along roads.

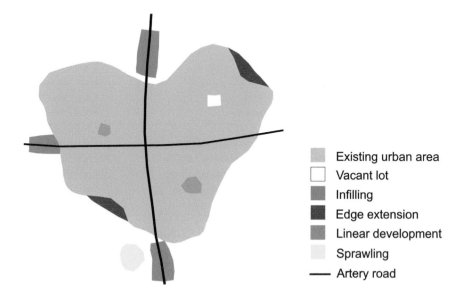

Existing urban area
Vacant lot
Infilling
Edge extension
Linear development
Sprawling
—— Artery road

FIGURE 7.11 Comparison of various modes of urban expansion. Sprawling is also known as outlying or leapfrog growth.

It is possible for a large city to expand in a mixture of several modes of expansion to varying degrees. Accurate simulation of urban expansion must take into account all possible modes of expansion. Of all these modes, organic growth (e.g., infilling and edge expansion) is the easiest to implement. In the CA environment, a cell cannot be turned into urban unless one of its neighbouring cells is urban already. The road-like mode can also be handled with relative ease by checking whether a cell lies within the specified threshold proximity to a major road via buffering. The most challenging mode to handle is leapfrog expansion, which can be accommodated by spatially allocating the suitable cells randomly within a band surrounding the existing urban limit.

7.5.2.2 Players in Urban Expansion

Urban expansion results from the intricate and mutual interactions of a plethora of demographic, socioeconomic, cultural, environmental, and physical variables (Table 7.8). Each of them may contain more sub-variables, all of which are just generic variables commonly considered in all urban expansion simulations. Other more specific variables, such as building density and proximity to transport

TABLE 7.8
Variables That Play a Role in Urban Expansion and Their Scope of Influence

Category	Variable	Scale of influence
Demography	Population growth	Global (driving)
	Net immigration	Global (driving)
	Age	Local
	Education level	Local (selective)
Socioeconomic status	Income	Local (selective)
	Family size	Local (driving)
	Kinship	Local (attractive)
Natural environment	Steep terrain	Local (inhibitive)
	Natural hazard	Local (exclusive)
	Land use	Local
Amenity	Proximity to commercial hubs	Local (attractive)
	Proximity to recreation facilities	Local (attractive)
	Proximity to education facilities	Local (attractive)
Accessibility	Distance to transport	Local (attractive)
	Distance to existing urban area	Local (attractive)
	Distance to main roads/motorway	Local (attractive)
Sense of community	Safety (crime rate)	Local (deterrent)
	Ethnic diversity	Local
	Religious affiliation	Local
Institution	Urban versus rural zoning	Global (inhibitive)
	Ecologically sensitive areas	Local (exclusive)
	Culturally important sites	Local (exclusive)

facilities, may need to be considered for special cities. All of these factors play either an inhibiting or driving role in urban expansion (Figure 7.12). The inhibiting factors attempt to contain urban expansion or confine it to certain parts of a city through zoning regulations. For instance, no development is permissible within landslide-prone or geo-hazardous areas. These factors are primarily intuitional, imposed by the local authority to prevent potentially costly damage and remediation later on. All the driving variables fall into two (global and local) types in terms of their scope of influence. Global drivers cause the overall urban extent to expand. They represent the demanding side of urban expansion. In contrast, local drivers cause the urban area to expand in only certain localities without any effect on the overall urbanised area. The most crucial global driver of urban expansion is population growth as a consequence of a higher birth rate and inward migration. Having more households equates to a larger urban area if population density is maintained spatially uniform. Environmental factors, such as amenity and topography, play a secondary role in affecting where new arrivals in a city choose to settle, and hence the spatial distribution of the newly expanded urban areas. For instance, topography affects urban expansion, in that steep terrains are less favoured for urban development due to the higher construction costs. It is such local variability that the bulk of urban growth simulation efforts are required to address, as the same local factor will not affect the preference of all residents equally, for example where to site a newly urbanised cell in the study area. For instance, high-income residents have a strong desire to live in an area close to all local amenities, while low-income residents may prefer to live in close proximity to transport facilities. In contrast, global drivers affect all residents equally regardless of their socioeconomic status and preferences. It is imperative to differentiate the role of these variables, as they must be treated differently in coming up with the transition rules for cell state.

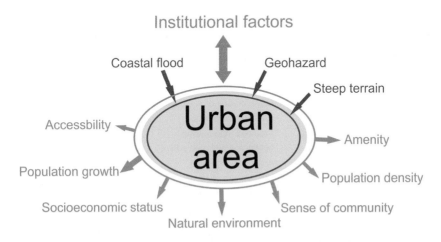

FIGURE 7.12 Commonly considered variables in simulating urban expansion and their different and contrasting roles in urban expansion simulation. Green arrows show drivers, red arrows show inhibitors, and orange arrows show policymakers.

7.5.2.3　Simulation Models of Urban Expansion

So far, several models have found applications in simulating urban expansion. In the early days, only individual models were used in isolation, such as logistic regression, CA, and agent-based models (Table 7.9). All of them have their own deficiencies, as they take into account only specific elements in the complex process of urbanisation. Understandably, the simulation accuracy is not satisfactory. One attempt to overcome the deficiency involves hybridising these traditional models with new machine learning algorithms, such as support vector machine, Random Forest, and ANN. These algorithms are ideally suited to establishing authentic relationships between urban expansion and its influencing factors because how exactly the latter affect the former remains unknown. They are particularly powerful at establishing more realistic rules regarding cell state transition. However, how many cells should transit their state remains unknown, an issue commonly resolved via the integration of CA with Markov chain.

When multiple models are involved in a simulation, the simulation process becomes increasingly complex, and the method of hybridisation also increases in sophistication. Figure 7.13 illustrates a way of integrating ANN with CA and MC. In the integration, CA, MC, and ANN fulfil a particular purpose in the simulation process (Table 7.9). MC predicts the expected number of cells to be transitioned into

---

TABLE 7.9

Comparison of Common Methods for Simulating Urban Expansion

Category	Model(s)	Pros	Cons
Singular	Logistic regression	Relationship between urban growth and drivers considered	Unable to spatially site new growth area
	CA	Simultaneous consideration of space and time; expansion simulated via micro-level cell state changes	Number of cells that should transit their state unknown; transition rules ambiguous
	Agent-based	Able to take into account interactions among cells; consideration of players in urban development	Agent rules and behaviour hard to articulate; neighbourhood effect ignored
Multiple	ANN+CA, RF+CA, SVM+CA ANN+ABM	Non-linear relationship between drivers and urban expansion exploited	Location of change uncertain
Hybrid	MC+CA	Total amount of change determined by MC	How cell state should transit unresolved
	CA+ABM	Agents' perception considered in a spatial context	Relationship between players and transition rules unclear
	MC+CA+ANN	ANN – transition rules; MC –total new growth area; CA – spatial allocation	Complex, binary (either change or no change)

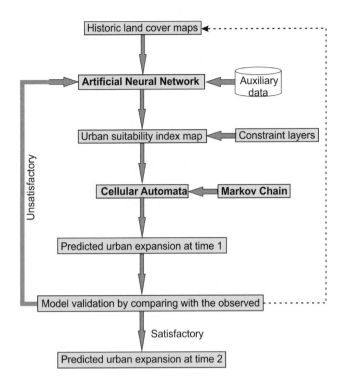

FIGURE 7.13   A possible method of integrating ANN with CA and MC for simulating urban expansion.

urban use at time t+1 (1 = one iteration, e.g., 5 or 10 years). ANN accepts training samples selected from historical land cover maps and produces the transition rules based on how the study area has been urbanised in the past. In the simulation, the ANN must be properly configured by optimising the relevant calibration parameters via repeated cross-validation. One of the most significant and primary outputs from the validated ANN is a map showing the spatial distribution of the urban suitability index, calculated as:

$$USI = f_{ann}\left(V_1,\ V_2,\ \dots\ ,\ V_i\right) \prod C \tag{7.2}$$

where f_{ann} represents the ANN function, V_i is the ith variable for predicting USI, and C denotes the constraint with a binary value of 0 (prohibited from urbanisation) or 1 (no restrictions). The USI is usually expressed on a continuous scale from 0 (utterly unsuitable) to 1 (perfectly suitable). The actual USI value of all cells is checked and ranked, and the cell with the largest USI will be given the highest priority of transition in state. This process is repeated in descending order until the expected number of cells to transit their state has been selected. It is possible to convert the continuous USI values to a few categories in order to reveal the spatial pattern of suitable candidate cells, but this is not related to the simulation itself.

After the suitability of all candidate cells has been derived and the total number of cells to be transitioned is determined, the spatial simulation problem evolves to how to allocate these new urban cells spatially, a task that is ideally accomplished via CA simulation. The exact allocation is subject to the mode of urban expansion. The urban expansion model is run multiple times, and all model runs are averaged to iron out randomness effects. The simulation will become rather messy or even unmanageable if multiple modes of expansion are considered, especially leapfrog (sprawl) growth. One possible solution is to run the simulation with a unique mode considered at a time. Then all modelled expansions are weighted by the proportion of expansion in different modes, and the final outcome is a weighted average of expansion in multiple modes.

Irrespective of the expansion mode, the transition of a given cell from one state to the next can be expressed as:

$$S^{t+1}_{ij} = f_{ca}\left(S^t_{ij}, A^{t+1}, USI^t, N\right) \tag{7.3}$$

where S^t_{ij} is the state of cell (i, j) at time t; S^{t+1} and A^{t+1} represent the cell state and the expected number of cells to be transitioned into urban use at time t+1 predicted by MC, and USI^t and N denote the CA transition rules combining urban expansion suitability from the ANN and the neighbourhood effect of cell (i, j).

7.5.2.4 A Case Study

The integrated ANN-MC-CA method illustrated in Figure 7.14 was applied to simulating urban growth of South Auckland by 2026, in which most of the relevant factors in Table 7.8 were considered, based on a cell size of 30 m by 30 m. This size was fully compatible with that of other geospatial data, such as DEM and historical land cover maps derived from Landsat satellite images. As shown in Figure 7.14, most of South Auckland is not suitable for urban development judging from the low USI values (<0.2) of predominant cells (approximately 90%). Those cells with an USI index value >0.6 are numbered relatively low at only 5.5%, all located next to current urban areas. In contrast, those cells with a higher USI value are widely scattered, with an exceptionally high concentration in the centre and north-west of the study area. These places will become the most likely hotspots of urban expansion in the next 10 years.

Figure 7.15 illustrates the predicted urban expansion in 2016 based on the historical land cover change from 1996 to 2006 using the integrated ANN-CA-MC model. The urban area is predicted to expand by 1,510.38 ha to reach 14,893.38 ha, against the actual area of 14,787.41 ha, leading to a maximum simulation accuracy of 93% (more accuracy details can be found in Table 7.1). One possible explanation for the discrepancy between the modelled and observed urban areas is the intensification and vertical development of residential areas, causing the newly urbanised areas to be smaller than predicted. This kind of over-prediction can be mitigated by incorporating the total floor area of buildings into the prediction model instead of the total urbanised area. Such information can be easily obtained from light detection and ranging (LiDAR) data. Another potential cause of the lower than expected expansion is the larger family size caused by children living with their parents for longer than in the past.

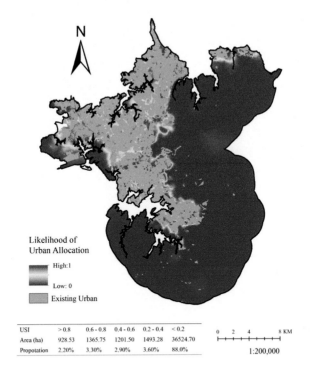

USI	> 0.8	0.6 - 0.8	0.4 - 0.6	0.2 - 0.4	< 0.2
Area (ha)	928.53	1365.75	1201.50	1493.28	36524.70
Propotation	2.20%	3.30%	2.90%	3.60%	88.0%

FIGURE 7.14 Spatial distribution of the urban suitability index of South Auckland generated from ANN and its statistics (source: Xu et al., 2019, with permission).

The validated ANN-CA-MC model predicts that the urban area of South Auckland will expand to 1340.55 ha by 2026. Most of the expansions will take place in new and rapidly developing suburbs in close proximity to the periphery of existing urban areas in the mode of edge expansion (Figure 7.15). In addition, some minor urban developments will occur in the mode of infilling expansion within open and bare land amid existent urban areas. Overlay of the predicted urban expansion map with the 2025 Master Plan of Auckland Council reveals that almost 70% of the predicted expansion areas coincide with the planned "New Growth" and "Residential" zones, and only a minority happen to be located in the "Business" (including industrial) zone. In addition, some urbanisation is predicted to take place in the "Country Living" zone adjoining current urban areas. Such a close spatial agreement between the simulated and the planned urban growth indicates that the ANN-CA-MC integrated model has produced a reasonable prediction of potential future urban development, and it can be applied to other parts of the city.

7.5.3 Simulation of Grassland Degradation

The Qinghai-Tibet Plateau in western China is home to millions of hectares of alpine grassland. This precious grazing resource sustains the livelihood of hundreds of thousands of pastoralists. In recent decades, the grassland has suffered degradation

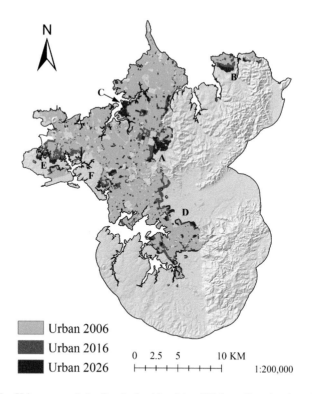

FIGURE 7.15 Urban growth in South Auckland by 2026 predicted using the integrated ANN-CA-MC model (source: Xu et al., 2019, with permission).

due to climate change and overgrazing. In the Three River (Yellow, Yangtze, and Lancang) Headwaters region, overgrazing averaged 67.88%, or 27.43 sheep units per km² above the sustainable level in 2010 (Zhang et al., 2014). How to achieve sustainable grazing under the joint influences of climate change and external disturbances requires an understanding of the effect of grazing intensity on grassland degradation. It refers to a gradual process in which rangeland quality is lowered to such a level to adversely impact its grazing value as a consequence of excessive external disturbances and/or unfavourable natural conditions (Li, 1997). This process may take years or even decades to complete, depending on the intensity of the external disturbances and the harshness of the natural conditions. In the short term, it manifests as a lowered biomass productivity, fragmentation of vegetation cover, the emergence of unpalatable and even toxic grass species, reduced soil fertility, and increased soil compaction. At the late stages when degradation reaches the severe level, the grassland surface can become completely denudated, a phenomenon locally known as *heitutan*, or "black soil beach", and the exposed bare ground is a site rife for severe soil erosion.

In addition to overgrazing, grassland degradation can also be triggered and aggravated by frequent population booms of small mammals, such as alpine pika

(*Ochotona curzoniae*), due to the tipped eco-equilibrium. In the long run, overgrazing creates a setting conducive to the invasion of pikas, thereby accelerating the process of degradation. Also, the harsh, frigid alpine climate makes the grassland atop the Qinghai-Tibet Plateau inherently vulnerable to degradation (Wang et al., 2001), even though it is impossible to disentangle climate effects from anthropogenic impacts. It is generally believed that the latter are the primary drivers of grassland degradation. Degraded frigid alpine meadows increased from 24.5% of the total grassland in the 1980s to 34.5% in the 1990s (Li et al., 2011). It is estimated that 16% of the 703,000 ha of alpine meadows in the Yellow River Headwaters zone alone have been severely degraded to form *heitutan*. Grassland degradation has posed a grave threat to the livelihood of pastoralists and has indirectly affected the fluvial environment in the downstream catchment.

It is crucial to simulate grassland degradation involving multiple plant functional types (PFTs) comprising the alpine grassland. Insights into and an understanding of how these plants compete and interact with each other dynamically, which cannot be gained from field studies, are prerequisites to improving protection and managing the alpine meadows sustainably. Through simulation, it is possible to gauge the timeframe of *heitutan* formation and the time it takes to restore it to productive use. Such simulation can answer a number of questions crucial to the sustainable utilisation of grassland resources. For instance, will overgrazing alter the plant composition of the grassland? Is it possible for the degraded grassland to recover naturally through seed germination if left undisturbed? How long will it take for the severely degraded grassland (e.g., bare ground) to recover if the grassland is grazed at a given intensity?

The answers to these questions can be obtained via agent-based spatiotemporal simulation. This simulation is highly challenging, as it involves an ecosystem comprising different PFTs (Figure 7.16). The areas' populations and habitats are dynamic, in that they compete against each other for living space, light, moisture, and nutrients. They produce new plants and die at a certain age. The consideration of grassland degradation in the emergence of bare patches complicates the simulation even further, in which the surface cover types (e.g., cell states) will rise from four PFTs (forbs, grasses, sedges, and weeds) to six (the two extra are related to ground cover, e.g., unoccupied and degraded). Unoccupied land refers to vacant ground created by the recent demise of healthy plants. It has the potential to be rapidly recolonised by other more competitive and tolerant PFTs, such as weeds. In comparison, degraded bare ground has lost its soil and nutrients to erosion. It is rather difficult for any plants to colonise this ground quickly unless they are extremely competitive and not choosy about their habitat. Apart from plants themselves, the grassland ecosystem is highly complex, in that it is actively grazed by livestock, during which occasional increases in the number of small mammals, especially alpine pika, take place. Consideration of these external disturbances further makes the simulation even more challenging and complex, as their effects can be either beneficial or destructive in the process of grassland degradation, depending on the intensity of disturbance.

In this example, the agent-based simulation was run as scenarios over a space of 100 m by 100 m. Given the spatial extent of the observed alpine grassland conditions in the field, this simulation space was considered sufficiently representative of the

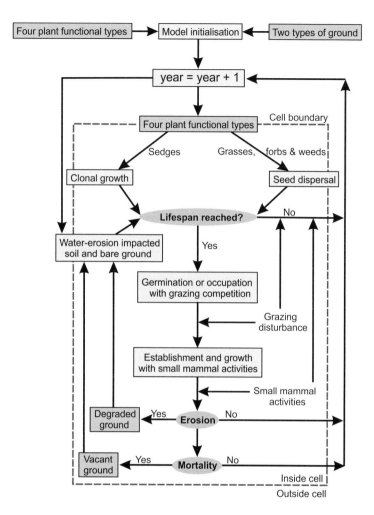

FIGURE 7.16 Flowchart of the spatially explicit PFTs simulation model for assessing grassland degradation and recovery potential under grazing and small mammal disturbance (source: modified from Li, 2012).

grassland diversity and conditions. The space was partitioned into 100 by 100 cells. The cell size of 1 m by 1 m was deemed suitable as it could adequately capture the spatial extent of clonal expansion and pika activities. In the simulation, the colonisation of a vacant (unoccupied) cell can be accomplished via two means: seed dispersal for all PFTs except sedges, and clonal growth for sedges (Figure 7.16). Once a cell is occupied by one of the PFTs, it is no longer possible for other PFTs to occupy it until after the current PFT has died. In order to yield a reasonable simulation, an extensive array of inputs were parameterised to run and validate the spatiotemporal simulation model, including grass life span, rate of forage consumption by livestock, forage reproduction rate, and recovery rate of degraded cells (Table 7.10). Other rates include the spread rate of vacant ground cells and the erosion rate of degraded ground

TABLE 7.10

Summary of Model Inputs Needed in Simulating Grassland Degradation and Recovery under Disturbance by Small Mammals

Variables	Patch/agent	Variables to be parameterised
Patches	PFTs	Lifespan, reproductive rate at different life stages (including growth rate, germination rate, survival rate of seedling, seeding rate per reproduction-tiller, death rate, and transition rate from single-tiller to composite-tiller and reproduction-tiller), number of pika burrows, grazing state
	Degraded	Probability and rate of erosion, erosion spread rate, recovery time, number of pika burrows, probability of transition to vegetative covers
	Vacant	Spread rate, recovery time, number of pika burrows, grazing state
Agents	Small mammals	Likelihood of an outbreak in relation to grazing intensity
	Livestock	Trampling effect on PFTs, and grazing effect on PFTs' reproduction rate

Source: Li (2012).

cells. Based on these parameters, rules were established for seed germination and establishment, the influence of human activities (e.g., grazing intensity and stocking rate), and disturbance by small mammals. The simulation was accomplished using four NetLogo modules for plant type competition and succession, grazing disturbance, and small mammal disturbance and mortality, and was augmented by two auxiliary modules (joint count statistics and result plotting).

All the rates and model inputs in Table 7.10 were carefully parameterised based on a combination of field-gathered evidence and reference to the literature. As with all simulation models, the initial conditions, such as the quantity of vacant and degraded ground, were set by randomly allocating them in the simulation space (Figure 7.17a) after their quantity has been determined in the field. This random distribution indicates a highly fragmented landscape that bears little resemblance to reality. Nevertheless, at the moderate grazing level, the same landscape is simulated to be dominated by sedges (green) and grasses (blue), with few denuded patches 100 years later (Figure 7.17b). However, if grazing rises to the intensive level, the portion of grass cells is drastically reduced. Instead, denuded cells will occupy a noticeable portion of the landscape (Figure 7.17c).

Dissimilar to urban growth simulation, grassland degradation simulation produces a result that cannot be verified because of the randomised initial inputs (conditions), so there is no reality against which the simulated outcome can be compared. The simulation outcome can be made more realistic by taking into consideration the environmental variability that affects plant seed germination. It is impossible to check the reasonableness of some parameterisations due to the lack of field data (e.g., the time it takes for a bare patch to become revegetated). Within the small area of 100 m by 100 m, it is reasonable to assume homogeneity in topography and climate. However, it is unlikely that the simulated outcome can be replicated at the broad

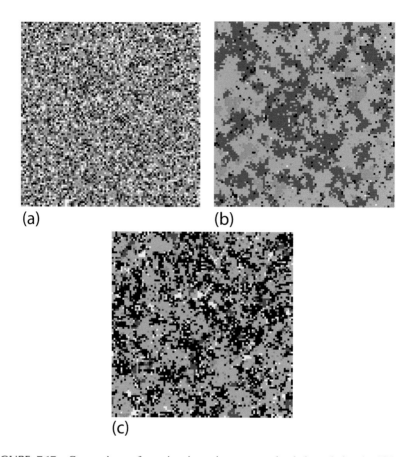

FIGURE 7.17 Comparison of grazing intensity on grassland degradation in 100 years within a landscape of 100 by 100 cells (cell size = 1 m), simulated using NetLogo: (a) initial state of degraded and vacant ground; (b) outcome of moderate grazing; and (c) outcome of intensive grazing. Black = degraded; grey = vacant; yellow = weeds; green = sedges; blue = grass; pink - forbs (source: Li, 2012).

landscape scale where topographic heterogeneity is the norm, so its influence on plant health and growth rate cannot be ignored. Topographic heterogeneity at a broad spatial scale can be taken into account via a DEM in future simulations. However, how climate will change in 100 years remains unknown, so its effect on the grass-land ecosystem cannot be factored into the simulations.

REVIEW QUESTIONS

1. Compare spatiotemporal dynamic simulation with spatial modelling, pay-ing particular attention to why the former is superior to the latter.
2. What are the main properties of CA in spatial simulation? How does each of them affect the implementation of CA-based simulation?

3. What are the main differences between the neighbourhood used in inverse distance weighting and in cellular automata simulation?
4. What are the pros and cons of CA in spatial simulation? How can the cons be minimized?
5. In which way is agent-based simulation superior to CA simulation?
6. In your view, what is the bottleneck in agent-based simulation?
7. It can be argued that the performance of agent-based simulation is as good as the data collected from the field. To what extent do you agree or disagree with this claim?
8. Compare and contrast agent-based simulation with CA simulation, paying particular attention to the treatment of the external environment in both simulations.
9. In which senses is NetLogo posed to be the ideal platform for running spatial dynamic simulation?
10. It can be argued that NetLogo is rather limited in performing spatial dynamic simulations. To what extent do you agree or disagree with this statement?
11. Compare and contrast wildfire simulation with urban growth simulation based on the edge expansion mode. Why is it much more challenging to accurately simulate urban growth than wildfire spread?
12. Discuss the extent to which the simulation of grassland degradation caused by livestock is CA simulation, as well as agent-based simulation.
13. Although time never appears directly in either CA or agent-based simulations, why are they still called spatio-temporal simulations?
14. What is sensitivity analysis? Is it possible to perform sensitivity analysis with cartographic modelling? Explain.

REFERENCES

Anderson, H. E. (1982) *Aids to Determining Fuel Models for Estimating Fire Behavior.* USDA Forest Service. General Technical Report INT-122.

Albini, F. (1976) *Estimating Wildfire Behavior and Effects.* USDA Forest Service. General Technical Report INT-30.

Angel, S., Parent, J., Civco, D., Blei, A., and Potere, D. (2011) The dimensions of global urban expansion: Estimates and projections for all countries, 2000–2050. *Progress in Planning,* 75: 53–107, doi: 10.1016/j.progress.2011.04.001.

Babiker, I. S., Mohamed, A. A., Hiyama, T., and Kato, K. (2005) A GIS-based DRASTIC model for assessing aquifer vulnerability in Kakamigahara Heights, Gifu Prefecture, central Japan. *Science of Total Environment,* 345: 127–40.

Balbi, J. H., Morandini, F., Silvani, X., Filippi, J. B., and Rinieri, F. A. (2009) Physical model for wildland fires. *Combustion and Flame,* 156: 2217–30.

Braun, W. J., and Woolford, D. G. (2013) Assessing a stochastic fire spread simulator. *Journal of Environmental Informatics,* 22(1): 1–12.

Brownlee, J. (2019) *Master Machine Learning Algorithms – Discover How They Work and Implement Them from Scratch.* , p. 163.

Clarke, K. C., Hoppen, S., and Gaydos, L. (1997) A self-modifying cellular automaton model of historical urbanization in the San Francisco Bay area. *Environment and Planning B: Planning and Design,* 24(2): 247–61.

Clarke, K. C. (2008) Mapping and modelling land use change: An application of the SLEUTH model. In *Landscape Analysis and Visualisation. Lecture Notes in Geoinformation and Cartography*, Pettit C., Cartwright W., Bishop I., Lowell K., Pullar D., and Duncan D. (eds.) Berlin, Heidelberg: Springer, doi: 10.1007/978-3-540-69168-6_17.

Coleman, J., and Sullivan, A. (1996) A real-time computer application for the prediction of fire spread across the Australian landscape. *Simulation*, 67(4): 230–40.

De Andrade, P. R. (2010) Game theory and agent-based modelling for the simulation of spatial phenomena. Ph.D. thesis, P. R. ANDRADE – São José dos Campos: INPE, p. 99.

De Angelis, D. L. and Yurek, S. (2017) Spatially explicit modelling in ecology: A review. *Ecosystems*, 20(2): 284–300, doi: 10.1007/s1002.

DeMers, M. (1997) *Fundamentals of Geographic Information Systems*, New York: John Wiley & Sons, p. 486.

Dietzel, C. and Clarke, K. C. (2007) Toward optimal calibration of the SLEUTH land use change model. *Transactions in GIS*, 11(1): 29–45.

Drissi, M. (2015) Modeling the spreading of large-scale wildland fires. In *Proceedings of the Large Wildland Fires Conference*, Keane, R. E., Jolly, M., Parsons, R., and Riley, K. (eds.). Fort Collins, CO: U.S. Department of Agriculture, Forest Service, Rocky Mountain Research Station. pp. 278–85.

Du, G., Shin, K. J., Yuan, L., and Managi, S. (2018) A comparative approach to modelling multiple urban land use changes using tree-based methods and cellular automata: The case of Greater Tokyo Area. *International Journal of Geographical Information Science*, 32(4): 757–82.

Finney, M. A. (2004) *FARSITE: Fire Area Simulator—Model Development and Evaluation*. US Department of Agriculture – Forest Service, Research Paper RMRS-RP-4 Revised, p. 47.

Freire, J. G., and Dacamara, C. C. (2019) Using cellular automata to simulate wildfire propagation and to assist in fire management. *Natural Hazards and Earth System Sciences*, 19(1): 169–79.

Ghisu, T., Arca, B., Pellizzaro, G., and Duce, P. (2015) An optimal Cellular Automata algorithm for simulating wildfire spread. *Environmental Modelling and Software*, 71: 1–14.

Hajian, M., Melachrinoudis, E., and Kubat, P. (2016) Modeling wildfire propagation with the stochastic shortest path: A fast simulation approach. *Environmental Modelling & Software*, 82: 73–88.

Heinsch, F. A., and Andrews, P. L. (2010) *BehavePlus Fire Modeling System (Version 5.0): Design and Features*. Gen. Tech. Rep. RMRS-GTR-249. Fort Collins, CO: U.S. Department of Agriculture, Forest Service, Rocky Mountain Research Station, p. 111.

Hertel, T. W. (2017) Land use in the 21st century: Contributing to the global public good. *Review of Development Economics*, 21(2): 213–36.

Jahdi, R., Salis, M., Darvishsefat, A. A., Mostafavi, M. A., Alcasena, F., Etemad, V., Lozano, O., Spano, D. (2015) Calibration of FARSITE simulator in northern Iranian forests. *Natural Hazards and Earth System Sciences*, 15(3): 443–59.

Knudby, A., Brenning, A., and LeDrew, E. (2010) New approaches to modelling fish-habitat relationships. *Ecological Modelling*, 221(3): 503–11, doi: 10.1016/j.ecolmodel.2009.11.008.

Koo, E., Pagni, P., Stephens, S., Huff, J., Woycheese, J., and Weise, D. (2005) A simple physical model for forest fire spread rate. *Fire Safety Science*, 8: 851–62, doi: 10.3801/IAFSS.FSS.8-851.

Li, B. (1997) The rangeland degradation in North China and its preventive strategy. *Scientia Agricultura Sinica*, 30(6): 1–9.

Li, X. (2012) The spatio-temporal dynamics of four plant-functional types (PFTs) in alpine meadow as affected by human disturbance, Sanjiangyuan region, China. Ph.D. thesis. New Zealand: University of Auckland, p. 211.

Li, X. L., Gao, J., Brierley, G., Qiao, Y. M., Zhang, J., and Yang, Y. W. (2011) Rangeland degradation on the Qinghai-Tibet Plateau: Implications for rehabilitation. *Land Degradation and Development*, 24(1): 72–80, doi: 10.1002/ldr.1108.

Macal, C. M., and North, M. J. (2009) Agent-based modeling and simulation. *Proceedings of the 2009 Winter Simulation Conference*, December 13-16, 2009, Austin, Texas, USA. M. D. Rossetti, R. R. Hill, B. Johansson, A. Dunkin and R. G. Ingalls (eds.), pp 86-98.

O'Regan, W. G., Kourtz, P., and Nozaki, S. (1976) Bias in the contagion analog to fire spread. *Forest Science*, 22(1): 61–8, doi: 10.1093/forestscience/22.1.61.

Parisien, M. A., Kafka, V. G., Hirsch, K. G., Todd, J. B., Lavoie, S. G., and Maczek, P. D. (2005) *Mapping Wildfire Susceptibility with the Burn-P3 Simulation Model*. Natural Resources Canada, Canadian Forest Service, Northern Forest Centre, Edmonton, Alberta. Information Report NOR-X-405, 36 p.

Rothermel, R. C. (1972) *A Mathematical Model for Predicting Fire Spread in Wildland Fuels*. Ogden, Utah: Intermountain Forest and Range Experiment Station. US Forest Service Research Paper INT 115, p. 40.

Smith, L., Beckman, R., Baggerly, K., Anson, D., and Williams, M. (1995) *TRANSIMS: TRansportation ANalysis and SIMulation System*, US Department of Transportation, Washington, DC, p. 10, doi: 10.2172/88648.

Torrens, P. M. (2006) Geosimulation and its application to urban growth modelling. In *Complex Artificial Environments*, J. Portugali (ed.). London: Springer-Verlag, pp. 119–134.

Tymstra, C., Bryce, R. W., Wotton, B. M., and Armitage, O. B. (2010) *Development and structure of Prometheus: The Canadian Wildland Fire Growth Simulation Model*. Inf. Rep. NOR-X-417. Natural Edmonton, AB, Canada: Resource Canada, Canada Forest Service, Northern Forest Centre.

Wang, G. W., Qian, J. Q., Cheng, G. C., and Lai, Y. L. (2001) Eco-environmental degradation and causal analysis in the source region of the Yellow River. *Environmental Geology*, 40(7): 884–90.

Xu, T., Gao, J., and Coco, G. (2019) Simulation of urban expansion via integrating artificial neural network with Markov chain – Cellular automata. *International Journal of Geographical Information Science*, 33(10): 1960–83, doi: 10.1080/13658816.2019.1600701.

Zhao, L., and Peng, Z. R. (2014) LandSys II: Agent-based land use–forecast model with artificial neural networks and multiagent model. *Journal of Urban Planning and Development*, 141(4), 04014045.

Zhang, J., Zhang, L., Liu, W., Qi, Y, and Wo, X. (2014) Livestock-carrying capacity and overgrazing status of alpine grassland in the Three-River Headwaters region, China. *Journal of Geographical Sciences*, 24: 303–12, doi: 10.1007/s11442-014-1089-z.

8 Time-explicit Spatial Analysis and Modelling

Data are not just numbers, they are numbers with a context.
In data analysis, context provides meaning.

(Cobb and Moore, 1997)

8.1 TIME

8.1.1 NATURE OF TIME IN GEOGRAPHY

Time is continuous and eternal. It never stops or ceases to exist. Time elapses linearly and unidirectionally from the past to the present and the future. It is irreversible. Time can be expressed in two ways. This takes into account the granularity of linear time (Figure 8.1a), which always increments towards the right. The starting time varies from case to case. The increase can have different granules, ranging from seconds (e.g., percolation rate), daily (e.g., evaporation), or monthly (e.g., temperature) to yearly (e.g., annual gross domestic product growth rate). The second way treats time as cyclic or periodic and having a limit. When it exceeds this limit, time reverts to where it started (Figure 8.1b). For instance, in daily time as commonly expressed, there are two cycles of 12 hours' duration each. After 12 hours, time reverts to 0, so 0 and 12 are identical. Similarly, both weekly and yearly time are cyclic: in weekly time, days repeat themselves every 7 days from Monday to Sunday, then return to Monday, and in annual time, months repeat every 12 months from January to December, then back to January.

Due to the duplex cycles of time in a day, it is cumbersome to know the exact time without additional adjectives to differentiate daytime from nighttime. This confusion can be avoided by adopting the 24-hour clock, which avoids the need to specify morning or afternoon. Similarly, the exact date of a day in a year is related to month length, which has a variable number of days. A better alternative is to make use of the Julian date. This refers to the count of cumulative days from the start of the Julian Period (e.g., 1 January) up to 365/366, with cyclic time ignored. Julian dates make it very convenient to calculate the number of days between two dates, and they are commonly used in astronomy and computer science.

Which granular unit of time should be adopted in spatial analysis and modelling is highly dependent on the phenomenon under study. The best time scale to study is subject to the timeframe of change and the speed at which the spatial entity of interest changes. If it varies considerably within a short period of time, the unit needs to be small enough to capture the variations. Conversely, if the subject under study varies slowly, then the duration can be increased accordingly. The adopted time duration

DOI: 10.1201/9781003220527-8

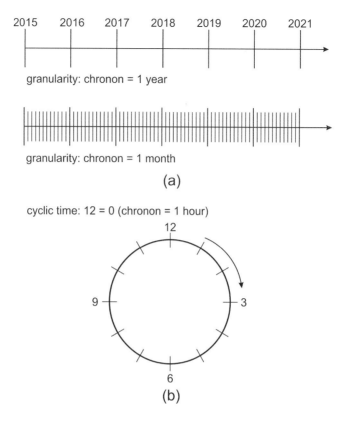

(a)

(b)

FIGURE 8.1 Comparison of linear time with cyclic time: (a) granularity of linear time can be annual or monthly; (b) cyclic time can return to its origin after 12 hours.

can be highly variable, ranging from seconds to millennia. Natural disasters such as landslides and tornados do damage within seconds. Bushfires can change the direction of their spread within minutes. On the other hand, other physical processes, such as the diversion of a river channel course to form an oxbow lake and the formation of an atoll island, may take place on the centennial or even millennial scale. Many other phenomena, such as the growth of crops and melting of icepacks, take place seasonally, a time scale that lies between these two extremes. Accordingly, the time interval adopted in spatiotemporal simulation can be as short as diurnal in studying king tide-induced coastal flooding, or as long as seasonal in studying plant phenology, or even annual in studying bird migration. The appropriate timeframe can be extended to decadal in simulating urban expansion, and centennial in studying landscape evolution.

8.1.2 SPACE, TIME, AND ATTRIBUTE

If combined with 3D location, all spatial entities have five dimensions: the three spatial components common to all spatial entities, plus time and attribute (Table 8.1). The three spatial components refer to easting, northing, and elevation (Equation 2.3).

TABLE 8.1

Five Dimensions of Spatial Entities

Dimension	Values		Type
First	X coordinate	Easting	Spatial (important to distance and area)
Second	Y coordinate	Northing	
Third	Z coordinate	Elevation	Spatial (volumetric information)
Fourth	A value	Attribute	Content or quality
Fifth	T value	Temporal	Time

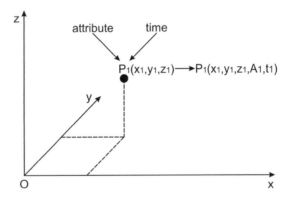

FIGURE 8.2 Five dimensions of point spatial entities in the 3D Cartesian space.

Conventionally, height means a disparity from the reference geodetic datum to the observed surface, and is commonly represented as a digital elevation model (DEM). However, it can also be used to represent attributes in a 2D space, such as the spatial distribution of air pollutants and soil pH at a certain depth. In both cases "height" represents the variation along the vertical axis (z). If the phenomenon is 3D in nature, such as air pollutant concentration or soil pH, then the attribute becomes the fourth dimension. No matter whether the studied space is 2D or 3D, all spatial entities must have the temporal dimension for spatiotemporal simulation. As demonstrated in the preceding chapters of this book, most spatial analyses and simulations concentrate only on the first four dimensions. In comparison, the temporal component is the least studied.

The presentation of the five dimensions in the traditional Cartesian coordinate system can be visualised as in Figure 8.2, in which the attributes can be colour-coded and time is presented by repeating the same representation multiple times.

The space, attribute, and time components of all spatial entities are intrinsically associated with each other. An attribute exists only under a given set of spatial and temporal circumstances. Of the five components, spatial and temporal components are inseparable from each other, and are the targets of spatiotemporal modelling. Naturally, both space and attribute can change with time. Attribute can also change over time at a given location. For instance, a beach can be submerged at high tide, but be exposed as a mudflat at low tide (Figure 8.3). This kind of interchange takes place periodically over a cycle of 12 hours. Similarly, the same land cover in a given geographic area can change from grassland to snow, depending on the season. It is grassland in summer, but can be covered by snow in winter. A vacant plot of land can be turned into a forest through tree planting. A mature forest can be converted to a wasteland after all the trees have been logged. In spatial analysis, space or location itself do not change with time. Rather, it is the spatial variation of the attribute that is the focus of spatial analysis, which is further compounded by its temporal variation.

In spatial analysis, space, attribute, and time can be translated to where, what, and when (Figure 8.4), with mutual interactions existing between any two of them. The relationship between attribute and time has been illustrated using the example of a forest. Here, the focus is on the relationship between the other two dimensions. If

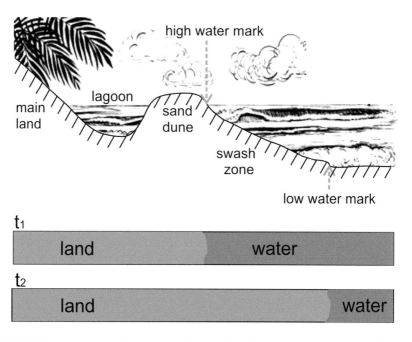

FIGURE 8.3   Interrelationship between time and attribute at the land–sea interface on a beach.

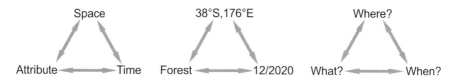

FIGURE 8.4   Interrelationship between spatial, thematic, and temporal components of geographic entities.

space is shifted to a nearby location, its attribute can change from forest to grassland. The grassland can be burned to bare ground after a wildfire, but can recover naturally to grassland after a certain period of time. It is this kind of interchange among space, attribute, and time that forms the backbone of spatiotemporal analysis and simulation, possibly under the influence of external disturbances.

8.1.3 Temporal Aggregation versus Discretisation

Since time is continuous, the attribute of an observed variable also varies continuously with time. Time-continuous attribute data are thorny to handle except in visualisation (see Section 8.5) because of the huge volume of data involved. In practice, time-continuous data are studied at certain crucial moments or over certain periods (see Section 8.5). For instance, beach erosion at a certain profile (linear spatial entities) during a hurricane may be monitored every 5 seconds, as it is impossible to maintain a continuous record. Besides, the temporal variation from one second to the next may not be sufficiently large to warrant data capture at a shorter temporal interval. In reality, it is either too troublesome or impractical to treat temporal phenomena as continuous, and the observed data are seldom logged incessantly. For instance, in studying beach morphology, the temporal interval can be prolonged from 5 seconds to 1 minute if the surface morphology does not vary much with time in the off-storm season. Similarly, daily temperature is continuous, but is recorded at certain intervals (e.g., every minute). In all these cases, it is sensible to discretise time, as the amount of variation from one moment to the next is so insignificant that continuous recording of all values does not add any value to the data collected.

With the increasing use of electronic gadgets, it is possible to continuously log data with time, such as temperature. However, it is not sensible to do so with areal data (e.g., data initially recorded at points but commonly enumerated over an area, such as rainfall). The study of the variation in rainfall spatial pattern with time requires summing up all the recorded rainfall readings over a duration, which is essentially temporal aggregation. The rainfall recorded at a gauge is taken as the total accumulated rain within a specified time frame (e.g., 24 hours). Apart from rainfall, other common examples of temporal aggregation include rush-hour traffic volume, daily cases of new infection, annual migration, and yearly channel discharge. As can be seen from these examples, the unit of temporal aggregation varies widely. It can be as short as seconds, but as long as years. For instance, the speed of ocean currents can be expressed as meters per second, but the erosion rate of a beach has to be annual (e.g., 5 cm per annum). In contrast, the number of vehicles travelling on a road can be aggregated to hours in studying rush-hour traffic. The sensible temporal interval of aggregation depends on the pace of mobility or change. There is no restriction on the temporal unit that can be adopted to aggregate the data if they are initially recorded continuously. It is possible to aggregate them over multiple time spans. For instance, migration data can be aggregated annually as well as monthly to study seasonal variations in tourism. However, if the data are collected periodically, such as census data every 5 years, then the temporal unit must be relaxed to exceed the temporal interval of data collection before any temporal aggregation is possible.

8.2 MODELS OF SPATIOTEMPORAL REPRESENTATION

It is rather challenging to represent spatiotemporal features clearly while also facilitating efficient data retrieval and analysis. It is even more difficult to represent spatiotemporal processes. Different methods of representation have been developed for this purpose. Common models of representation include time-location path, gazetteer, snapshot, space-time composite, event-based, and amendment vector (Table 8.2). Each of these has its own strengths and special applicability to certain types of data.

8.2.1 TIME-LOCATION PATH MODEL

The *time-location path model* is the best for representing the movement of point features within a limited spatial extent in vector format. Point data of a certain feature can be mobile spatially, and its position at two consecutive times is linked with a straight line segment. The same point event can occur or be captured at different times. To keep the representation legible, the timeframe is usually restricted to a diurnal period (Figure 8.5). A longer time period is possible by splitting the data into smaller temporal fragments. Both time and location can be continuous over space, but the attribute does not change. It can be a tracked animal roaming in a forest, or a student moving around in his or her daily routine. The exact location can be shown at specific times. Such point data can be easily collected using a Global Positioning System (GPS)–enabled device (e.g., a smartphone). The exact location of the tracked entity (e.g., vehicles or animals) is available at any given moment. In epidemiology, this kind of spatiotemporal tracking is crucial to identifying close contacts and the potential spread caused by a person infected with a communicable virus, such as COVID-19.

This model shows only the locations of point features, which vary with time, so time is implicit in the representation. Usually, the exact time is annotated at critical and representative moments (Figure 8.5a). The annotated time must have a coarse resolution to keep the representation legible. In order to maintain a high degree of legibility, the number of indicated locations (times) has to be limited. More precise representation at a finer temporal scale is possible by assigning a more suggestive annotation scheme to the temporal component. For instance, the line segment linking two indicated time spots can be coded as a continuum of colour or shade, even though it is rather imprecise. This can be done even with 3D time-space path diagrams, in which the location is represented in a 2D space (perspective), and time is shown in the third dimension (Figure 8.5b). This time-space path can indicate the general location at a rough time, as well as the duration at specific locations (e.g., a dentist's office). The representation will become messy quickly as the time duration prolongs.

With time-space data, a number of variables can be derived to describe the movement. Change between any two points in time can be easily calculated from their start and end coordinates. Such results as distance or range of travel (useful for studying animal behaviour), direction, and speed of movement can be derived from the duration between two observations. Apart from spatial extent, the representation can also show the speed of movement, although not directly. It is also possible to temporally sequence the movement, which can be described statistically.

This spatiotemporal model is suited to representing objects at discrete times, such as a moving vehicle. Common applications of this model include animal tracking

TABLE 8.2

Comparison of Major Ways of Representing Spatiotemporal Data

Models	Main features	Pros	Cons	Author(s)
Location-time/time-space path	Attribute constant, position changes with time, can be plane or 3D (z-axis reserved for time)	Good for point data, continuous movement possible	Ill-suited for polygon data or showing processes	Peuquet (1994)
Gazetteer	Linkage of spatial features with their names and location	Able to search data using names that may have changed	Unable to show spatial change	Hill (2000)
Snapshot	Snapshots at periodic times	Both raster and vector data possible	Hard to perceive change, inefficient	Armstrong (1988)
Space-time composite	Division of space into homogeneous units, and tables list attributes at different times	Good at showing temporal change, no data redundancy, concise	Unable to show changes across space, history of transition unknown	Segev and Shoshani (1993)
Spatiotemporal object	Similar to space-time composites with geo-atoms	Location of extent of change clearly visible	Not suitable for time-series raster data	Goodchild et al. (2007)
Event-based	Time-oriented, location-indexed	Good at detecting changes	Applicable to raster data only	Peuquet and Duan (1995)
Amendment vector	Boundaries at different times are stored	Able to show change with time	With vector data only	Langran (1992)

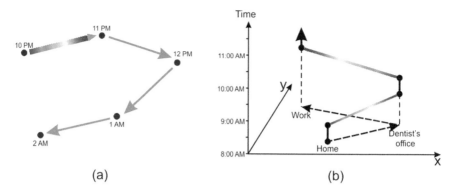

FIGURE 8.5   Two ways of representing time-location features in which the position of the same entity at multiple times is shown to indicate mobility: (a) time annotation; (b) 3D time-space path coded with colour to show time more precisely.

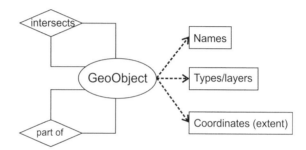

FIGURE 8.6   The gazetteer model of representing spatiotemporal features by linking place names that may have changed over time to their location (source: modified from Bandholtz, 2003).

and fleet management, in which the attribute is constant, but location changes with time. The target of tracking should usually be limited to a few points, as should the period of tracking, otherwise the tracked movement will become illegible, and the readability of the time-space diagram will deteriorate quickly when too many targets are being tracked simultaneously. This method of representation is not suitable for representing a changing scene or phenomena over a geographic area. In this case, the gazetteer and snapshot methods are better suited.

8.2.2   GAZETTEER METHOD OF REPRESENTATION

The *gazetteer method* can be defined as a geographic index. In representing spatio-temporal features, a gazetteer can be considered as a geospatial dictionary of place names. Each entity is linked to its various names, spatial extent, and thematic components (Figure 8.6). All the possible linkages between the names of spatial entities and their thematic attributes, as well as their spatial representation, are explicitly spelled out beforehand (Hill, 2000). This mode of representation requires defining spatial objects and associating their names to the spatial and thematic components, which are the same for all geospatial entities. Only the names of geographic objects (e.g., points, lines, and polygons) are allowed to change. The names are typically expressed in plain language. Time is embodied in the representation, in that place names may have changed with time, so both the original and variant names must be stored, though with no information about when the change took place. As a method of indirectly representing spatiotemporal data, a gazetteer can fulfil the following two functions:

(1) *Indirect geo-referencing*: The location and spatial extent (for polygons) of an entity can be automatically determined by just specifying its name via its linkage to the geographic coordinates. This linkage also allows the easy retrieval of data using their names.

(2) *Vertical data integration*: The linkage between location and the thematic components in a gazetteer allows vertical integration of the attribute data at the same location easily. Gazetteer is good for integrating time-series attributes at the same location, and retrieval and integration of online information.

The gazetteer model of representation is particularly suited to representing census data, where the name of some census tracks or units may have changed over the years, together with their enumeration boundaries. In this application, a large quantity of names can be stored, and all the stored data can be efficiently retrieved. The gazetteer representation is best at representing point and polygon entities with their own names. It is popular in online searches for point and polygon features via their names, such as searching for a restaurant that may have changed its name. Naturally, these entities must be geo-coded. As such, this model is not suited to representing linear data (e.g., highways), although it is possible. However, it is not good at showing changes to the spatial extent of the entity being represented, only a change in its name.

8.2.3 Snapshot Model

In the *snapshot model*, the world is represented by a stack of time-stamped layers, either raster or vector, all registered to the same coordinate system, each capturing the world at a particular static moment (snap) (Figure 8.7). All the layers must encompass the same geographic extent, but can be composed of cells of the same grid size for raster layers. The snapshot model is able to illustrate continuous changes in the scene being covered if the time interval between snaps is kept sufficiently short. In raster format, the change at a given cell can be examined longitudinally. However, it is impossible to identify the spatial changes that have occurred at locations between the given points in time as they are not explicit, nor does this model show the change directly and quantitatively. The temporal interval between snapshots is not necessarily fixed. For seasonally changing phenomena such as grassland, all the snapshots must be captured at a similar time of the year to avoid seasonal variations. In this representation, every event has a binary code of 0 or 1, signifying in or out of the event. As such, it is particularly useful in illustrating urban expansion in a city over a

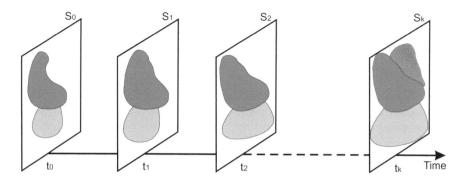

FIGURE 8.7   The snapshot approach of representing spatiotemporal changes of vegetation from time t_0 to time t_k.

period, in which land cover at each time is represented as a single layer, and changes in an urban area over time (0 = no change, 1 = change) can be visually identified from the continuous updating of all the layers.

A slight variation of this representation is called the *past in present snapshot*, in which the dataset is divided into multiple layers, each having its own time. The birth and death of entities in the dataset can be shown only by comparing multiple layers. This comparison is vital to determining the age of certain entities.

This layer-oriented model of representation is straightforward and easy to comprehend. It is intuitive and direct, but limited, in that the temporal information, such as timestamps and temporal intervals, is not obvious or explicitly represented. Instead, such information is recorded in a separate metadata or header file. There is also a lack of protocols for standardising the encoding of temporal information. Also, the interval between snapshots is not always uniform, which may project a misleading impression about the pace of change. The other limitation is its low efficiency of representation, especially when the phenomena being depicted have not experienced much change, because the same spatial entities are repeated in all the snapshots. Lastly, the time interval must be kept very coarse, otherwise the enormous volume of data (especially in raster format) will quickly become unmanageable.

8.2.4   SPACE-TIME COMPOSITES MODEL

In the *space-time composites model*, temporal information about an object is composited with its spatial features and treated as an additional attribute (Segev and Shoshani, 1993). In this model, space is partitioned into a maximum number of homogeneous units or polygons (Figure 8.8), then a table is created listing all the units and their attributes at each of the time specified. Thus, it is not compromised by the spatial and temporal resolutions of the data. The addition of new time-series layers can be accommodated by simply adding extra columns to the table. However, only the final state at each time is recorded, while the history of transition remains unknown. Hence, the exact time at which a change takes place is uncertain. The period of change can be detected from the temporal overlay of snapshot layers. This

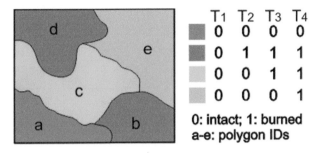

FIGURE 8.8   The space-time composites model of representing spatiotemporal data (source: modified from Yuan 1996).

model is good at representing changes in the spatial extent of objects over time (Nadi and Delavar, 2003), but incapable of representing changes in attributes across space (e.g., movement over space). Also, the entire spatial database must be reorganised whenever the geometrical or topological relationships among the represented homogeneous units have changed over time, together with the attribute table.

8.2.5   Spatiotemporal Object Model

In the *spatiotemporal object model*, spatial entities are treated as a set of discrete objects comprising spatiotemporal atoms by adding a temporal dimension perpendicular to the 2D planimetric space (Nadi and Delavar, 2003), identical to Figure 8.5b, except position being replaced with objects. Both vector and raster data can be represented in the spatiotemporal atom model. Each spatiotemporal atom is the largest homogeneous unit whose spatial and temporal properties are stored, together with abrupt changes in the temporal and spatial dimensions, similar to snapshots and space-time composites. However, it is impossible to model gradual changes in space or time using this representation as the spatiotemporal atoms have a discrete structure. Thus, this model is unable to show transitions, spatial processes, or movements.

This limitation can be effectively overcome with the spatiotemporal object model depicted in Figure 8.9, which is very similar to the space-time composites model, in that the temporal information is added to the 2D space as the third dimension, and the object is marked as a geo-atom (Goodchild et al., 2007). The time scale can be fine or coarse. The change history can be retrieved only via an overlay of multiple spatiotemporal atoms corresponding to the parcel or cell of interest. This model is good at representing objects with the spatiotemporal dimension, but not suitable for representing time-series raster data.

This representation clearly illustrates the location and spatial extent of change, but the map can be difficult to read if the change is shown many times. This difficulty can be overcome with two variants of the initial spatiotemporal model: amendment vector and least common geometries. The *amendment vector* variant is applicable to data represented in vector format, in which changes at different times are shown in different legends (Figure 8.10a). It extends the classical vector means of representation by showing the same geographic feature (e.g., a polygon) at multiple times

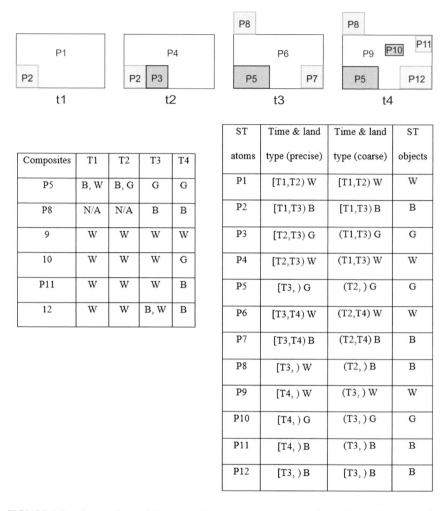

FIGURE 8.9 Comparison of the space-time composites model (left table) with the spatiotemporal object model (right table) of three types of land covers (blue, green, and white) at four times from *t1* to *t4* (source: modified from An and Brown, 2008, with permission).

(Langran, 1992). Central to this feature-oriented representation is the amendment of features whose spatial extent has changed over time (e.g., shift in polygon boundary). It is able to preserve the integrity of spatial entities such as lakes and urban areas. Similar to the snapshot concept, the boundaries captured at specific times are represented, together with the original position, thus avoiding the data redundancy issue that plagues the original model of representation. While this variant of representation is able to illustrate the spatial expansion (successive increment) of the spatial entity graphically from one time to the next (Figure 8.10b), it is difficult to perceive the change if the spatial entity does not have a consistent trend of change (e.g., expansion during one period of time, but contraction during the next). Also, time is not

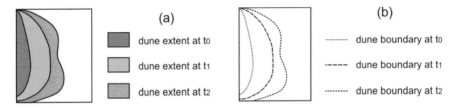

FIGURE 8.10 The amendment approach of representing spatiotemporal change of a coastal sand dune in the vector format: (a) temporal composite of areal change of the sand dune extent; (b) temporal composite of areal changes based on dune boundaries.

explicitly obvious in the representation as it is embodied as an amendment to the attribute. Similar to the snapshot model, there is no guarantee that the time interval is uniform. This form of variant representation offers a high degree of legibility as the changes of only one feature are shown, so it is possible to perceive the pace and location of change. Nevertheless, this model of representation applies to vector data only. With raster data, it is impossible to show multiple attributes at different times using the same grid layer.

The *least common geometries* variant overcomes the limitations of existing methods of representation as it is able to cover occurring changes continuously. This variant is spatially and temporally reliable as it covers not only time slices, but also continual records of changes. It can generate boundaries for each given time step as the model is object-oriented. Thus, points, lines, and polygons can all be represented.

8.2.6 Event-based Spatiotemporal Model

The *Event-based Spatiotemporal Data Model* was proposed by Peuquet & Duan (1995). This raster-based model overcomes the representation inefficiency endemic to the snapshot model by using a collection of time-stamped layers. Only the spatial component at the initial time t_0 (the base map) is stored, and each cell, identifiable via its row and column coordinates in the spatial layer, is then indexed temporally (Figure 8.11). In this time-based representation, a new time slice is handled simply by adding a column to the timeline or temporal vector. This timeline lists all the events in a temporally ordered progression to locations of change, so it is able to cope with as fine a temporal resolution as necessary. More importantly, this model allows the detection of change easily by comparing the attributes at different times in the tabular form. It can perform three temporally based retrieval tasks:

(1) Retrieval of locations that have changed to a given value at a given time span;
(2) Retrieval of locations that have changed to a given attribute value over a given time span;
(3) Calculation of the total area that has changed to a given value over a given time span.

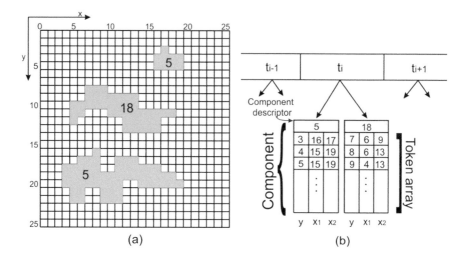

FIGURE 8.11   The event-based spatiotemporal model: (a) spatial change at time *t* rendered as a simplified map; (b) the corresponding event components (source: modified from Peuquet and Duan, 1995, with permission).

It must be noted that this model was devised to represent only raster data. How to adapt it to represent vector maps is an issue yet to be resolved.

One downside of this approach of representation is that the repetition of the (x,y,v) triplets (v = a new value at each location at time *t*) is related to the spatial extent of the change between t_i and t_{i+1}. This triplet can increase when the spatial extent is large. One possible way of overcoming this repetition is to group the cells. Such a value-specific group can be stored as a single sub-structure called a "component", as shown in Figure 8.11. Owing to this grouping, the changed value is stored only once per event instead of repeating many times for different locations. Each component has two elements, a component descriptor and an array of locations called a "token". A token encompasses three entries of *x* (the row number), y_1 (the first column), and y_2 (the last column) of the same row of consecutive cells having the same value *v* (Peuquet and Duan, 1995), in a manner remarkably resembling run-length encoding.

None of the aforementioned models for representing spatiotemporal data is applicable to transport over road/street networks, which must incorporate travel behaviour from origin to destination. This difficulty can be overcome with the model proposed by Frihida et al. (2002). In this model, chief entities are represented as highly structured classes. Time is embodied in the model via an encapsulated memory of time-bound connections and states. Although the prototype of the model is able to navigate through a nested hierarchy of objects and yield descriptive information on an individual's travel behaviour over space and time, it is unable to explicitly display the object's histories, such as object space-time paths. In brief, it cannot perform temporal reasoning.

8.2.7 ORGANISATION AND STORAGE OF SPATIOTEMPORAL DATA

In order to facilitate efficient spatiotemporal analysis and modelling, spatial data (especially raster data) must be organised strategically in a geographic information system (GIS) database. Time-series raster data can be stored in database form or file-based form. The former is exemplified by Esri's file geodatabase for raster spatial data. The latter can be implemented as either multiple single-band raster files or a single raster file with multiple bands (Table 8.3) (Song et al., 2016). Single-band raster files simply store all the spatial data at different time stamps as a set of standalone files, all conforming to a uniform naming convention for the convenience of data look-up. All the standalone files are merged into a single file, with each time-stamp file as a band. This file is accompanied by two elements, "TimeCoverage" and "TimeSchema". The former defines the start (t_1) and end dates (t_2), while the latter prescribes the time interval (i) between adjacent bands.

The untiled, unstacked method of organisation is identical to the space-time composites model, except that the stored spatial layers must be in raster format only. In the tiled, unstacked organisation method, all the raster layers are divided into equal and non-overlapping tiles, each being stored as a record, instead of the entire scene. All the tiles can be indexed to enable rapid searching. They can also be indexed temporally.

No matter which format of storage is used, it must facilitate spatiotemporal analysis in a GIS. At present, no format can do this.

8.3 SPATIOTEMPORAL DATA ANALYSIS

8.3.1 TIME-EXPLICIT VERSUS TIME-IMPLICIT ANALYSIS

Similar to the snapshot spatiotemporal data model, *time-explicit spatiotemporal analysis* is also a snapshot approach to handling the temporal dimension of spatiotemporal data. Space and time are treated separately, with a sole focus on time. It is commonly implemented in the form of time-series analysis that is repeated at a given temporal increment (e.g., annual). The same spatial analysis is undertaken for data

---

TABLE 8.3

Main Methods of Organising Spatiotemporal Data in a GIS Database

Method	Main features
File system	Multi single-band raster files
	Single raster file with multiple bands
Database	Untiled, unstacked raster files
	Tiled, unstacked raster files
	Tiled and stacked raster files

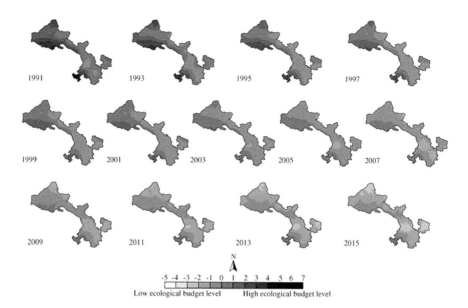

FIGURE 8.12   Spatiotemporal variation of per capita ecological budget for Gansu, China, 1991–2015 (source: Yue et al., 2006, with permission).

collected at a particular time. Such time-series results illustrate how the phenomenon under study evolves over the study period.

In time-series spatial analysis, time is handled in two ways. The first is time slicing to a fixed interval (e.g., 2 years), in which time is regarded as discrete (Figure 8.12). The second way is to treat time as continuous via animation of discrete frames of results, or video. This treatment is commonly adopted to study and illustrate temporal processes, such as the continuous erosion of a beach. The beach profile can be captured using a video camera every 5 minutes. If the results are obtained from aerial photographs, the time interval can be decadal, with a very irregular interval between temporally adjacent frames.

In *time-implicit spatial analysis*, time is treated as continuous with a fixed increment, as in dynamic spatiotemporal simulation. The same phenomenon is simulated multiple times, and the results at the intermediate times are not the main concern, but may be conjectured from the preceding and subsequent time-series results. It is the temporal variation of the variable under study that is the focus. The results can be visualised in particular snapshots. Time-implicit analysis is commonly used to study spatiotemporal processes by incrementing time iteratively. The same transition rules apply in all iterations, but neighbours and their conditions may change from one iteration to the next (see Section 7.1.2).

8.3.2 SPATIOTEMPORAL ASSOCIATION

In order to understand spatiotemporal auto-correlation, it is imperative to first examine temporal association. This refers to the correlation of the same variable at different times, such as the volume of vehicle traffic on a road between morning rush hours (X_T) and evening rush hours (Y_T). Temporal trend proximity can be studied using the temporal correlation coefficient of two time-series data, and is commonly used to study human mobility (Gao et al., 2019). It can be calculated using the first-order temporal correlation coefficient, modified from Equation (2.14) as:

$$CORT(X_T, Y_T) = \frac{\sum_t^{T-1}(X_{t+1} - X_t)(Y_{t+1} - Y_t)}{\sqrt{\sum_t^{T-1}(X_{t+1} - X_t)^2}\sqrt{\sum_t^{T-1}(Y_{t+1} - Y_t)^2}} \tag{8.1}$$

where X_T and Y_T denote two time-series variables with a total of T observations, and X_t and X_{t+1} represent the observations at time t and $t+1$, respectively. As with ordinary correlation, CORT(X_T, Y_T) ranges from -1 to 1 in value, with -1 meaning that X_T and Y_T have the same rate of change, but in the opposite direction. A correlation of 0 means that they are stochastically and linearly independent of each other because they behave differently time-wise.

The aforementioned temporal association can be further extended to the spatio-temporal domain. *Spatiotemporal auto-correlation* is an extension of bivariate spatial auto-correlation to the temporal domain. It refers to the correlation of a variable with itself over space and time. It can be estimated using the usual univariate and bivariate Moran's I equation with some modifications. Let $\{X_i^T\}$ (i=1, 2, …, N) be a set of spatiotemporal series data at N locations, X_i^T is a time-series dataset at location i, and $X_i^T = \{X_{it}^T\}$ (t=1, 2, …, T). All X_i^T have the same duration of T. The correlation coefficient is calculated as:

$$I^T = \frac{N}{\sum_i^N\sum_j^N W_{ij}} \frac{\sum_i^N\sum_{ij}^N W_{ij}Z_i^T Z_j^T}{\sum_i^N Z_i^{T^2}} \tag{8.2}$$

where W denotes the weight matrix based on the spatial proximity of neighbouring observations. It may need to be standardised. Commonly used means for its determination include inverse distance and inverse distance square, identical to the distance-decay weighting function used in inverse distance weighted interpolation; Z_i^T stands for the deviation between the time series at location i and the mean series $\bar{X}^T$. It is calculated as:

$$Z_i^T = \emptyset\left[CORT\left(X_i^T, \bar{X}^T\right)\right]\cdot\left(V_i - \bar{V}\right) \tag{8.3}$$

where ϕ is an exponential adaptive tuning function in the form of $\emptyset(x) = \dfrac{2}{1 + e^{2x}}$, Vi represents the accumulative of X_i^T, and $\bar{V}$ refers to the accumulative of $\bar{X}^T$, the mean of $\{X_i^T\}$. It shows a series of mean values at each time, or $\bar{X}^T = \left\{\dfrac{1}{N}\sum_i^N X_{it}^T\right\}$.

This measure integrates the temporal variance with the value deviation at a location to reveal the spatial pattern. It is able to indicate global and local spatiotemporal patterns. In the case of transport, it can reveal the spatial auto-correlation between different roads, but is unable to show the correlation at different times. Also, the global Moran's I results are heavily influenced by spatial clustering of the time-series data, as in the case of traffic crashes (e.g., certain spots are more prone to crashes than elsewhere). It may be good at studying human mobility, but not at identifying hotspots.

The above limitations can be overcome by emphasising the temporal aspect of the auto-correlation. The coefficient can be calculated by modifying Moran's I (Equation 3.7). The bivariate Moran's I, which is good at revealing the spatial auto-correlation between the same variable separated by a spatial lag, can be adapted as separation by a time lag between t and t'. Let z_t and $z_{t'}$ be the same variable observed at time t and t', respectively. The global Moran's I space-time autocorrelation statistic ($I_{t,t'}$) can be defined as:

$$I_{t,t'} = \frac{z_{t'} \cdot w \cdot z_t}{n} \tag{8.4}$$

where z must be standardised by the mean and standard deviation, or $z_t = \left| x_t - \bar{X}_t \right| / \sigma_t$. This statistic measures the influence induced by a change in a spatial variable z at time t' and location $i(z_t)$ on its neighbours at time t (Matkan et al., 2013). This global measure may not be able to identify the spatial auto-correlation at local spots. The alternative is to use bivariate LISA:

$$I_{t,t'}^i = z_t^i \sum_j w_{ij} z_{t'}^j \tag{8.5}$$

where i and j refer to the location. As before, z_t and $z_{t'}$ must be standardised prior to calculating I.

The main applications of spatiotemporal auto-correlation are detection of spatiotemporal dependencies of crashes (Matkan et al., 2013), and in studying collective human mobility (Yang et al., 2019).

8.3.3 TIME SLICING OVERLAY ANALYSIS

Time slicing is a way of representing features over a specific temporal range in a linear reference system. In transport, time slicing routes are an efficient way of representing road segments that have changed their shape (e.g., realignment of the route), and surface sealing. *Time slicing overlay analysis* involves at least two layers in the input. Both of them must cover an identical ground area, but are captured at different times. If in raster format, their grid cells must be of the same size. As with ordinary overlay analysis, time slicing overlay analysis is a way of identifying changes that have occurred in the interim. Changes are identified through an overlay of territorial areal units at different times. It must be noted that time slicing overlay analysis is not just an ordinary way of detecting changes. It focuses on the spatiotemporal change of the same phenomenon whose spatial extent has changed (e.g., urban sprawl).

8.4 TIME-EXPLICIT SPATIOTEMPORAL MODELLING

All the spatiotemporal simulation models presented in Chapter 7, be they cellular automata or agent-based, do not treat time explicitly. Time is considered in the sense that some variables change with time, which is realised through temporal increment at a fixed interval. This section concentrates on time-explicit models. The temporal component can be studied by focusing on the study area or phenomenon of interest at different times (Figure 8.13). There are four types of time-explicit spatial models: extended semi-variogram models, diffusion models, state and transition models, and modified Susceptible, Infectious, Removal (SIR) models.

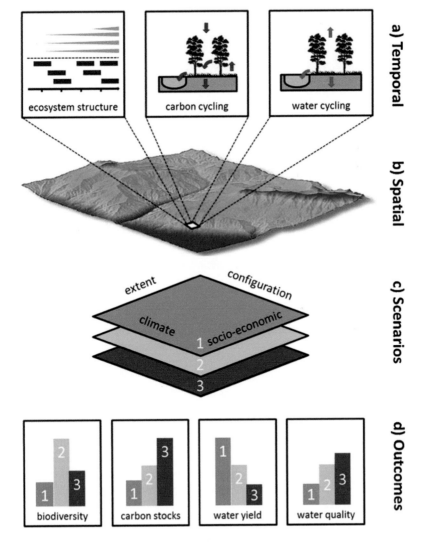

FIGURE 8.13 An example of spatiotemporal modelling in which the same study area is captured at multiple times (source: Cunningham et al., 2015, with permission).

8.4.1 EXTENDED SEMI-VARIOGRAM MODEL

The variable y in space (s) and time (t) or y(s, t) to be quantitatively modelled can be expressed as a sum of two components, $u(s, t)$ and $v(s, t)$, or:

$$y(s, t) = u(s, t) + v(s, t) \tag{8.6}$$

where $u(s, t)$ denotes the structured mean field and $v(s, t)$ is essentially the random space-time residual field. The mean field $u(s, t)$ can be modelled as:

$$u(s,t) = \sum_{l=1}^{L} \gamma_l M_l(s,t) + \sum_{i=1}^{m} \beta_i(s) f_i(t) \tag{8.7}$$

where $M_l(s, t)$ stand for the spatiotemporal covariates, γ_l denote their coefficients, and $\beta_i(s)$ are the spatially varying coefficients for the temporal functions. Equation (8.6) is virtually the same as the conventional semi-variogram expressed in Equation (5.1), except that all the terms are a function of time, with the exception of coefficients $\beta_i(s)$. They are time-invariant, and are handled with universal kriging, allowing the temporal structure to vary with locations:

$$\beta_i(s) \in N\left(X_i \alpha_i, \sum_{\beta_i}(\theta_i) \right) \quad \text{for } i = 1, 2, 3, \ldots, m \tag{8.8}$$

where X_i represent $n \times p_i$ matrices of regression coefficients (n = number of observations), often containing geographical covariates, and $\sum_{\beta_i}(\theta_i)$ are $n \times n$ covariance matrices. The $\beta_i(s)$ fields can have different covariates and covariance structures, all assumed to be a priori independent of each other.

8.4.2 DIFFUSION MODELS

Certain spatial phenomena, such as fire and disease spread, are a function of both location (x,y) and time (t). They can be represented as z(x,y,t). Such spatial processes are ideally studied using *diffusion models*. These are mathematical equations that depict the spatial-temporal variation of a phenomenon over time and across space. Diffusion models are commonly used to study the gradual spread and dispersal of a spatial feature or phenomenon from a single source, such as the spread of communicable diseases from known cases, and of wildfires from the ignition point. They are able to reveal spatial diffusion, spatial spread, and spatial change. Some diffusion processes, such as the diffusion of air pollutants, are inherently 3D in nature, and 3D diffusion is complex, and hence will not be covered here. Instead, the discussion will be confined to 2D diffusion. The generic equation for describing 2D diffusion can be expressed as:

$$\frac{\partial Z}{\partial t} = D\left(\frac{\partial^2 Z}{\partial x^2} + \frac{\partial^2 Z}{\partial y^2} \right) \tag{8.9}$$

where D is the diffusion coefficient. It dictates the speed at which Z changes with time. The above model is underpinned by two assumptions:

(1) All individuals disperse simultaneously in all directions in a spatially homogeneous domain with no directional preference.
(2) All individuals have exactly the same ability of dispersion, and there is no change in population (e.g., mortality and reproduction are ignored).

These assumptions can considerably simplify 2D spatiotemporal modelling, but they also make the model in Equation (8.9) inapplicable to the dispersion of plants in a landscape. This process is more complex and dynamic as it involves mortality and population growth in a heterogeneous environment (e.g., a rugged terrain). Spatial heterogeneity can be taken into consideration via a friction layer served by a DEM in spatiotemporal modelling, but not population dynamics.

If the entire population concentrates at a single point at t=0, then its spatial distribution in a 2D space will be normal, or:

$$Z(x,y,t) = \frac{Z(0,0,0,)}{4\pi Dt} \exp\left(-\frac{x^2 + y^2}{4Dt}\right) \tag{8.10}$$

What this model does is to predict the asymptotic rate of expansion of the Z population front. The rate of diffusion is defined as the distance between sites with equal population densities at two successive times. D is usually estimated in independent experiments, such as mark-recapture experiments. If marked animals are set free within a uniform grid of traps, D can be estimated as:

$$D = \frac{2M(t)^2}{\pi t} \tag{8.11}$$

where $M(t)$ refers to the mean displacement of animals recaptured t times after their releases (Skellam, 1973).

8.4.3 State and Transition Model

The *state and transition model* is a framework for modelling the successional dynamics of ecosystems. In order to be studied using this model, an ecosystem must exist in a set of discrete *states*, between which *transition* can take place. All states are transitional and interchangeable, depending on the rules governing how one state can change to another (Plant, 2019). However, the transition between two states is not always reciprocal, but may be unidirectional. As shown in Figure 8.14, there are four states and seven transition rules among them. The transitions between wasteland and grassland, and between shrubland and woodland are both reciprocal, but not between wasteland and woodland directly. A woodland can be turned into a wasteland immediately following clear-cut logging. A wasteland cannot be turned into woodland naturally via self-recovery. It must become grassland first, and later shrubland, to reach the woodland state.

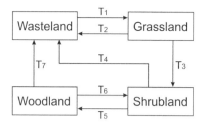

FIGURE 8.14   A schematic representation of a state and transition model illustrating how four states of land cover can transit between them under different rules.

State and transition models are the best at studying the temporal change between a limited number of states at discrete times. Time is not explicitly set in the transition. For instance, the time it takes for wasteland to transit to grassland, or from grassland to shrubland remains unknown. This kind of model is highly limited in its application. It is totally unsuitable for dynamic spatial-temporal modelling in which the spatial extent and change are closely related to time, such as fire and disease spread. For this reason, epidemiologic modelling must rely on the modified SIR model.

8.4.4   Modified SIR Model

Endemic-epidemic modelling of the gradual spread of a communicable disease in a community from infectious individuals is so complex that none of the aforementioned time-explicit models is able to fulfil this task. This modelling is highly dynamic, as the infected persons can be removed from the population (e.g., recovered or perished). Just like in the state and transition models, an individual can have one of three possible states: Susceptible (S), Infectious (I), and Removal (R) caused by recovery, hence the expression SIR model. It is able to compute the theoretical number of people to be infected with a contagious disease in a closed population over time. However, there is no information on the spatial distribution of this subpopulation. Strictly speaking, it is only temporal modelling. The simplest version of SIR is the Kermack-McKendrick model, because it treats the population as closed and stagnant (i.e., there is no birth and death, either natural or caused by the disease itself). Other assumptions of the model include:

(1) An instantaneous incubation period of the infectious agent;
(2) The infectious period is the same as the length of the disease
(3) A completely homogeneous population without any variation in age, social status, or spatiality.

The simplest SIR model comprises three coupled non-linear ordinary differential equations:

$$\frac{dS}{dt} = -\beta SI \tag{8.12}$$

$$\frac{dI}{dt} = \beta SI - \gamma I \qquad (8.13)$$

$$\frac{dR}{dt} = \gamma I \qquad (8.14)$$

where $S(t)$ stands for the number of people susceptible to infection at time t, $I(t)$ represents the number of infected people, $R(t)$ denotes the number of people who have recovered and who are immune to further infection, and β and γ refer to the infection rate and the recovery rate, respectively. The ratio of βS to γ is known as the epidemiological threshold, or the number of people infected by a single primary source. This model has been implemented in the surveillance program in R (Meyer et al., 2015). Apart from this simple SIR model, more complex models involving far fewer assumptions about the population and the infectious period have been developed for the same purpose. Needless to say, they are far more complex mathematically than the highly simplified SIR model.

These four types of temporal models fulfil quite different objectives, each suitable for a unique and specific domain of applications (Table 8.4), so it is relatively easy to decide which one to use. Caution still needs to be exercised in selecting the appropriate model. In general, the models will become more complex if they are underpinned by fewer assumptions about the spatial phenomenon being modelled. Of the four models, the state and transition model is the simplest, but it also yields the least amount of information about change. In contrast, the *modified SIR model* is the most complex. The complexity will be much higher than that shown in Equations

---

TABLE 8.4

Comparison of Four Time-explicit Spatiotemporal Models and Their Best Uses

Model	Main features	Best uses
Extended semi-variogram	Variable under study varies spatially and temporally	Describing the spatial and temporal behaviour of spatial phenomena
Diffusion (2D)	The spread of a phenomenon is a function of time and speed of propagation in a homogeneous space	Studying how the same attribute changes spatially with time (e.g., disease and fire spread)
State and transition	The allowable transitions between multiple states are governed by rules, time-implicit, duration of change unknown	Modelling of successional dynamics of ecosystems
Modified SIR	The spread of infectious cases in a closed population over time is modelled based on infection and recovery rates, but there is no information on their spatiality	Endemic-epidemic modelling of communicable diseases

(8.12–8.14) if fewer assumptions are made about the population. This is the only model that does not have a spatial component, so one possible area of improvement is incorporation of the spatial component. One way of adding this dimension to the modelled outcome is through the use of census tracks. The same spread process from the infected individuals to the population can be shown to vary spatially based on the demography in each census tract.

8.4.5 PACKAGES FOR SPATIOTEMPORAL ANALYSIS AND MODELLING

At present, there are no commercially available computer systems designed specifically for undertaking generic spatiotemporal analysis and modelling. However, researchers have developed niche packages for performing some of these tasks, and have made them publicly available with generous resources for support. Of these packages, two merits a brief introduction. The first is the *R surveillance* package. This open-source system is excellent for temporal and spatiotemporal modelling and monitoring of epidemic events. This regression-based modelling framework offers three classes of spatiotemporal analytical functions (Table 8.5): `twinstim`, `twinSIR`, and `hhh4`. `twinstim` is a two-component, endemic and epidemic-superposed, spatiotemporal intensity model. This point process model is designed to analyse the spatiotemporal patterns of geo-referenced and time-stamped cases. This kind of analysis is useful in continuously monitoring infective events (Meyer et al., 2015). The sporadic events caused by unobserved sources of infection (i.e., the endemic component) is modelled by a piecewise constant log-linear predictor that is spatially and temporally varying. The epidemic component is observation-driven, and models the force of infection from a log-linear predictor associated with the point patterns of past events, in which the infection pressure decays with increasing spatial and temporal distance from the infective events.

TABLE 8.5

Spatiotemporal Endemic-epidemic Models and Corresponding Data Classes in the R Surveillance Package

Package	`twinstim`	`twinSIR`	`hhh4`
Data class	epidataCS	Epidata	sts
Resolution	Individual events in continuous space-time	Individual SIR event history of a closed population	Geographically and temporally aggregated event counts
Example	Crash patterns	COVID-19 infection cases	Daily infection of COVID-19 in a city
Model	Spatiotemporal point process	Multivariate temporal point process	Multivariate time series (Poisson or NegBin)

Source: Modified from Meyer et al. (2015).

The second class of functions in **twinSIR** traces the event history over time for a select group of units using the spatial SIR model that is represented as a multivariate temporal point process. Both classes of analysis focus on individual cases. The last class of analysis concentrates on spatially aggregated time-series data fitted with the multivariate negative binomial time-series model **hhh4**. Its format is almost identical to the land use regression (LUR) model (Section 6.6.2), except that the independent variables (predictors) have a temporal component and are log-linear. Similar to the point process models of **twinstim** and **twinSIR**, its mean can be decomposed additively into endemic and epidemic components.

The second is the *spatiotemporal R package*, a system of R functions for analysing NO_x data recorded at the same location over time, either along a national network or at fixed sites (Bergen and Lindström, 2019). It accepts time-aggregated data (e.g., monthly or weekly) that may need transformation prior to analysis. Predictions at times and locations where no measurements are available can be made from a set of spatial and/or spatiotemporal covariates. They can be geographical (e.g., distance to roads and the coast, or buffer width from a population centre) (Yang et al., 2020). It has a number of functions for data import, analysis, display, and plotting, as well as model validation. The core to this spatiotemporal R package is the **createSTmodel()** module, which allows the user to specify spatial-temporal covariates (if any), and the type of spatial covariance model for each β-field and v-field, as well as define the covariates used for each of the β-fields. This means it is able to provide accurate spatiotemporal predictions of ambient air pollutants at small scales (Lindström et al., 2014).

8.5 VISUALISATION OF SPATIOTEMPORAL DATA AND PROCESSES

Visualisation is a graphic means of communicating spatiotemporal data and/or processes and exchanging knowledge with the reader or stakeholder. Through visualisation, we can gain a better understanding of spatial entities and their relationships. *Spatiotemporal visualisation* is concerned with the graphic display of spatial entities to demonstrate their changing behaviour in space and time. It is able to reveal the overall tendencies and movement patterns and provide a global perspective of progression, from which evolutionary trends and patterns of change can be detected. Effective visualisation relies on dynamic cartography, which plays an increasingly important role in providing insights into the complex spatiotemporal relationships between spatial phenomena and models, and in understanding spatial processes (Mitas et al., 1997). If properly designed, visualisation can enhance the interactivity of exploration, and visual analytics can make the hidden patterns more visible. Thus, visualisation has been widely used to illustrate spatiotemporal analysis results and processes. Of these two, spatiotemporal processes are much more challenging to visualise because their visualisation must be 3D (2D space plus the changing attribute). The difficulty rises disproportionately if the spatiotemporal processes also involve movement. Visualisation can be achieved using a number of methods in the map format, via animation in post-modelling and analysis, and via simulation during modelling (Table 8.6). Of the three means, maps are the most common and easiest to produce, so they will be covered first.

TABLE 8.6
Comparison of Three Means of Visualising Spatiotemporal Processes and Their Main Features

Means	Formats	Main features	Best use
Map display	Single map, space-time cube	Static maps capturing a moment; clever use of cartographic elements to show movements	When only a few chronological frames are available; good at showing simple movements with time
Animation	Time-series maps, video clips	Continuous display of time series maps to illustrate spatial changes; general impressions only; not quantitative	When a large number of frames are available; good at showing the spatial change of a mono-variable (e.g., urban sprawl)
Simulation	NetLogo graphic onscreen display	Display of spatial dynamic process onscreen as simulation runs over the entire simulated space; quantitative information of change available; unable to show movement	When it is desirable to view the spatial pattern of change at multiple moments/ locations, or changes in cell state involving a large number of factors

8.5.1 Simple Map Display

A simple display can be created using *static maps* in the analogue or digital format. On a static map, movements can be visualised through the innovative and creative use of various cartographic symbols. For instance, in visualising mobility, arrows can be used to show the direction of movement, arrow width can be made proportional to the quantity of movement, arrows can be colour-coded to indicate the time of movement, the horizontal and vertical positions can denote the origin of the movement, while the environment of the place (e.g., temperature) can be shown in a graph along the timeline. Needless to say, these symbols must conform to the existing cartographic conventions and be intuitive. It must be emphasised that static maps can only visualise simple movements (e.g., movements of the same entities in certain directions). If the movements are made by multiple entities that move haphazardly in the visualised space, then the effectiveness of visualisation diminishes drastically. However, this problem can be lessened with digital maps.

In digital media, more map elements can be introduced to the cartographic symbols, such as the use of appearance duration or flashing intervals to represent dynamic phenomena (Slocum et al., 2013). In addition to arrows and lines, charts and diagrams can be inserted into a map to indicate state changes. However, this is still a baseline approach that indicates time and dynamics through the use of varying arrow sizes (e.g., for speeds in a vector field). It is unable to indicate the absolute state of a variable at a given point in time. Therefore, baselines are ill suited for showing relationships between discrete events.

The second means of *static display* is via the space-time cube, in which the third dimension is reserved for time (Figure 8.15). This cube is able to show the whole scene in one frame, but only one scene at a time. At a given time, only the top scene is visible. The space-time cube is good at illustrating spatial processes (e.g., transition in state). In addition, more information can be overlaid onto the scene or individual events. It is possible to interactively query the 3D environment. If delivered online, the degree of interactivity can be higher, but Internet-based interactive visualisation requires the translation of the geo-referenced data to the Virtual Reality Modeling Language format. Compared to analogue maps, digital maps offer much more benefits, such as supporting the exploration and presentation of spatiotemporal data (Mitas et al., 1997), and increased flexibility of visualisation. For instance, it is possible to change the viewing perspective, temporal focus, and dynamic linkage of maps with corresponding symbols. However, the space-time cube is comparably hard to understand for those not well trained in reading perspective maps. Critically, this display is unable to illustrate movements from one time to the next, as only one scene can be viewed at a time (Figure 8.15). Also, no quantitative information about temporal changes or movements is available.

An excellent alternative to space-time cube visualisation is GeoTime, a mature commercial software package that can improve perception and understanding of

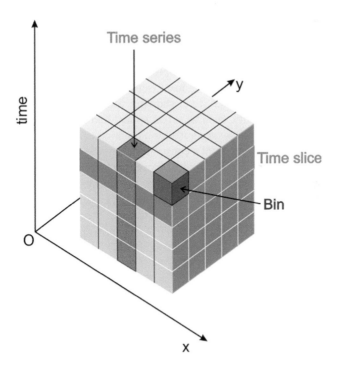

FIGURE 8.15 A space-time cube illustrating spatial variability along the vertical axis (time = orange). The data can be analysed at a given time (time-sliced, green scene). "Bin" refers to a grid cell in the spatiotemporal domain.

entity movements, events, relationships, and interactions in the spatiotemporal context much more effectively than a static display. It makes use of the time-location model presented in Figure 8.5b, and animates events in time through a slider bar. All the visualised events and their connections in 3D format are easy to read and understand. GeoTime is best used for visualising the mobility of the same spatial entities, such as tracked animals and crime suspects.

8.5.2 Animation

In cartography, *animation* refers to the visualisation of a changing phenomenon through the continuous and sequential display of time-series maps, usually on a computer screen. Animation involves many more frames of maps than simple display, so it is able to show spatiotemporally dynamic processes or progression vividly, which is achieved by displaying the static maps in succession at a fixed speed. All the maps in an animation show the same entity at a particular moment. The display of a series of maps one after another sequentially is known as time-series animation, in which each frame captures one moment in time. When a cursor moves along the timeline, the maps change accordingly. A timeline animation can be enhanced with interactive operations, such as zoom in, zoom out, pan, or focus on selected event types (Zhong et al., 2012). If the animation involves more than tens of frames, then it can be turned into a video that can be viewed conveniently on-screen. There are two strategies for animating a process: baseline and real-time. *Baseline animation* refers to animation starting from a common time, with all subsequent frames showing changes based on what is contained in the first frame. *Real-time animation* means that animation takes place concurrently with the simulation itself (see Section 8.5.3).

Time-series animation is effective at showing simple and gradual expansion or contraction of a spatial entity, such as urban sprawl and landslides. If constructed from time-series satellite images, time-series animation is also effective at illustrating the spatial dynamic processes of a constantly changing phenomenon, such as the spatiotemporal variations of harmful algal blooms. The effectiveness of visualising spatial-temporal changes has been enhanced by the advances in digital technologies that enable only the changed parts of a frame to be constantly updated. In this way, the viewer's attention is directed exclusively towards them, rather than the entire frame. Time-series animation can be 2D or 3D. 3D animation has gained vast popularity in recent years. It can be further extended to 4D by draping the frames onto a surface represented by a DEM.

Animation has been widely used to visualise the output of mobility and transportation simulations, such as depicting activities and tendencies of movements over a period of time. The quality of an animation is affected by a number of factors, such as the number of data frames used, the optimal display speed, the number of intermediate frames to create between the known frames, and the media for the created animation (e.g., computer screen, video, or online) (Acevedo and Masuoka, 1997). These factors require careful consideration in order to produce an animation of the ultimate quality. If delivered online, image size, file size, and colours to be used in the animation will be more restrictive than if animated locally. The spatial and temporal accuracy of the spatiotemporal process being visualised is subject to the number of original map frames. The number of intermediate frames affects the visual perception and smoothness of the animation.

Finally, it must be acknowledged that animation is able to illustrate only the general pattern of spatial variation, but not the quantity of change visually, nor the spatial extent of change quantitatively unless additional cartographic displays are added to it. These limitations are lessened if the spatial process being animated is simple and unidirectional, such as urban expansion. Since the nature of the spatial process (urban growth) is known already, it is much easier to appreciate the pace of spatial spread visually from the animation than otherwise.

8.5.3 SIMULATION

Some dynamic spatial processes, such as the spatial-temporal degradation of grassland quality under excessive grazing and strong external disturbances, are better visualised simultaneously as they are being simulated, because of the absence of maps. Simulation-based visualisation is good at illustrating spatiotemporal changes graphically and showing dynamic processes in which the state of numerous cells is changing at the same iteration. It is particularly good at showing how the state of a given cell changes with time. It is also able to illustrate movements or spatial spread of an event (e.g., fire), depending on the phenomenon being simulated. The spatial change is shown graphically on a computer screen locally (Figure 7.8). The onscreen display is updated at a default interval specified by the simulator. If the display is too fast, or critical frames need to be preserved permanently, they can be captured as snapshots with the "step" button in NetLogo. Apart from the general impression of spatial changes, simulation-based visualisation enables the quantity of change to be displayed on the computer screen using additional graphic outputs (e.g., line charts). Ledges may be inserted into the interface of the modelling codes to halt the visualisation temporally, and the speed of updating the display may be altered manually by manipulating an onscreen slider. In both cases, only the quantitative change with time is visualised, while no quantitative information on the spatial distribution or pattern of the visualised phenomenon is displayed. In fact, such information is seldom generated in the simulation.

REVIEW QUESTIONS

1. While some geographic phenomena change so fast that time must be considered in the simulation. In contrast, some other geographic entities may not change much even within years. Identify a few examples of each in real life, paying particular attention to the latter whose attributes can be regarded as time-invariant at a fine temporal scale of spatiotemporal simulation.
2. In spatial analysis, the scale of analysis exerts a significant impact on the results. The conclusions based on them can be erroneous if extended to another scale of study. Is this also the case with the temporal scale? Identify two examples in which the temporal granularity of aggregation has little effect on the results, but can have huge impacts on the results of temporal analysis.

3. Compare and contrast the strengths and limitations of the following spatiotemporal representation: temporal composite and space-time composite. What is going to happen to their limitations if the data to be represented are in the raster format?

4. In spatial-temporal analysis, which element is featured more prominently, spatial or temporal? Why?

5. In your view, how does the spatial-temporal autocorrelation of traffic induced by a traffic accident vary spatially from the accident spot at the time of the accident and five hours after it?

6. What is the relationship between timeless and time-extended semivariograms?

7. The diffusion model depicted in Equation (8.9) is applicable to the homogeneous 2D space. How should it be modified if it is applied to model the spread of chemicals leaked into the soil from a traffic accident in a heterogeneous environment?

8. In the state and transition model shown in Figure 8.14, all the states are mutually transitional under certain conditions that must be spelled out in any simulation. Under what conditions can wasteland be transitioned to grassland, and vice versa? What is the role of the temporal scale in the transition?

9. It can be argued that the modified SIR model is just a temporal model. Suggest two changes to it to make it a spatial-temporal model.

10. Compare and contrast the strengths and limitations of simply display with animation in visualising spatial-temporal data/processes. What is the most important message an animation must convey irrespective of its particular format?

REFERENCES

Acevedo, W., and Masuoka, P. (1997) Time-series animation techniques for visualizing urban growth. *Computers & Geosciences*, 23(4): 423–35.

An, L., and Brown, D. (2008) Survival analysis in land change science: Integrating with GIScience to address temporal complexities. *Annals of the Association of American Geographers*, 98: 323–44, doi: 10.1080/00045600701879045.

Armstrong, M. P. (1988) Temporality in spatial databases. *Proceedings of GIS/LIS'88*, vol. 2. Bethesda, MD: American Congress of Surveying and Mapping, pp. 880–89.

Bandholtz, T. (2003) Sharing ontology by web services: Implementation of a semantic network service in the context of the German Environmental Information Network (gein). *Proceedings of the First International Conference on Semantic Web and Databases, September 2003*, pp. 177–89.

Bergen, S., and Lindström, J. (2019) Comprehensive Tutorial for the Spatio-Temporal R-package, retrieved from https://rdrr.io/cran/SpatioTemporal/f/inst/doc/ST_tutorial.pdf.

Cobb, G., & Moore, D. (1997) Mathematics, statistics, and teaching. *The American Mathematical Monthly*, 104(9): 801–823, doi: 10.2307/2975286.

Cunningham, S., Marc, N., Baker, P., Cavagnaro, T., Beringer, J., Thomson, J. R., and Thompson, R. (2015) Balancing the environmental benefits of reforestation in agricultural regions. *Perspectives in Plant Ecology, Evolution and Systematics*, 17(4): 301–317, doi: 10.1016/j.ppees.2015.06.001.

Frihida, A., Marceau, D. J., and Thériault, M. (2002) Spatio-temporal object-oriented data model for disaggregate travel behavior. *Transactions in GIS*, 6(3): 277–94, doi: 10.1111/1467-9671.00111.

Gao, Y., Cheng, J., Meng, H., and Liu, Y. (2019) Measuring spatio-temporal autocorrelation in time series data of collective human mobility. *Geo-spatial Information Science*, 22(3), 166–73, doi: 10.1080/10095020.2019.1643609.

Goodchild, M. F., Yuan, M., and Cova, T. J. (2007) Towards a general theory of geographic representation in GIS. *International Journal of Geographical Information Science*, 21: 239–60.

Hill, L. L. (2000) Core elements of digital gazetteers: Place names, categories, and footprints. In *ECDL 2000, Research and Advanced Technology for Digital Libraries*, Borbinha, J. and Baker, T. (eds.), Lisbon, Portugal, pp. 280–90.

Langran, G. (1992) *Time in Geographic Information Systems*. London: Taylor & Francis, p. 180.

Lindström, J., Szpiro, A., Sampson, P. D., Sheppard, L., Oron, A., Richards, M., and Larson, T. (2014) A flexible spatio-temporal model for air pollution with spatial and spatio-temporal covariates. *Environmental and Ecological Statistics*, 21(3): 411–33, doi: 10.1007/s10651-013-0261-4.

Matkan, A. A., Mohaymany, A. S., Shahri, M., and Mirbagheri, B. (2013) Detecting the spatial–temporal autocorrelation among crash frequencies in urban areas. *Canadian Journal of Civil Engineering*, 40: 195–203.

Meyer, S., Held, L., and Höhle, M. (2015) Spatio-temporal analysis of epidemic phenomena using the R package surveillance. *Journal of Statistical Software*, 77(11), doi: 10.18637/jss.v077.i11.

Mitas, L., Brown, W., and Mitasova, H. (1997) Role of dynamic cartography in simulations of landscape processes based on multivariate fields. *Computers and Geosciences*, 23(4): 437–46.

Nadi, S., and Delavar, M.R. (2003) Spatio-temporal modeling of dynamic phenomena in GIS. *ScanGIS*, pp. 215–225.

Plant, R. E. (2019) *Spatial Data Analysis in Ecology and Agriculture Using R* (2nd edition), Boca Raton, FL: CRC Press, p. 684.

Peuquet, D. J. (1994) It's about time: A conceptual framework for the representation of temporal dynamics in geographic information systems. *Annals of the Association of American Geographers*, 84(3), 441–61, doi: 10.1111/j.1467-8306.1994.tb01869.x.

Peuquet, D. J., and Duan, N. (1995) An event-based spatiotemporal data model (ESTDM) for temporal analysis of geographical data. *International Journal of Geographical Information Systems*, 9(1): 7–24, doi: 10.1080/02693799508902022.

Segev, A., and Shoshani, A. (1993) A temporal data model based on time sequences. In *A.U. Tansel, J. Clifford, S. K. Gadia, A. Segev, R. T. Snodgrass (Eds.): Temporal Databases: Theory, Design, and Implementation.* Benjamin-Cummings Pub Co, pp. 248–70.

Skellam, J. G. (1973) *The Formulation and Interpretation of Mathematical and Diffusionary Models in Population Biology*. In The Mathematical Theory of the Dynamics of Biological Populations, edited by M.S. Bartlett and R.W. Hiorns. New York: Academic Press, pp. 63–85.

Slocum, T. A., McMaster, R. B., Kessler, F. C., and Howard, H. H. (2013) *Thematic Cartography and Geovisualization: Pearson New International Edition*, p. 624.

Song, M., Li, W., Zhou, B., and Lei, T. (2016) Spatiotemporal data representation and its effect on the performance of spatial analysis in a cyberinfrastructure environment - A case study with raster zonal analysis. *Computers & Geosciences*, 87: 11–21, doi: 10.1016/j.cageo.2015.11.005.

Yang, C., Chan, Y., Liu, J., and Lou, B. S. (2020) An implementation of cloud-based platform with R packages for spatiotemporal analysis of air pollution. *The Journal of Supercomputing*, 76: 1416–37, doi: 10.1007/s11227-017-2189-1.

Yang, J., Sun, Y., Shang, B., Wang, L., and Zhu, J. (2019) Understanding collective human mobility spatiotemporal patterns on weekdays from taxi origin-destination point data. *Sensors (Basel)*, 19(12): 2812, doi: 10.3390/s19122812.

Yue, D., Xu, X., Li, Z., Hui, C., Li, W., Yang, H., and Ge, J. (2006) Spatiotemporal analysis of ecological footprint and biological capacity of Gansu, China 1991–2015. *Ecological Economics*, 58(2): 393–406, doi: 10.1016/j.ecolecon.2005.07.029.

Yuan, M. (1996) *Temporal GIS and Spatio-Temporal Modeling*. NCGIA, SANTA_FE_CD-ROM.

Zhong, C., Wang, T., Zeng, W., and Arisona, S. (2012) Spatiotemporal visualisation: A survey and outlook. In *Digital Urban Modelling and Simulation*, S. M. Arisona, G. Aschwanden, J. Halatsch, and P. Wonka (eds.) Berlin: Springer-Verlag, pp. 299–317, doi: 10.1007/978-3-642-29758-8_16.

Index